Earth and Life

Earth and Life

A History of Four Billion Years

ANNE NÉDÉLEC

translated with the participation of
GALEN HALVERSON

Great Clarendon Street, Oxford, OX2 6DP,
United Kingdom

Oxford University Press is a department of the University of Oxford.
It furthers the University's objective of excellence in research, scholarship,
and education by publishing worldwide. Oxford is a registered trade mark of
Oxford University Press in the UK and in certain other countries

© Originally published in French La terre et la vie – une histoire de 4 milliards d'années by Anne Nédélec

© Odile Jacob, 2022

English translation © Oxford University Press, 2025

The moral rights of the author have been asserted

All rights reserved. No part of this publication may be reproduced, stored in a retrieval system, transmitted, used for text and data mining, or used for training artificial intelligence, in any form or by any means, without the prior permission in writing of Oxford University Press, or as expressly permitted by law, by licence or under terms agreed with the appropriate reprographics rights organization. Enquiries concerning reproduction outside the scope of the above should be sent to the Rights Department, Oxford University Press, at the address above.

You must not circulate this work in any other form
and you must impose this same condition on any acquirer.

Published in the United States of America by Oxford University Press
198 Madison Avenue, New York, NY 10016, United States of America

British Library Cataloguing in Publication Data
Data available

Library of Congress Control Number: 2024950431

ISBN 9780198945413
ISBN 9780198945420 (pbk.)

DOI: 10.1093/9780198945451.001.0001

Printed and bound by
CPI Group (UK) Ltd, Croydon, CR0 4YY

Links to third party websites are provided by Oxford in good faith and
for information only. Oxford disclaims any responsibility for the materials
contained in any third party website referenced in this work.

The manufacturer's authorised representative in the EU for product safety is Oxford University Press España S.A. of El Parque Empresarial San Fernando de Henares, Avenida de Castilla, 2 – 28830 Madrid (www.oup.es/en or product.safety@oup.com). OUP España S.A. also acts as importer into Spain of products made by the manufacturer.

Preface

After thirty years of geological research dedicated to the ancient rocks and past climates of the Earth in Africa and Brazil, I was eager to write the history of our planet—a history that long precedes the appearance of humans and explains how the world formed and evolved to become what it is today. This is a long history, full of drama and cataclysm, during which the Earth suffered meteorite impacts, episodic and massive outpourings of lava, and deep freezes under glacial ice.

This book recounts four billion years of the history of the Earth and life. It follows a chronological thread and allocates text in proportion to geological time, such that the emphasis appears to be on the long expanse of Precambrian time. Special focus is given to the interactions among the many forms of life on Earth, for these feedbacks not only resulted in many upheavals in the Earth's history, but have also helped to stabilize a habitable environment.

The book attempts to answer fundamental questions. What can we learn from the oldest minerals on Earth? Where are the oldest rocks? Why did life begin in the ocean? What do the oldest fossils look like? We'll discover that life on our planet gradually diversified, beginning with tiny microbes in the ocean depths, followed by shallow-water reefs, the first algae, then the first marine animals, and, finally, plants and vertebrates that colonized the continents. Ocean was everywhere at the beginning of Earth's history. Land masses emerged slowly, but with enormous consequences for climatic and biogeochemical cycles. Plate tectonics constructed and destroyed supercontinents, and the Earth's geography constantly changed.

The most important change occurred more than two billion years ago, when the atmosphere and surface ocean became oxygenated due to the growth and burial in sediments of photosynthetic bacteria. Meanwhile, giant mineral deposits (iron, manganese, uranium) formed as a result of this rise in oxygen. The surface of the continents changed in colour from grey to red at low latitudes after the oxygenation of the atmosphere, and finally to green, when, beginning four hundred million years ago, forests formed on land.

The book is based on recent scientific work, but it is meant to be accessible to readers who did not study geology or biology at university. This approach required the topics to be highly selective and for me to impose my own interpretations in some cases. My colleagues may deplore certain omissions or disagree with my interpretations, and I assume responsibility for these choices. I hope that reader will enjoy this story of Earth and life, and gain new understanding of the planet we inhabit—hopefully for a better future!

Acknowledgments

Throughout my career, I have worked with colleagues and students from France and other countries, some of whom became friends. I thank them for all the good times shared in the field in Cameroon, Madagascar, and Brazil, and for all the lively discussions that enlightened our ideas. I cannot overlook the role played by the drivers, guides, and farmers we met in distant places, or the technicians at GET (Geosciences Environnement Toulouse) in the Midi-Pyrénées Observatory, all of whom contributed to the collective research endeavour. A scientific adventure is always a human one...

My warmest thanks go to Christiane Cavaré-Hester and Anne-Marie Cousin, illustrators at GET, for their talent and constant good humour. My partner, Jean-Luc Bouchez, my colleague, Sébastien Fabre, and my sister Claudine were kind enough to read the first versions in French and suggested some improvements. Many friends supported me with their interest in this book. Several colleagues kindly sent me their own photos: Tomaso Bontognali, Hervé Diot, Abderrazak El Albani, Galen Halverson, Paul Hoffman, Morgane Ledevin, John Purso, Robert Rainbird, Pierre Thomas, Sam Wilde, and the late Grant Young. Brigitte Meyer-Berthaud edited the paleobotany section of the last chapter.

Odile Jacob, the publisher in Paris, was receptive to my initial proposal, which was defended with conviction by Gilles Ramstein. Marie-Lorraine Colas encouraged me to broaden the subject by developing the history of life. Héloïse Mahé edited the final version in French.

Thanks to Sonke Adlung, commissioning editor at Oxford University Press, who agreed to publish this English translation. It has benefitted from a few additions, thanks to my first readers and to reviewers' comments. My translation was greatly improved by my friend and colleague, Galen Halverson, Professor at McGill University in Montreal (Canada), who also provided many relevant suggestions and to whom I am very grateful. Final book production was due to Adam Breivik (OUP) and Rajeswari Azayecoche (Integra).

Anne Nédélec
Plougonvelin, April 2025

Contents

Introduction ... 1

1. Formation of the Earth ... 5
   Earth's place in the solar system ... 5
   Formation of the solar system ... 6
   Meteorites: witnesses of planet formation ... 8
   Structure and composition of the Earth ... 10
   A bit of chemistry: isotopes ... 12
   A bit of physics: radioactive decay ... 13
   How old is the Earth? ... 14
   Formation of the Earth's core and mantle ... 16
   Formation of the Moon ... 20

2. The mysteries of the first 500 million years ... 25
   The oldest minerals in the world ... 25
   What can we learn from the Hadean zircons? ... 27
   The oldest lunar rocks ... 29
   Comparison of the Moon and the Earth ... 32
   Asteroid impacts ... 34
   Evolution of the number of impacts over time ... 38
   The contribution of late impacts ... 39

3. Origin of the atmosphere and ocean ... 41
   Composition of the current atmosphere ... 41
   Volatile components in the protosolar nebula ... 43
   Origin of terrestrial water ... 44
   Degassing of the magma ocean ... 46
   Information provided by noble gases ... 47
   Formation of the ocean ... 48
   The atmosphere of other terrestrial planets ... 49
   The greenhouse effect ... 51
   The faint young Sun paradox ... 53

4. The oldest rocks on Earth ... 55
   Acasta gneisses (Canada) ... 55
   Isua (Greenland) ... 56
   Barberton (South Africa) ... 57
   Greenstones ... 58
   Komatiites ... 61
   TTGs ... 62
   The terrestrial heat flux ... 64

5. Earth's internal cooling through time — 67
   - Mode of internal heat transfer — 67
   - Rising plumes and hot spots — 69
   - Plate tectonics — 71
   - Magma genesis due to plate tectonics — 74
   - Characteristics of the Archean magmatism — 76
   - When did plate tectonics begin? — 78
   - Before plate tectonics — 80
   - The Hadean–Archean transition — 82

6. The Archean ocean — 85
   - An Archean beach in the moonlight … — 85
   - Little land mass emerged — 86
   - Submarine hydrothermalism and the origin of cherts — 89
   - Temperature of the Archean oceans — 94
   - Sulphates or sulphides? A consequence of the composition of the atmosphere — 94
   - What happened to dissolved iron in the ocean? — 95
   - Giant iron ore mines — 97

7. Origin of life — 101
   - The molecules of life — 101
   - The genetic code — 102
   - An experimental approach to prebiotic chemistry — 103
   - Organic matter in meteorites and comets — 106
   - In search of the oldest fossils — 108
   - The phylogenetic approach: the tree of life — 113
   - Archea in oceanic hydrothermal vents and the methane cycle — 115
   - Who was LUCA, the ancestor of all living beings? — 118
   - Before LUCA: from prebiotic chemistry to the first cell — 119
   - The RNA world hypothesis — 120

8. Everything changes on Earth — 123
   - Continental growth and long-term climate change — 123
   - Gradual oxygenation of the ocean and formation of BIFs — 127
   - The Great Oxygenation Event of the atmosphere — 129
   - Ongoing rise of oxygen in the ocean and formation of giant manganese deposits — 132
   - Major disturbances in the carbon cycle — 133
   - First oils — 135

9. Columbia: The first supercontinent — 137
   - Cratons at the heart of the continents — 137
   - The growth of cratons — 139
   - Paleomagnetism: a paleogeographic tool — 141
   - Reconstructing the first supercontinent — 144
   - Sedimentation on the surface of Columbia — 144
   - Thermal effect of the supercontinent — 147
   - Formation of giant uranium deposits — 148

10. Diversification of life in the Proterozoic	151
Multiplicity and diversity of stromatolites	151
The Franceville Basin in Gabon: oil, biomarkers, and enigmatic fossils	154
Acritarchs: the first fossil eukaryotes	158
The origin of eukaryotes	159
The emergence of algae	163
First marine animals: the Ediacara fauna	165
11. From Columbia to Gondwana: The ballet of the continents	169
The Wilson cycle	169
The role of hotspots in the breaking of supercontinents	171
In search of lost oceans	173
The evolution and fragmentation of Columbia	174
The Grenville orogeny and the formation of Rodinia	176
Fragmentation of Rodinia	177
The Pan-African orogeny and the formation of Gondwana	178
The Cadomian chain on the northern edge of Gondwana	180
12. The Snowball Earth	183
The causes of the last ice age	183
Origin of Quaternary climate oscillations	184
Ice core data	188
The geological record of the late Precambrian glaciations	190
Evidence of global glaciations	193
The comeback of BIFs	195
Causes of the Snowball Earth	196
What happened to life during the Neoproterozoic ice ages?	201
13. Thawing of the Snowball Earth	203
How did Earth recover from global glaciation?	203
Evidence of global warming	205
How were post-glacial cap dolostones formed?	205
The post-glacial ecosystem	208
Landscapes of the Snowball Earth aftermath	210
Duration of deglaciation	213
14. Life invades the continents	219
The Cambrian explosion: optical illusion or reality?	219
The first land plants	221
Consequences of the greening of continents	224
The first land animals	229
Under threat from volcanoes and other hazards: the great extinctions	229
First hominins	232
The expansion of *Homo sapiens*	235
Epilogue	237
References	239
Glossary	247
Index	257

Introduction

The Earth is about four and a half billion years old. If people are asked about the major events in this long history, some of them will likely mention the existence and disappearance of the now-extinct dinosaurs, which captivate the imaginations of children and adults alike. Because of a blockbuster film, many of them will associate dinosaurs with a special time, the Jurassic period, which belongs to the Mesozoic era. The Quaternary period—the last period in the geologic timescale—began two and a half million years ago, and continues today; it was originally thought to encompass the appearance of humans, though it is now thought humans evolved earlier.[1] The Quaternary period formerly had the status of an era, following the Tertiary era in the geological timescale, but both are now included in the Cenozoic era, which spanned the formation of the Alps, Pyrenees, and Himalayas, as well as the diversification of modern mammals. The preceding Mesozoic era is commonly referred to as the 'age of the dinosaurs' and came to a dramatic end some 66 million years ago. The Paleozoic era, which began some 539 million years ago, is referred to as the 'age of invertebrates', due to the prominence of marine molluscs, brachiopods, and arthropods, among other invertebrates. Perhaps the most emblematic of the Paleozoic invertebrates were the trilobites—crustacean-like animals that disappeared at the end of the Paleozoic, some 252 million years ago, widely sought by paleontologists and amateur fossil collectors alike.

The Paleozoic, Mesozoic, and Cenozoic eras together comprise the Phanerozoic eon. The preceding four billion years constitute the **Precambrian**[2] (i.e. all of Earth's history preceding the **Cambrian**). The Cambrian is the first period of the Phanerozoic eon (and by extension, the first period of the Paleozoic era). What do we know about the expansive four billion years of Precambrian Earth? Both little and much, as we shall see. Progress in geological knowledge over the last few decades has enabled us to describe Precambrian landscapes in broad terms, while new details are regularly emerging. Earth as seen from space did not always look as we know it today; it was only during the Paleozoic era that the continents became green, with the colonization of land by plants and animals. The distribution of the oceans and continents also changed. Early in the Precambrian, there was little if any emerged land. Life appeared very early in the oceans and remained aquatic for over three billion years. During this long interval, major changes occurred to Earth's environment, including

---

[1] Anthropologists today place the origin of humans at the end of the Tertiary period. Modern humans, *Homo sapiens*, appeared about 300,000 years ago
[2] Terms printed in bold when they first appear in this book are defined in the glossary

the appearance of oxygen in the atmosphere nearly two and a half billion years ago, and dramatic fluctuations in climate.

To tell the story of the Precambrian, which occupies most of this book, a few milestones must be set. The Precambrian is divided into three parts or **eons** (Figure 0.1). The **Hadean** eon lasted about 500 million years: it began with the formation of the Earth and ended four billion years ago. It was followed by the **Archean,** which lasted for one and a half billion years, between 4,000 and 2,500 million years ago. Note that the name 'Archean' in the geological sense does not refer to a branch of prokaryotic life, yet life appeared early in the Archean eon.

Geologists often use the notation **Ma** (from the prefix 'mega') to represent ages in millions of years and **Ga** (from the prefix 'giga') for ages in billions of years. Thus 1 Ga equals 1,000 Ma. Usually, geological ages are never preceded by a minus sign.

The youngest eon of the Precambrian is the **Proterozoic** (etymologically the period of primitive animals), which lasted nearly two billion years. It began at 2,500 Ma and ended at 539 Ma with the beginning of the Paleozoic Era. It is divided into three eras: the Paleoproterozoic, the Mesoproterozoic, and the Neoproterozoic, which lasted 900, 600, and 461 million years respectively.

It may be surprising that the limits of the Precambrian eras and eons correspond to figures rounded to the nearest 100 million years, whereas the Phanerozoic eon began precisely at 539 Ma. Indeed, very precise geological landmarks define the geological boundaries in the Phanerozoic, and these have been dated relatively exactly. This is not the case for the Precambrian. Only the Ediacaran period, at the end of the Neoproterozoic, is well defined, and its base dated precisely at 635 Ma. For all the other Precambrian periods, the proposed boundaries simply correspond to ages that are approaching major events in Earth's history and are easy to remember. Efforts are being made to formalize these boundaries as for the Phanerozoic, but in the absence of widespread fossils, this objective is especially challenging.

Large regions of the Earth are formed by rocks of Archean or Proterozoic age. Precambrian rocks are therefore quite common. This is fortunate for geologists, because the Precambrian represents almost 90% of the total history of our planet! This fact may come as a surprise to, for example, French readers, for there are no Archean rocks in metropolitan France, where the oldest rocks, dating back barely 2 Ga, only constitute very small areas in northern Brittany. In contrast, South African, Australian, Canadian, or Brazilian geologists can survey entire provinces made of rocks of Archean age. Moreover, very important mineral resources are found in Precambrian terrains and justify their detailed exploration.

The work of the geologist begins in the field. Several photos of characteristic landscapes and other geological features are presented in this book. These photos illustrate how geology takes us not just back in time, but also to far away places, combining science with the delights (and challenges!) of travel. Fieldwork is typically followed by analytical work on collected samples in the laboratory, the drafting

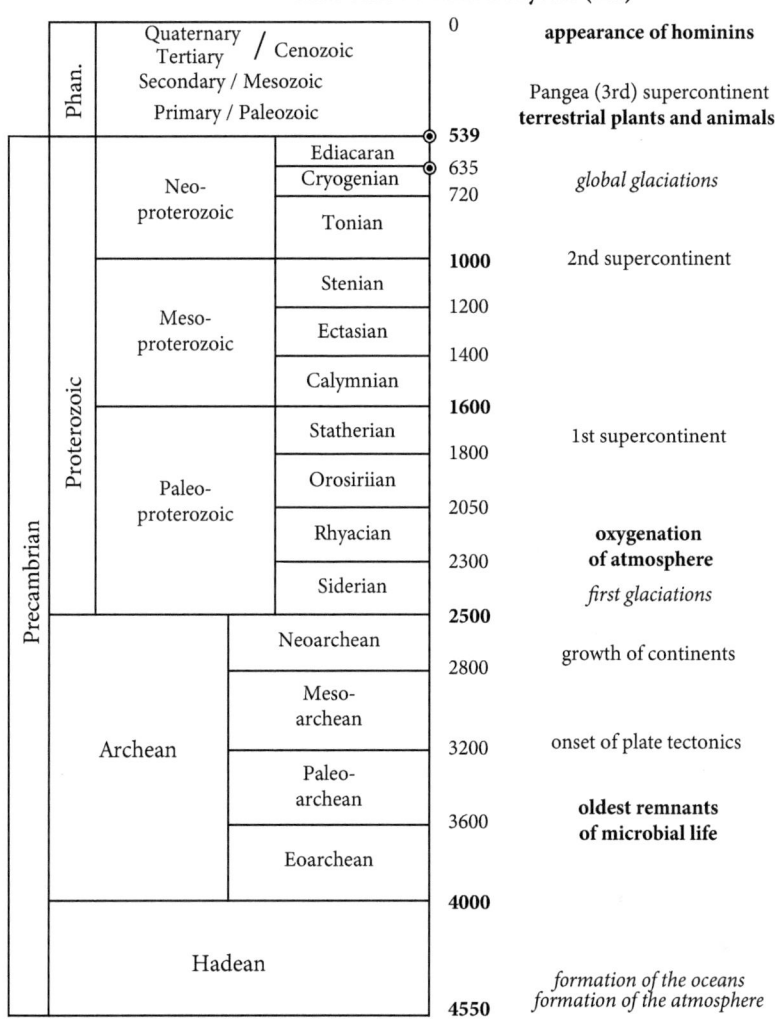

Figure 0.1 The major divisions of the Earth's history, with their boundaries in millions of years (Ma). The main tectonic and environmental events are also indicated, as well as some stages in the evolution of life. Phan.: Phanerozoic. The Quaternary period (only 2.5 million years!), now included in the Cenozoic era, should be represented only in the thickness of the line at the top of the figure.

of maps and diagrams, and numerical modelling. As the scientific questions posed by geology are increasingly interdisciplinary, discussions and collaborations with astronomers, climatologists, biologists, and other specialists have become routine. These wide-ranging efforts are leading to an increasingly detailed history of the Earth, of the life that has inhabited our planet, and crucially, the interactions between the two. This history provides an explanation of how the modern Earth and its inhabitants originated. One need not be a geologist or even a scientist to appreciate this history and the perspective it provides for the world we live in. A few basics of chemistry or physics are enough. They will be briefly reviewed throughout the book where needed to help understand its content.

> Most of the Earth's history falls into the Precambrian, which preceded the Paleozoic era and lasted nearly four billion years. The Precambrian is subdivided into three eons: the Hadean, the Archean, and the Proterozoic. The subsequent Phanerozoic eon is aptly named to indicate the visible record of animal life.

1
Formation of the Earth

Before our great journey through time, we need to locate the Earth in space. The Earth is just a small planet orbiting the Sun, a medium-sized star in the Milky Way. The Milky Way, our galaxy, so beautiful on summer nights, consists of several hundred billion stars. It formed shortly after the birth of the universe 13.8 billion years ago. Its most recent stars were born yesterday or the day before, while the oldest are nearly 13 billion years old, putting the impressive numbers listed in the introduction into perspective. . . . Now, let's forget about outer space and come back to the Earth within the solar system.

Earth's place in the solar system

The Earth is one of the four rocky inner planets of the solar system (Figure 1.1). It is about 150 million kilometres from the Sun. This distance defines the astronomical unit (AU). Beyond Mars (the fourth rocky planet), a multitude of small rocky bodies orbit the Sun in sometimes very eccentric orbits; they form the asteroid belt.

Next are the four outer giant planets, starting with the largest, Jupiter. Despite its distance (5 AU), Jupiter can be seen with the naked eye. Jupiter and Saturn are mostly made of hydrogen (gaseous at the surface, liquid at depth), with some ice and rocky

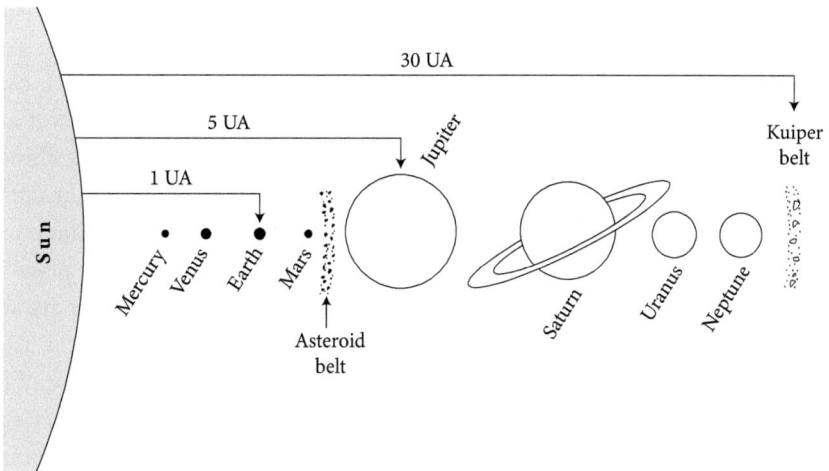

Figure 1.1 The solar system. Distance scale is not respected, and the Oort cloud is not shown as it is too far away.

Earth and Life. Anne Nédélec, Oxford University Press. © Originally published in French La terre et la vie - une histoire de 4 milliards d'années by Anne Nédélec © Odile Jacob, MMXXII; English translation © Oxford University Press (2025).
DOI: 10.1093/9780198945451.003.0002

material at their cores. Uranus and Neptune also have a small solid rocky core but are mostly icy with a smaller proportion of gas.

Beyond Neptune, at 30 AU, is the Kuiper belt, which contains many small icy bodies and some dwarf planets such as Pluto. Even further out, beyond 20,000 AU, a group of very poorly known objects make up the Oort cloud. The Oort cloud extends to more than 100,000 AU, the limit of the solar system, defined as the furthest distance where objects subject to the Sun's attraction can still be found. However, the solar wind (i.e. the flow of particles emitted by the Sun) decreases much closer, at around 100 AU.

Formation of the solar system

All the planets and other objects in the solar system are thought to have formed at the same time, just over four and a half billion years ago, from a **nebula**—a large cold cloud of gas and dust. This gradually condensed into a protostar at its centre, the future Sun, surrounded by a protoplanetary disc. The solar system is much younger than the universe.

In 2014, the giant ALMA radio telescope[1] in northern Chile obtained the first image of a protoplanetary disc around a very young star (Figure 1.2). Before this image, only a theoretical idea of planet formation was available. ALMA has made it possible to observe this process around the star HL Tauri, located 450 light years from Earth and only one million years old. Within the bright disc surrounding this young star, dark rings correspond to areas where dust has already agglomerated to form larger objects, which will become planets. Indeed, as a planet forms, it sweeps up anything nearby in its orbit. Since 2014, ALMA has provided even more precise images of another solar system forming only 150 light years away from our own. Knowing that one light year is equivalent to 9,461 billion kilometres, one can appreciate the high performance of the telescope.

Dusts in orbit around a young star gradually agglomerate. The larger objects capture the smaller ones in their vicinity under the effect of **gravitation**, a force of attraction called 'gravity' on Earth. This **planetary accretion** mechanism forms clusters of blocks, then small bodies, and finally protoplanets (dwarf planets) when the diameter of the object reaches 1,000 km. The asteroid belt between Mars and Jupiter contains more than 520,000 objects of varying size that have failed to form a single planet. The Itokawa asteroid is one of these small, sub-kilometre-sized objects, which are simply clumps of boulders and debris (Figure 1.3). Ceres, the largest object in the asteroid belt, has a diameter close to 1,000 km and is truly spherical in shape, resembling a small planet. Vesta, the second-largest asteroid, with a diameter half that of Ceres, is not perfectly spherical (Figure 1.4). Both asteroids have impact marks from encounters with smaller objects.

---

[1] ALMA stands for Atacama Large Millimeter(/submillimeter) Array.

Figure 1.2 (a) **The large ALMA antenna array inaugurated in 2013 on a 5,000 m elevation plateau in the Atacama Desert (Chile).** Used together, the 66 antennas function as a single telescope. The largest are 12 m in diameter. ALMA is designed to observe gas clouds where stars are born. These clouds emit radiation in the millimetre to submillimetre wavelengths. **(b) ALMA image of a planetary disc forming around the young star HL Tauri.** The dark rings indicate forming planets.
Source: ESO (European Southern Observatory): https://www.eso.org/public/images/archive/category/alma.

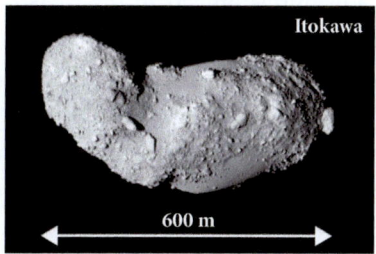

Figure 1.3 **Itokawa asteroid photographed by the Japanese Hayabusa mission in 2005.** After several attempts, the probe managed to collect a few grammes of dust and returned to Earth in 2010. This first *in situ* sample shows that the surface of Itokawa has a composition of ordinary chondrite (a common type of meteorite).
Source: JAXA (Japan Aerospace Exploration Agency).

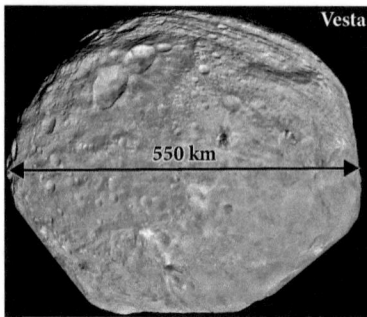

Figure 1.4 The asteroid Vesta. Composition of several images taken by the American probe Dawn between 2011 and 2012.
Source: NASA/JPL/Caltech.

Meteorites: witnesses of planet formation

Planetary accretion by gravitational capture of objects is a violent process. Collisions with smaller bodies leave scars on the surface of a planet or asteroid; these are the impact craters that can be seen on all rocky bodies in the solar system. During the impact, the impactor is often broken into smaller debris, which may even be ejected back into space along variable trajectories where they may encounter other orbiting planets. These fragments that sometimes reach the Earth's surface are called meteorites. The flow of meteorites was very large early in Earth's history. The meteorite flux has decreased but is still estimated to be equivalent to more than 10 tonnes per day. Most of these fragments are very small and remain unnoticed. A special environment, such as the pristine ice of Antarctica, is needed to find these micro-meteorites. Larger objects leave a trail visible at night when they enter the atmosphere: these are known as shooting stars. Fortunately, objects that can cause significant damage are rare and the probability of an encounter with the Earth decreases with their increasing size.

Meteorites represent the primitive objects of the solar system, which means they help us to understand how our planet was formed. More than 90% of meteorites are **chondrites**, containing millimetre-sized spherical structures called chondrules (Figure 1.5). Chondrules are thought to have formed during the condensation of the protosolar nebula. Ordinary chondrites consist of silicate minerals, notably olivine, with variable proportions of metallic iron. Silicates and iron (metal) are the main materials from which the inner planets of the solar system were formed.

Other meteorites are formed exclusively of iron or, more precisely, of an alloy of iron and nickel: these are the iron meteorites (or **siderites**). The meteorite responsible for the formation of Meteor Crater in Arizona (Figure 1.6) about 50,000 years ago was a siderite, of which many fragments have been found near Devil's Canyon, hence the name Canyon Diablo meteorite. The largest fragment weighs over 600 kg. A 360 kg fragment is in the Museum of Natural History in Paris. These impressive masses result from the very high density of metallic iron: about 5 (g/cm^3). For comparison, the density of water is equal to 1 and that of granite is around 2.7. Other smaller fragments

Figure 1.5 Different types of meteorites. (a) Fragment of the Allende meteorite, a carbonaceous chondrite. (b) Microscopic image of a chondrule under crossed polars. (c) Fragment of the Canyon Diablo iron meteorite, a siderite. (d) Polished section of a pallasite.

Source: https://planet-terre.ens-lyon.fr and Anne Nédélec for the siderite photo, courtesy of the Ries Krater Museum (Nördlingen, Germany).

Figure 1.6 Meteor Crater (Arizona). A single crater 1.2 km in diameter and 190 m deep.
Source: USGS (United States Geological Survey).

of the Canyon Diablo meteorite can be found in many museums around the world. The total number of fragments, although over 30 tonnes, is only a part of the meteorite, as the rest was vaporized during the impact. Siderites are thought to represent fragments of the iron core of ancient planets that had grown large enough for differentiation to occur (i.e. iron had separated from the silicate minerals to form the core). These fragments were subsequently blasted apart during collisions with a larger planet. Chondrites, which contain both silicates and metallic iron, are thought to represent fragments of small bodies or small planets that never differentiated.

Structure and composition of the Earth

A planet like the Earth or its neighbours Venus and Mars consists of two very different parts: an iron core at the centre and a silicate mantle at the periphery (Figure 1.7). The Earth's core cannot be sampled directly. Its existence was first suspected at the end of the 18th century, when the English scientist Henry Cavendish calculated the density of the Earth and deduced that the interior must be made of a material five times denser than water—a remarkable result for the time. However, it was not until the end of the 19th century that the idea of an Earth with an iron core[2] surrounded by a rocky mantle was accepted by all.

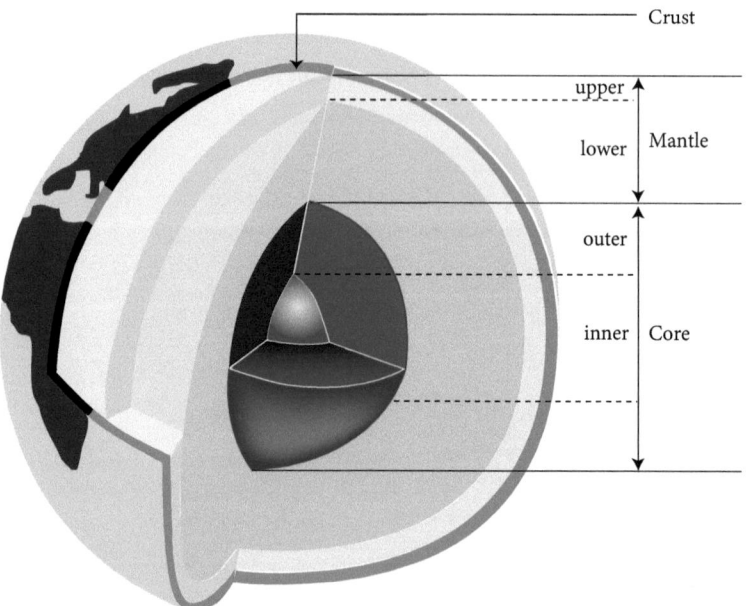

Figure 1.7 **Structure of the Earth.**

[2] There are other elements in the core, including nickel.

At the beginning of the 20th century, the technique of recording seismic waves was developing. Seismic waves are generated at the focus of an earthquake and propagate through the interior of the Earth at speeds dictated by its physical properties. Seismographs record their arrival at different places on the planet. In 1912, the German physicist Beno Gutenberg discovered an area where the arrival of seismic waves was never recorded, because their path was refracted (i.e. deflected) when passing between two layers with different density. One notices the refraction phenomenon when bathing or swimming: the immersed part of our legs or arms appears to form an unnatural angle with the remainder of the limb due to the refraction of light rays in the water. The discontinuity highlighted by Gutenberg defines the mantle–core boundary; it separates the dense iron–nickel core from the mainly silicate mantle. It is located at a depth of 2,885 km. Furthermore, some seismic waves are stopped at this interface: these are **shear waves** that cannot pass through liquids. The lack of shear wave propagation proves that the upper core is liquid. With a radius of 3,486 km, the core represents one third of the Earth's mass, but only one sixth of its volume.

In 1936, the Danish seismologist Inge Lehmann discovered another discontinuity, this time within the core, separating the outer core from the inner core (Figure 1.7). The inner core is solid, made of iron that has crystallised out of the ferrous liquid of the outer core. The inner core has a radius of 1,218 km: it is still growing slowly because of the progressive cooling of our planet. Crystallization started at depth and continues very gradually towards the surface, even though the temperatures at the centre of the Earth are higher. This somewhat counter-intuitive property is due to the very high pressures in the inner core; it will factor in again in relation to the mantle. The surface of the inner core is therefore a crystallizing surface that slowly migrates outwards from Earth's centre. We do not know when the crystallization of the inner core began, meaning the age of the inner core remains an unanswered question! The pressure and temperature at its surface are also difficult to estimate. These pressures cannot be attained experimentally, though a temperature of around 5,200 °C is proposed for the surface of the inner core today.

Convection currents of liquid iron in the outer core are responsible for Earth's magnetic field. In a very long time, when the Earth's core has completely crystallized, the magnetic field will disappear. Mars, which is smaller than the Earth, has cooled down much faster and no longer has an internal magnetic field, likely because its core is fully solid. It should be noted in passing that the iron in the outer core is not pure. The metallic liquid contains about 85% iron, 5% nickel, and up to 10% lighter elements based on precise density calculations. While we assume that the other elements comprise mainly sulphur (S), silicon (Si), and oxygen (O), we have few experimental constraints on their proportions or the abundance of other light elements.

It is easier to know the composition of the mantle because mantle rocks outcrop in certain places on the Earth's surface. This is the case, for example, in the Ariège Pyrenees close to Lers (or Lherz) Lake, near the village of Massat. The name for typical mantle rock, **lherzolite**, derives from this toponym. Lherzolites belong to the **peridotite** family, which are rocks rich in the mineral **olivine**. Olivine is a silicate of magnesium (Mg) and iron (Fe). Mantle olivine contains much more magnesium than iron,[3] meaning that the mantle contains a lot of magnesium, in addition to silicon and oxygen, while the core contains mostly iron.

A bit of chemistry: isotopes

The Russian chemist Dmitri Mendeleev attempted a classification of the known chemical elements in 1869. The principle of this classification was to align elements with similar properties in columns. The classification table now displayed in all high school chemistry classrooms follows this approach (Figure 1.8). The elements are listed in ascending order of atomic number from number 1, hydrogen (H), to number 92, uranium (U), then 93, and 94 (plutonium). The number of each element corresponds to the number of protons (Z) present in the nucleus of the atom concerned. Protons are positively charged. According to Rutherford's atomic model, small negatively charged particles orbit the nucleus: electrons. Their number is equal to that of the protons within the nucleus. Finally, uncharged particles, the neutrons, complete the nucleus.

The number of protons is always invariant for each element, but some elements have a variable number of neutrons (N). These elements are said to have multiple isotopes. An **isotope** of an element is therefore distinguished from other isotopes of that element by its mass number, which is the sum of protons and neutrons. For example, the

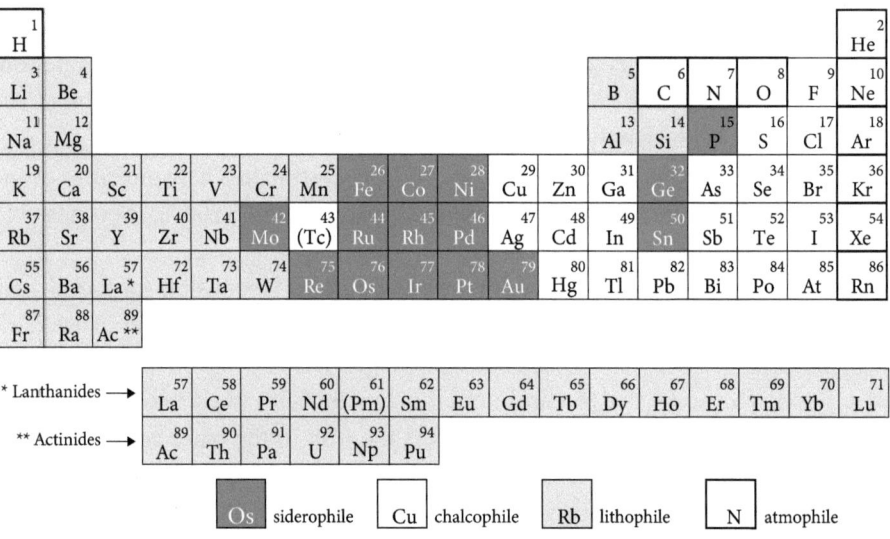

Figure 1.8 The periodic table of chemical elements. Four families of elements are identified: atmophiles, which are light elements present in the atmosphere, such as nitrogen (N); lithophiles, which are abundant in rocks (*lithos* in Greek), such as silicon (Si); siderophiles, which have similar properties to iron; and chalcophiles, which have a strong affinity with sulphur.

---

[3] The chemical formula of magnesian olivine, known as forsterite, is Mg_2SiO_4.

familiar element carbon has three known isotopes. The most common isotope of carbon has a mass number of 12 and is denoted as ^{12}C; its nucleus contains 6 protons and 6 neutrons. Another isotope of carbon, ^{13}C, has 6 protons and 7 neutrons. Finally, ^{14}C has 6 protons and 8 neutrons. Another example is uranium 238 (^{238}U), whose nucleus contains 92 protons and 146 neutrons; its sister isotope ^{235}U has 92 protons and 143 neutrons. Isotopes will be referred to frequently in this book because they are widely used to date rocks and to trace geological and biological processes.

A bit of physics: radioactive decay

Many isotopes are stable, meaning they will always retain their allotment of both protons and neutrons. Radioactive isotopes, on the other hand, have unstable atomic nuclei that are not stable.[4]. These isotopes tend to transform, or decay, into another isotope, at a rate determined by a probability specific to each radioactive isotope. When we say that an element is highly radioactive, it means that this probability is relatively high.

Radioactive decay always produces energy, accompanied by the emission of radiation. The isotope resulting from this decay is referred to as '**radiogenic**'. In more familiar terms, the radioactive isotope is commonly known as the parent isotope, and the derived radiogenic isotope is called the daughter isotope. Sometimes the daughter isotope is itself radioactive and will decay in turn. A succession of multiple linked radioactive isotopes is known as a decay chain. For example, radioactive uranium transforms into radiogenic lead through a chain of about 10 radioactive intermediates. The final lead daughter product is itself a stable isotope.

The law of radioactive decay, first elaborated by the New Zealand physicist Ernest Rutherford, states that the number of radioactive atoms generated by an initial stock of radioactive atoms (N) is governed by time and a constant of proportionality (λ), known as the decay constant.[5] The graphical representation of this law is an exponential decay curve (Figure 1.9). The decay is very slow in the case of weakly radioactive isotopes, which have a correspondingly small decay constant. This is the case for both uranium-238 and uranium-235, though decay of the former is especially slow. Both ^{238}U and ^{235}U generate radiogenic lead, ^{206}Pb and ^{207}Pb, respectively, the abundance of which therefore increases with time at the expense of uranium. The most common lead isotope on Earth is ^{204}Pb, which is neither radioactive nor radiogenic.

The length of time after which the number of radioactive atoms present will have decreased by half is called the **half-life** of the radioactive isotope.[6] The radioactive decay curve shows that, after ten radioactive half-lives, the number of radioactive atoms will have become about one thousand times smaller. After even more time, the

---

[4] Radioactivity was discovered in 1896 by the French physicist Henri Becquerel.
[5] This law is written: $\Delta N = -\lambda N \Delta t$.
[6] Decay constant (λ) and half-life ($t_{1/2}$) are related by the relation: $\lambda = \ln 2 / t_{1/2}$.

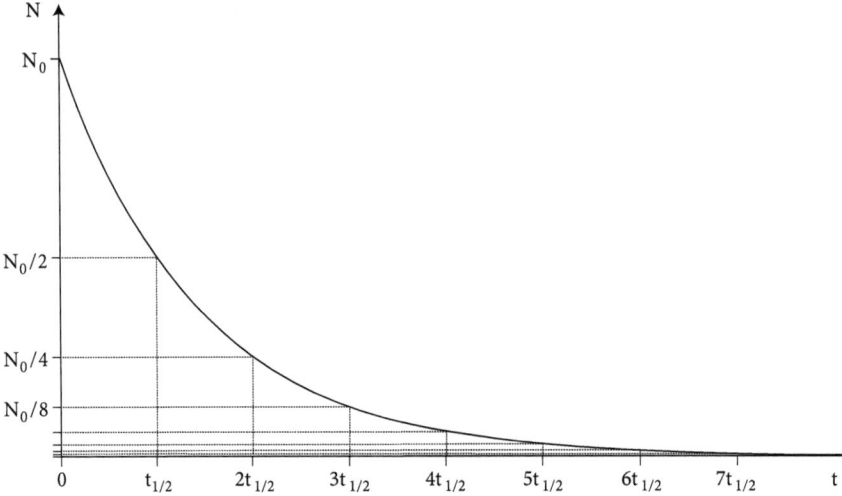

Figure 1.9 Graphical representation of the radioactive decay of a stock of radioactive isotopes (N) as a function of time (t). The quantity of parent isotope decreases by half after a time $t_{1/2}$ which is known as the half-life of this isotope.

quantity of radioactive atoms remaining will be so small that the original radioactive isotope can be regarded as extinct. This is the case for very short-lived isotopes that decayed rapidly after the formation of the Earth. Short-lived isotopes are useful for inferring the timescale of processes that occurred early in the history of the solar system.

How old is the Earth?

The determination of radioactive and radiogenic isotopes is a reliable chronometer, provided that the decay constants are known. Today, the instruments and techniques available for measuring the abundances and ratios of isotopes are so powerful that it is possible to determine geological ages with remarkable precision. The role of the geologist remains fundamental in selecting appropriate and undisturbed samples and interpreting the significance of the **radiometric ages.**

Very long-lived radioactive isotopes are used to date distant events in the Earth's history. This is the case for uranium-238, which is the most abundant uranium isotope today (more than 99% of natural uranium) and has a half-life of 4,467 million years. Uranium-235, which is much less abundant (less than 1% of natural uranium), has a shorter half-life (704 million years), which explains its low abundance today.

In 1956, the American geochemist Clair Patterson dated a set of five meteorites—three chondrites and two iron meteorites, including the Canyon Diablo meteorite—by determining their abundance of radiogenic lead isotopes formed by the decay of the two uranium isotopes. The results, when plotted with each of the radiogenic isotopes divided by the abundance of stable ^{204}Pb isotope, define a straight line (Figure 1.10),

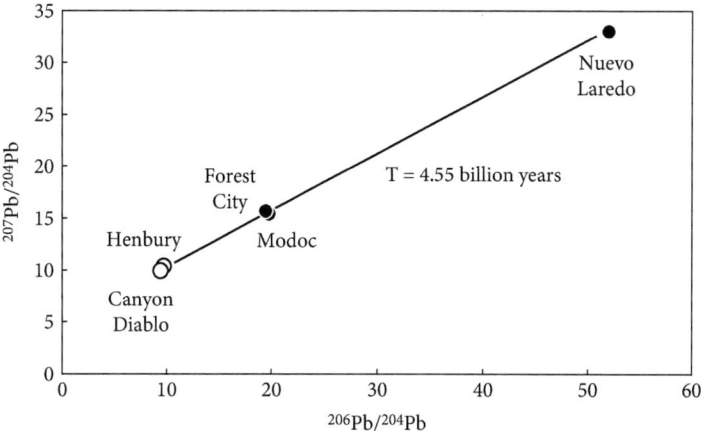

Figure 1.10 The age of the solar system according to Patterson (1956). The analysed samples come from five different meteorites (Canyon Diablo, Forest City, Henbury, Modoc, and Nuevo Laredo) and are aligned along a straight line (the so-called isochron), because they have the same age and originate from the same source, namely the protoplanetary nebula. The age is calculated from the slope of the isochron, a result derived from the law of radioactive decay of uranium isotopes.
Source: Adapted from Allègre et al. (1995).

proving that all the analysed meteorites have the same age and originate from the same source material, namely the protoplanetary disc. This line is called an 'isochron', and according to the law of radioactive decay applied to uranium isotopes, the age can be calculated based on the slope of the line.[7]

Patterson made the logical assumption that all meteorites were formed at the same time, along with Earth, at the very beginning of the formation of the solar system. Patterson's isochron yields an age of 4.55 (± 0.07) billion years (Ga). This is the age of the meteorites, and inferentially, the age of the Earth as well. To fully appreciate the importance of this first accurate determination of the age of the Earth, it should be remembered that, at the beginning of the 20th century, the age of the Earth was estimated to be about 100 million years, based on calculations made by Lord Kelvin. Early attempts to date rocks based on Rutherford's law of radioactivity produced ages older than one 1 billion years, which indicated that the Earth must have been at least that old, many times older than Kelvin's estimate. Patterson's work went further, and remarkably, his estimate of the age of the Earth has endured the test of time, despite massive technological advances in radiometric dating. The inquisitive reader may wonder why geologists have not simply dated the age of the Earth directly. The reason is that no terrestrial rocks are known to date back to the formation of our planet.

---

[7] The equation of the isochron line results from the integration of the relationship mentioned in the note above. The equations from the decay of each of the uranium isotopes are combined, so that the initial lead contents (which are unknown) are not taken into account. The slope of the isochron line is: $(1/138) \times [(e^{\lambda^{235U}} T-1) / (e^{\lambda^{238U}} T-1)]$, where T is the age of interest and 1/138 is the current ratio of the uranium isotopes $^{238}U / ^{235}U$.

Since Patterson's pioneering work, knowledge about meteorites has deepened. A small number of meteorites (less than 5%) are found to contain carbon, the building block of life on Earth. The first discovery of organic matter in a meteorite dates back to the analysis of the Orgueil chondrite, which fell in 1864 in the village of Orgueil in Tarn-et-Garonne in south-western France. Over the years that followed, an intense discussion on the origin (biological or not) of this organic matter ensued. On 8 February 1969, a very large meteorite was observed entering the atmosphere as a fireball in northern Mexico, between Chihuahua and Durango. The bolide fragmented in the atmosphere, and the pieces were scattered over an area of 300 km² around the small village of Allende. Many fragments of the meteorite were collected from this fortuitously semi-arid region (which made it easier to find them), resulting in a harvest of more than two tonnes of meteorite, much of it destined for geochemical study. The Allende meteorite is the largest carbonaceous chondrite identified to date. It contains whitish, multi-millimetre mineral inclusions (Figure 1.5). These inclusions are formed by **refractory** minerals rich in calcium and aluminium. Isolation of these inclusions and radiometric dating using the uranium–lead system generated ages of around 4.55 Ga and slightly older. These fragments of the Allende meteorite are the oldest known objects in the solar system, and the oldest of them (4,567 Ma) provides the best estimate for the age of the solar system, which is now thought to have preceded the formation of the Earth by about 12 million years (Amelin et al., 2010).

Formation of the Earth's core and mantle

Let's return to Earth and try to find out more information about the beginnings of our planet, and in particular the stage of **planetary differentiation** when the core separated from the mantle. How did this process occur, and how long did it take?

The formation of a core in a planetary embryo depends on temperature, as the process requires melting. The heating of a growing planet has two main sources: impacts with other objects and radioactive decay. Bolides that hit a planetary body have varying masses, and speeds of between 10 and 20 km/second. The corresponding energy, which is almost entirely converted into heat on impact, is very high.[8] In the early solar system, impacts were relatively frequent and could involve objects like Vesta or even larger. It was therefore easy to achieve partial or complete melting of the surface of incipient planets after a large impact. However, the effect was transient, and the surface of the forming planet would rapidly solidify, until the next impact.

More important was the role played by radioactive decay, and in particular the decay of long-lost, short-lived isotopes such as iron-60 (^{60}Fe) or aluminium-26 (^{26}Al), that have half-lives of less than one million years. They differ from the common isotopes of iron and aluminium, ^{56}Fe and ^{27}Al. Most elements are formed in the stars by fusion reactions of hydrogen and helium, which are the main constituents of stars: this

---

[8] The relationship between the kinetic energy E, the mass m, and the velocity V of the object is: $E = 1/2\, m\, V^2$.

process is called **nucleo-synthesis**. Stars resembling our sun mainly produce light elements such as carbon and nitrogen. The heavier elements are the product of explosive nucleosynthesis; in other words, they are formed in massive stars (at least ten times more massive than the sun) at the end of their lives, when they explode spectacularly—to the point of sometimes being visible in broad daylight in the sky.[9] These exploding stars are called supernovas. The testament to this vanished early radioactivity is in the radiogenic isotopes they produced, namely nickel-60 and magnesium-26, the respective daughter isotopes of ^{60}Fe and ^{26}Al, that were present in the protosolar nebula. In addition to the heat released by the decay of the short-lived isotopes, the longer-lived isotopes still present today such as ^{235}U were much more abundant at the very beginning of the Earth's history. In short, radioactive decay is the main source of internal heat in our planet.

Due to the heat produced by impacts and radioactive decay, a planet as small as 1,000 km in diameter would start to melt. What is the melting temperature of a young planet with a chondritic composition (i.e. composed of silicates and metallic iron)? To answer this question, we must bear in mind that a rock, which is an assemblage of minerals, begins to melt at a lower temperature than each of its constituent minerals considered individually. This property is well-known by glassmakers, who add fluxes (additional compounds) to melt silica sand at a lower temperature than 1,713 °C, the melting temperature of pure silica at atmospheric pressure. Furthermore, the melting of a mineral mixture takes place over a wide temperature range. For example, the silicate fraction of chondrites, consisting of olivine and some other minerals, begins to melt at around 1,300 °C at atmospheric pressure, but does not melt completely until around 2,000 °C. Pure iron melts at 1,538 °C, but the iron in meteorites, which contains a significant proportion of nickel and some sulphur, has a lower melting point. At about 1,500 °C on the surface of a forming planet, both iron and silicates begin to melt, producing two very different liquids that do not mix.

The denser ferrous liquid will migrate towards the core of the planet (Figure 1.11). This is the initiation of **planetary differentiation**, a process that leads to the formation of the planet's iron core. The silicate liquid, which is called a **magma** (i.e. a bath of molten silicates that may contain unmelted crystals), remains on the surface, where it forms a magma ocean. The dense ferrous liquid accumulates rapidly at the base of the magma ocean and then continues to advance towards the interior of the planet by a process of slow percolation. The outer part of the Earth's core is still made of liquid iron. The inner core, which has crystallized out of the liquid iron, is solid, even though it is the hottest part of the planet. Complete solidification of the core will occur in the distant future.

By contrast, solidification of the silicate magma ocean began very early and proceeded from the bottom (i.e. the core–mantle boundary) to the top (i.e. the surface of the Earth). Indeed, the melting curves of silicate mixtures show a pronounced slope

---

[9] This is the case of the Crab supernova, observed by Chinese astronomers in the year 1054. They reported that the light remained visible for three weeks in daylight, and for two more years at night. The remains of the explosion are now the Crab Nebula.

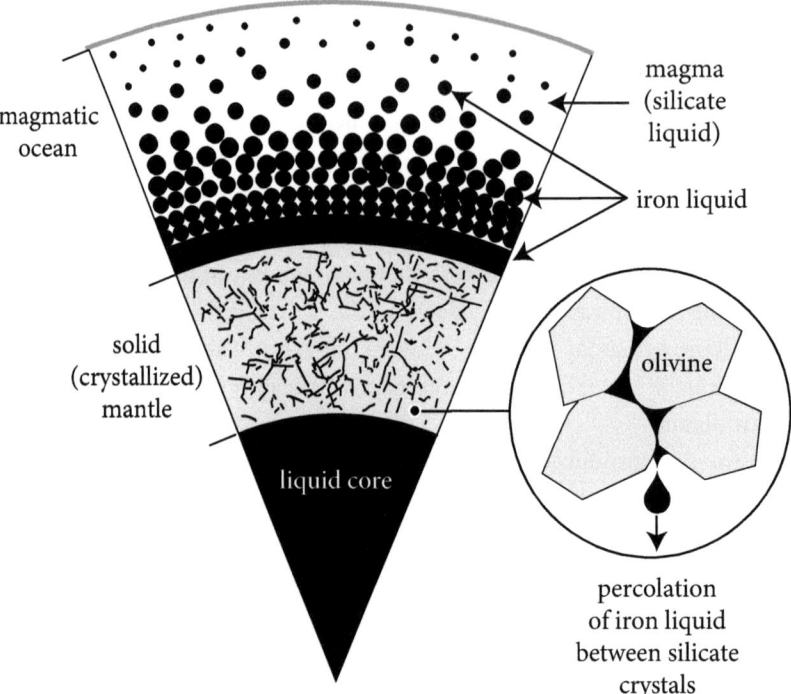

Figure 1.11 Separation of the Earth's core in a partially molten Earth. Note that at this early stage, the core is entirely liquid; solidification of the inner core began later, though exactly when remains unknown.
Source: Adapted from Shi et al. (2013).

because the effect of pressure (depth) counteracts the effect of temperature on melting, as we have already learned. In other words, the higher the pressure, the higher the temperature required to achieve melting. For peridotites, the typical mantle rocks, as for all silicate rocks, melting and crystallization occur over a range of temperatures. Two distinct curves, called '**solidus**' and '**liquidus**', are recognized. The solidus corresponds to the beginning of melting (or the end of crystallization) and the liquidus to the end of melting (or beginning of crystallization), depending on whether temperature increases (or decreases). At temperature and pressure conditions falling between these two curves, the mantle consists of a mixture of crystals and melt (Figure 1.12).

Complete melting of magnesian silicates requires a higher temperature than melting of metallic iron. In the interior of a differentiating planet, iron is therefore still liquid when olivine begins to crystallize in the magma ocean. **Pallasites**, arguably the most visually striking of meteorites, illustrate this situation (Figure 1.5). They contain olivine crystals that appear to float in (originally molten) metallic iron. A pallasite thus captures an image of the core–mantle boundary in a differentiating planet.

How long did it take to separate the iron core from the silicate mantle of our planet? Isotopes (again!) will provide information on the age and time of core formation. Siderophile elements such as nickel, cobalt, and tungsten (Figure 1.8) display great

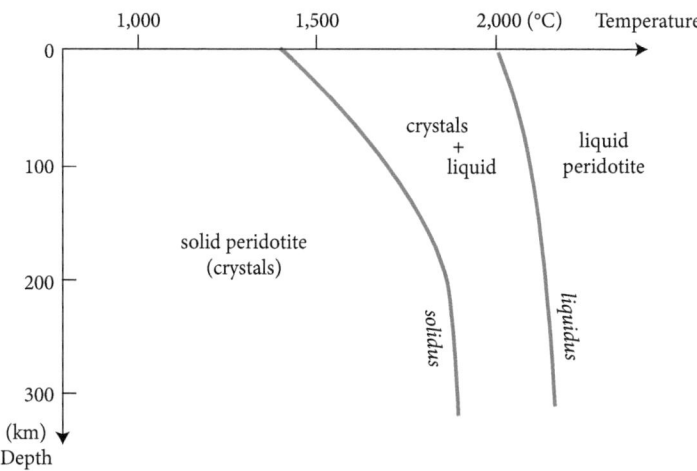

Figure 1.12 **Melting curves of a peridotite**, a rock formed of magnesian olivine and pyroxene, typical of the mantle. The solidus and liquidus frame a domain where melting is incomplete and where silicate liquid and crystals coexist.

affinities with iron. The lithophile elements, on the other hand, prefer silicates. The lithophile elements uranium, zirconium, and **hafnium** are present in very small quantities in minerals and rocks; for this reason they are called **trace elements**. Trace elements were distributed according to their respective affinity with ferrous liquid or silicate liquid during the differentiation process of the young planet. Hafnium (Hf) remained in the silicate liquid, while tungsten (whose elemental symbol, W, derives from the less-used term 'wolfram') migrated to the core with the ferrous liquid. However, an isotope of hafnium, ^{182}Hf, is radioactive and produces radiogenic tungsten, ^{182}W, an element whose affinities are very different from its parent isotope. The half-life of ^{182}Hf is only 9 million years, making it ideally suited to test the age of the core.

Due to its highly lithophilic nature, radioactive hafnium remained entirely in the mantle during mantle–core differentiation, but does not exist anymore due to its short half-life. Radiogenic tungsten, its daughter element, could have migrated into the molten iron as long as the formation of the core was not complete. If the core segregated very quickly, say in less than a million years, only a small amount of radioactive ^{182}Hf would have decayed at that time and most radiogenic ^{182}W would be preserved in the mantle. On the other hand, if core segregation exceeded 90 million years (more than ten times the half-life of ^{182}Hf), all radiogenic tungsten resulting from the decay of ^{182}Hf would have migrated with the ferrous liquid into the core and none would be found in the mantle. Analysis of terrestrial mantle rocks and comparison with undifferentiated chondritic meteorites provide an intermediate result between these two extreme cases. The mantle ^{182}W content (i.e. the so-called tungsten anomaly) indicates that the core was formed 30 million years after chondrite formation (i.e. 30 million years after the formation of the solar system), because hafnium, a refractory element, was contained in the early refractory inclusions of meteorites (Kleine et al., 2009). Planetary differentiation of the Earth would therefore have ended at around

4.53 Ga. Although useful, this calculation is somewhat simplistic as it considers that core formation occurred as a single event. Actually, the Earth's core grew at each addition of metal by impacts with large bodies.

Formation of the Moon

An unlikely event occurred early in Earth's history: the formation of the Moon. The Moon, with a diameter of around 3,500 km, is very different from the two small satellites of Mars, Phobos and Deimos, whose diameters are 22 and 12 km, respectively. Venus and Mercury have no satellites. The Earth–Moon pair is therefore unique. It turns out that the Moon plays an important role in stabilizing the Earth.

The Moon orbits the Earth at an average distance of 384,000 km (Figure 1.13), while rotating around its own axis in slightly over 27 days, always presenting the same face to the Earth. At the same time, it is slowly drifting away from the Earth, at a rate of a few centimetres per year. Its orbit is at an angle of about 5° to the plane of the ecliptic (the plane of the Earth's orbit around the Sun). At the time humans visited the Moon, between 1969 and 1972, its origin was being debated, with no consensus on how our satellite was formed. The hypotheses put forward up to that point, such as separation from the Earth or capture by the Earth, could not explain the Moon's size and orbit.

Lunar samples returned from the Apollo missions have shown that the Moon and Earth are closely related. Specifically, the stable oxygen isotopes ^{17}O and ^{18}O are present in the same proportions, in contrast to other bodies in the solar system, such as Mars (Figure 1.14). In addition, lunar rocks are totally anhydrous, meaning that they lack hydrated minerals and are depleted in highly **volatile** trace elements.

Another unique aspect of the Moon is that it is deficient in iron, relative to its size. Its small core is only 200 km in diameter. All the Moon's features can be explained by a hypothesis that that has emerged as the consensus since a conference on the origin of the Moon, which took place in Hawaii in 1984. The accepted scenario is that of a giant impact—that is, a collision between the Earth and a planet the size of Mars, which has been given the name Theia (Figure 1.15).

This scenario was proposed independently by two different teams who published within a year of each other, Hartman and Davis in 1975 and Cameron and Ward in 1976. But it took nine years for the scientific community to accept the giant impact

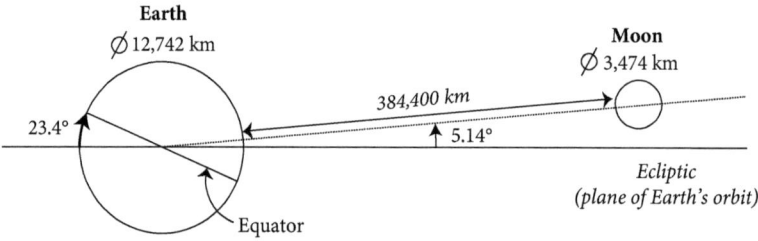

Figure 1.13 **Current characteristics of the Moon and its orbit.**

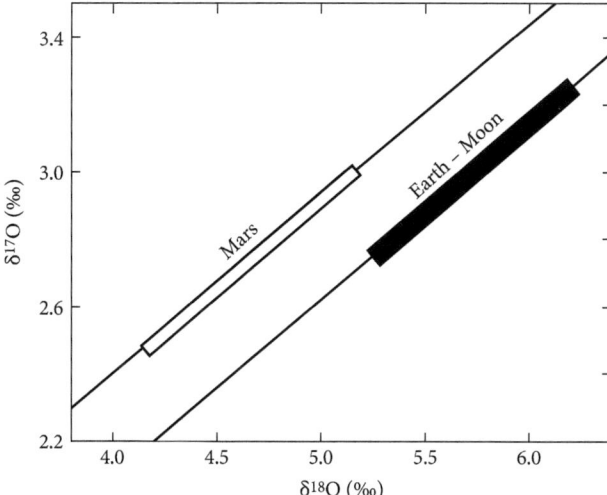

Figure 1.14 Comparative isotopic proportions of ^{17}O and ^{18}O. Terrestrial and lunar samples have the same isotopic signature, but differ from Martian meteorites; the notation *delta* (δ) indicates the deviation from a reference standard.
Source: Adapted from Pahlevan and Stevenson, 2007.

Figure 1.15 Drawing of the giant impact that would have created the Moon.
Source: NASA.

hypothesis. Although the model of planetary accretion by the addition of objects during impacts had long been accepted, it had not been thought that it could involve a body as large as another planet, probably because a planet in formation is expected to sweep its orbit and agglomerate all the other material before a competing planet could grow.

It has been calculated that this giant impact would have generated temperatures of at least 4,500 °C, thus creating a magma ocean over the whole surface of the Earth. A large part of the impactor would have volatilized, creating a cloud of solid fragments, magma droplets, and silicate vapour around the Earth. This cloud contained material mainly derived from Theia, but also from a fraction of the Earth's mantle. Alastair Cameron, a Canadian astrophysicist and professor at Harvard at the time, developed the first numerical model of the collision between the Earth and the impactor Theia (Figure 1.16).

The two planets that collided had already differentiated into mantle and core. The model predicts that the impactor would have almost entirely pulverized and/or vaporized, but that part of its core was incorporated into the Earth's mantle. The transit's timescale for the impactor metal to reach the Earth's growing core depends on whether the Earth's mantle was entirely molten or not. An accretionary disc of the pulverized

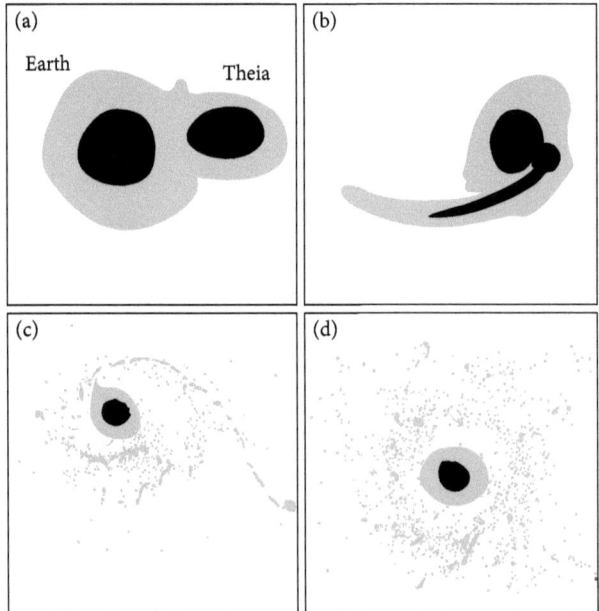

Figure 1.16 **Example of a numerical model of a collision between the Earth and a slightly smaller planet, Theia.** (a) The planets' mantles are in grey and the cores in black. (b) The impactor Theia is pulverized and partially vaporized by the impact. Some of the Earth's mantle is also expelled as debris, which joins the debris from Theia to form a disc orbiting the Earth—(c) and (d). The Earth's core is not changed. Note that the different diagrams are not to the same scale and that the Moon has not yet formed in (d).
Source: Adapted from Cameron, in Canup and Righter, 2000.

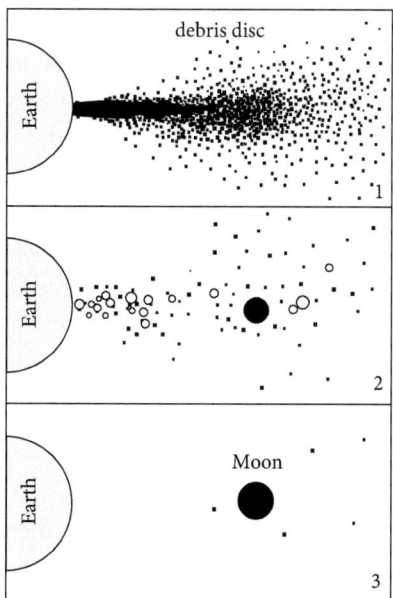

Figure 1.17 **Evolution of the debris disc orbiting the Earth after impact.** The Moon forms from debris at some distance from the Earth. The closest debris fall back to Earth.

and vaporized material formed around the Earth, from which the Moon was born (Figure 1.17).

The Moon is therefore not Theia. It is formed from fragments of Theia, but also from some terrestrial constituents. This model neatly explains the close kinship between the Earth and the Moon, but also some of the differences, in particular the case of the missing volatile elements on the Moon, which were lost in space just after impact because they were too light to be retained in the accretionary disc by Earth's gravity. The model also predicts that the Moon formed mainly from fragments of silicate (mantle) composition and contains relatively little ferrous material from the core of the impactor. Hence, the giant impact model explains all the observations concerning the Moon and lunar rocks.

A final key component of this model is that it implies that the Moon formed closer to the Earth than its present position, though with no precision in terms of how close. Only a minimum distance can be indicated, corresponding to the Roche limit, which is the distance below which a satellite would break up (or not be able to form) due to the tidal forces exerted by the Earth. This distance is about three Earth radii, or about 18,000 km. Since its formation, the Moon has been slowly inching away from the Earth. Precambrian nights must have been illuminated by a Moon that appeared much larger than it does in the sky today.

When did this giant impact occur? The oldest lunar rocks cluster around 4.36 Ga, which provides the minimum age of the Moon (Borg and Carlson, 2023). However, the age of formation of the Moon may be older than the crystallization of its oldest rocks. The duration of preservation of a magma ocean on the Moon is debated and ranges from 20 to 150 million yrs. Thiemens et al. (2019) suggest a Moon formation age of about 4.51 Ga from the analyses of the ^{182}Hf/^{182}W system in various Moon samples.

We will leave the question open, and return to the geology of the Moon in the next chapter. For now, let us consider the state of the Earth after the giant impact. All numerical models of the impact imply partial or total melting of the Earth's mantle as it shed heat delivered from the impact. The magma ocean which then blanketed the Earth gradually crystallized. With the exception of an early formed upper crust, solidification progressed from the inside to the outside due to the slope of the peridotite solidus, as shown earlier (Figure 1.12). The end of crystallization of the post-impact magma ocean marked the end of our planet's formation, but the Hadean—the first geological period of the Precambrian (between 4.5 and 4 Ga)—was just beginning.

> The Earth was formed 4.55 billion years ago by the accumulation of chondritic meteorites consisting of silicates and iron metal. Over the next twenty to fifty million years, the iron core and silicate mantle separated through extensive melting processes due to numerous impacts, culminating in the giant impact that created the Moon.

2
The mysteries of the first 500 million years

The earliest period in Earth's history (between 4.5 and 4 Ga) is the Hadean, whose name is derived from Hades, the god of the underworld in Greek mythology. Such a name suggests a period of hell and fire. Little is known about the Hadean because no rocks still exist that are this old. Nevertheless, there are reasons to believe that this period was not as hot and inhospitable as first imagined.

The oldest minerals in the world

Fortunately, a succession of extraordinary discoveries has lifted a corner of the veil on Earth's most ancient history. In 1983, Australian researchers published dates for isolated minerals, **zircons**, which are 4.18 billion years old (Froude et al., 1983). Zircon is a common mineral in granitic rocks, where it forms small crystals that are easily recognized under the microscope. It is a zirconium silicate with the formula $ZrSiO_4$, not to be confused with zirconium oxide or zirconia, ZrO_2, a fake synthetic diamond used in jewellery. In 1986, again in Australia, zircons 4.28 Ga were reported (Compston and Pidgeon, 1986). Finally, in 2001, Sam Wilde and colleagues dated zircons approaching 4.4 Ga. These precious witnesses to the Earth's most distant past all come from a small desolate region of Western Australia called the Jack Hills (Figures 2.1a, b).

How were these minerals dated? Zircon is unique in that it can accommodate some uranium in its crystal lattice. As it never initially contains lead, all the lead that forms in zircon therefore comes from the decay of uranium. The determination of the radiogenic lead isotopes contained in zircons provides the age of these crystals and by extension the age of the enclosing magmatic rock, where they crystallized. This is the principle of uranium–lead (U/Pb) dating on zircon, the most reliable and widely used method for determining the age of magmatic rocks.

The success of the Australians in the hunt for Hadean zircons can be explained by their technological advance in the field of isotope dating in the late 1970s, when they developed an instrument that allowed *in situ* analyses within a single crystal. Previously, analyses were carried out on samples consisting in a number of zircons that first had to be dissolved. The age obtained by this method then corresponded to an average age of the pooled zircons, potentially masking part of the individual history of the zircons.

The instrument invented by the Australians has an easy-to-remember name: SHRIMP, which stands for Sensitive High-Resolution Ion MicroProbe. In the

26 THE MYSTERIES OF THE FIRST 500 MILLION YEARS

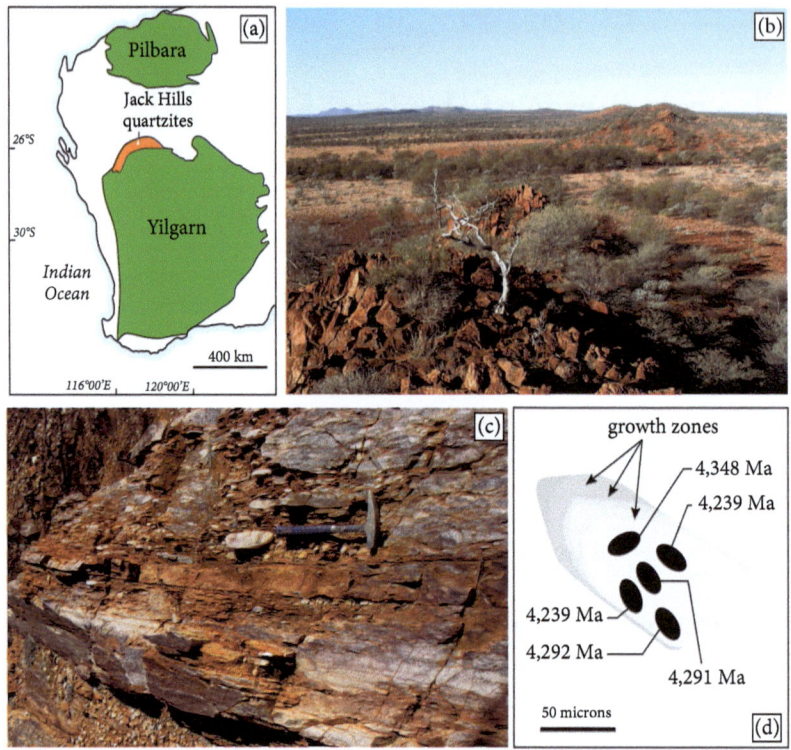

Figure 2.1 Jack Hills and the oldest zircons. (a) Location of the Jack Hills quartzites in Western Australia; the Pilbara and Yilgarn cratons (ancient terrains, shown in green). (b) Landscape of the Jack Hills. (c) An outcrop of quartzite, an ancient consolidated and metamorphosed sand, where Hadean detrital zircons were found (d) Section of a detrital zircon; black ellipses are SHRIMP-analysed zones in the core of the mineral and the *in situ* determined ages in millions of years (Ma); growth zones generally yield younger ages. This zircon has a history dating back to 4,348 Ma.
Source: (b, c) Sam Wilde. (d) Adapted from Cavosie et al. (2004).

SHRIMP, an ion beam strikes the sample, in this case an isolated zircon, and extracts a minute amount of material that is then analysed for its isotopes. The area analysed is between 5 and 30 microns in diameter. As the average zircon is generally just a few tenths of a millimetre long, several analyses are possible on a single zircon grain. This method can reveal very old ages in the core of some grains (Figure 2.1d).

Where did these zircons come from? They initially formed in granitic rock resulting from the crystallization of **magma** at depth. Much later, when a granite comes to the surface as a result of erosion, it is weathered and fragmented. Some minerals are chemically altered, such as **feldspars**, which become clays, while others, such as quartz, are preserved and can be transported further without modification. In addition to countless grains of quartz, a granite also contains a few grains of zircon, which is so resistant that it is commonly preserved even after the rock in which it formed has been destroyed by weathering. The zircon crystals inherited from the granite are called 'detrital zircons'. They retain the original age of the granite, even though the sedimentary rock that contains them is much younger. For instance, today's beaches of Brittany are an accumulation of quartz grains derived from the disintegration of granites that crystallized just over 300 million years ago.

The Hadean zircons from the Jack Hills are detrital zircons found in a sandstone. The sand derived from on older magmatic rock was first buried and turned into a **sandstone**, that was metamorphosed through heat and pressure to form another rock called a **quartzite**. Much later, erosion brought these quartzites to the surface, where they are easily identified by their pale colour (Figure 2.1c). Over a period of 20 years, more than 100,000 detrital zircon grains have been isolated from the Jack Hills quartzites and dated using SHRIMP. About 10 per cent of these grains have yielded Hadean ages, though the magmatic rocks where these zircons originally formed have never been found in outcrop. Detrital Hadean zircons have also been found in other parts of the world (Canada, China, Greenland, southern Africa), but the Australian database remains the largest in the world.

What can we learn from the Hadean zircons?

As the only evidence of the vanished Hadean crust, Hadean zircons attracted the attention of researchers. These minerals crystallized in magmatic rocks, but which ones? On Earth, there are two main types of magma: **basaltic** magmas, resulting from the partial melting of the mantle, and granitic magmas, whose origin often involves the partial melting of pre-existing crust. These two types of magmas correspond to the two types of crust found today: basaltic oceanic crust and granitic continental crust. Zircons are known to be more common in granitic magmas than in basaltic magmas. It is tempting to think that the same was true for the Hadean zircons, which would imply that they record an early granitic continental crust.

Isotopic analyses can identify the source, or **protolith**, of a magmatic rock, that is, the nature of the material that melted to form the magma. Magmatic processes do not alter the isotopic proportions of a given element, such that the **isotopic signature** of the source of the magma is conserved. Isotopic signatures are therefore tracers of the protolith. Two isotope systems are commonly used as protolith tracers in zircons: oxygen isotopes and hafnium isotopes.

The element oxygen is abundant in all silicates, including zircon, whose formula is $ZrSiO_4$. The most common oxygen isotope is ^{16}O. It accounts for 99.8% of Earth's oxygen. Two other isotopes, ^{17}O and ^{18}O, are present in very small quantities. The proportions of isotopes 18 and 16 are compared to a standard that serves as a reference. By convention, this standard is the actual sea water for which the proportions of $H_2^{16}O$ and $H_2^{18}O$ molecules are known; it is referred to as SMOW (Standard Mean

Ocean Water). Deviations from this reference composition, which are always small, are expressed in ‰ (per thousand) preceded by the notation delta ($\delta^{18}O$)[1]. They can be positive or negative, that is, the studied sample may contain oxygen that is slightly heavier (richer in ^{18}O) or slightly lighter (poorer in ^{18}O) than the standard. Peridotites in the Earth's mantle have a $\delta^{18}O$ of +5 to +6 ‰. The same is true for the basalts in the oceanic crust that are derived from them. In contrast, granitic rocks have oxygen isotope signatures between +6 and +12 ‰, showing that these magmas generally do not originate from the Earth's mantle, or at least not directly, unlike basalts. The highest values correspond to granites derived from the partial melting of ancient sediments. Most of the $\delta^{18}O$ values of Hadean zircons are between +6 and +7 ‰, although a few values as high as +9 ‰ have been found (Whitehouse et al., 2017). These $\delta^{18}O$ values demonstrate that the Hadean zircons crystallized in magmas that were not exactly basalts, nor granites either! Furthermore, the somewhat high values suggest that the source rocks of these magmas were in contact with liquid water. However, the question of the nature of the Hadean crust has not yet been resolved and is still a matter of debate. We'll come back to this point later.

The isotopes of hafnium provide additional information. Hafnium (Hf) is a very rare element on Earth and its isotopes have been measured only recently. Hafnium is found just below zirconium (Zr) in the periodic table (Figure 1.8). It has the same chemical properties as zirconium and replaces it locally in the zircon crystal. The very small amount of hafnium present in a rock is entirely contained in its zircon crystals. There are many isotopes of hafnium. One was mentioned in the previous chapter: ^{182}Hf, a radioactive isotope that is now extinct. Here, we consider the radiogenic isotope ^{176}Hf, which is the product of the radioactive decay of lutetium 176 (^{176}Lu). Lutetium is an extremely rare element from the group of so-called rare-earth elements or lanthanides i.e., elements from lanthane (La) to lutetium (Lu), see Figure 1.8, known to the general public since some of them are used in the manufacture of mobile phones. Lutetium, however, has no industrial utility and is the rarest of the rare-earth elements. It is so scarce that precise measurement of its isotopes has only been possible since the 1990s.

Let's go back to hafnium in Hadean zircons. The isotopic ratio $^{176}Hf/^{177}Hf$ is a tracer of the **mantle** or **crustal** origin of the magmas (Blichert-Toft and Albarède, 1997). All but the oldest Hadean zircons have a crustal signature, as if the Hadean crust had undergone repeated episodes of melting after its initial extraction from the mantle, without significant input of new mantle magmas, or so-called juvenile (i.e. mainly basaltic) magmas. This evolution dominated by **crustal recycling** is interrupted at the beginning of the Archean, when magmas with a mantle signature reappear in abundance (Figure 2.2).

This is very surprising as it suggests that during the Hadean, there was no magma production directly from the mantle. This observation, based initially on the analysis of

---

[1] Definition: $\delta^{18}O$ signature of the sample = {[($^{18}O/^{16}O$)sample / ($^{18}O/^{16}O$)SMOW] -1} × 1000. As the measured deviations are very small, multiplying by a thousand has the advantage of giving figures of a few units (instead of a few thousandths), which are easier to handle and remember.

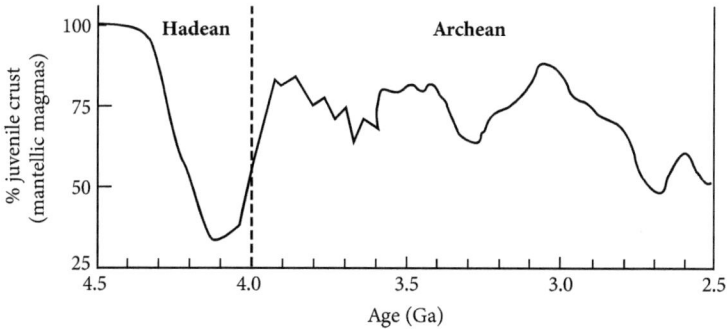

Figure 2.2 **Percentage of zircons indicating a juvenile hafnium isotopic signature**, that is, crystallized in magmas of mantle origin, compared to zircons indicating crustal recycling. There is a break and change just after 4 Ga, suggesting the disappearance of the Hadean crust and the contribution of mantle-derived magmas in the early Archean.
Source: After Belousova et al. (2010).

Australian zircons, suggested a sampling bias. However, the non-Australian examples analysed subsequently follow the same pattern. Therefore, it seems that a global change involving the Earth's mantle occurred at the Hadean–Archean transition some 4.0 Ga ago; an adequate explanation remains to be found.

These isotopic data raise several questions. The first concerns the composition of the Hadean crust. The second addresses the nature of the crustal recycling process that produced the magmas in which Hadean zircons crystallized. The third considers why juvenile magmas were absent during the Hadean. And finally, why did this Hadean crust disappear? That is a lot of enigmas for the Hadean. To gain additional clues, we will now turn our attention to the Moon, as our satellite has not suffered the tectonic modifications and surface weathering that recycle rocks on Earth, and as a consequence, it still preserves its oldest rocks.

The oldest lunar rocks

While rocks of Hadean age are unknown on Earth, this is not the case for the Moon. Viewed with the naked eye or binoculars, the visible side of the Moon shows two types of terrain: light areas and dark areas. The dark regions were called 'mare', plural 'maria' (from the Latin word for sea), which is a misnomer because there is no water on the Moon (Figure 2.3). The lighter areas are highlands, which loom above the maria. The entire lunar crust is heavily cratered, the highlands even more so than the maria, which means that the maria are younger. Note that the far side of the Moon has no mare. Between 1969 and 1972 (the golden age of lunar exploration), the Apollo missions enabled astronauts to walk on the Moon. Like good geologists, they returned about 400 kg of lunar rocks to Earth, and these have been the subject of extensive studies.

Two types of Moon rock have been identified: basalts and **anorthosites**. Mare basalts are the result of massive outpourings of lava; they attain thicknesses of up

Figure 2.3 The visible face of the Moon with the location of the American Apollo missions (which brought back about 400 kg of samples) and Russian Luna missions (which brought back only a few grammes).
Source: NASA.

to 5 kilometres. Anorthosites are found in the highlands. They are pale grey magmatic rocks that are very rich in **plagioclase feldspar**. Plagioclases are very common silicate minerals, whose composition varies from one endmember, albite, that is rich in sodium, to another endmember, anorthite, that is rich in calcium. Logically, an anorthosite is rich in calcic plagioclase. Anorthosites are also known on Earth, though they are relatively rare and mostly Proterozoic in age. By contrast, anorthosites are the most abundant rocks on the Moon. And they happen to be Hadean in age.

The Moon is covered by a blanket of dust and rock fragments of various sizes, due to the action of asteroid impacts throughout its history. Without drilling equipment, American astronauts only collected rock samples easily available on the lunar surface. Thus, all samples brought back by Apollo missions were rock fragments scattered by previous meteorite impacts. Many of them are brecciated. This is the case of the rock shown in Figure 2.4, known as sample number 60,025. This piece of anorthosite is the oldest lunar rock dated so far: 4.367 Ma old (Connelly et al., 2011).

Later, an age of 4.420 Ma was obtained from an isolated zircon from another rock sample dated at 4.331 Ma (Nemchin et al., 2009). In contrast, the mare basalts have ages ranging from 3.8 to 3.3 Ga. No younger volcanic rocks have been identified. Thus, the lunar crust consists of Hadean anorthosite highlands and Archean mare basalts.

The Apollo astronauts also installed seismometers that revealed the interior structure of the Moon (Figure 2.5). The seismic recordings showed that the lunar core is

Figure 2.4 Lunar oldest rock. (a) Photo taken in 1972 by the Apollo 16 crew in the Descartes region; the rock fragment indicated by the arrow next to the exploration vehicle track is anorthosite 60,025. (b) The same anorthosite sample photographed upon arrival in a laboratory in Houston, Texas.
Source: NASA—Lunar Sample and Photo Catalog.

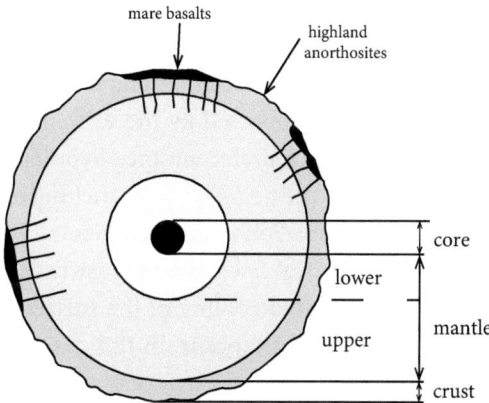

Figure 2.5 Cross-section showing the structure of the Moon, with a thick anorthositic crust and a small core. Lunar radius: 1,737 km.

small, whereas the lunar crust is thicker than the Earth's crust. Finally, the lunar mantle has a rigid upper part and a more deformable lower part.

The thick anorthositic crust is thought to have crystallized from the magma ocean that formed following the accretion of the Moon. The first mineral to crystallize out of the lunar magma ocean would have been olivine, which sank and accumulated at the base of the magma ocean because it was denser than the magma. Olivine can contain magnesium and iron, but the early crystallized olivine is always rich in magnesium. As a consequence, the remaining magma becomes progressively richer in iron and lower in magnesium, hence denser. Plagioclase that crystallized later was less dense than this magma, such that floating plagioclase crystals accumulated at the surface of the crystallizing magma ocean, where they formed the anorthosite crust of the highlands. Lunar anorthosites are therefore direct evidence of the crystallization of the initial magma ocean, but they formed only after 70–80% of crystallization of the lunar magma ocean, whose initial age remains undated. More specifically, they are evidence of a process called **fractional crystallization** which separated minerals formed during the progressive crystallization of the magma. In the case of the Moon, a unique type of fractional crystallization occurred, with sinking of dense olivine and pyroxene and floating of late plagioclase.

Comparison of the Moon and the Earth

It is tempting to apply the model of the crystallization of the lunar magma ocean to the Earth magma ocean. Unfortunately, this is not possible, as the pressure conditions were different. Earth is larger than the Moon, such that the sequence of crystallization and fractionation was also different. In the Earth's magma ocean, plagioclase could not have formed a thick anorthositic crust comparable to the Moon's crust. Therefore, the crust of the lunar highlands is not an analogue of the primitive Hadean crust on Earth.

In the absence of Hadean terrestrial rocks, we are left to speculate on the nature of the Earth early crust, and there is no consensus among the experts. Many colleagues favour a basaltic primitive crust, as supported by the existence of a minor group of meteorites, the basaltic **achondrites**. I prefer another hypothesis. Since the Earth's early magma ocean corresponded to the near-complete melting of the mantle, its composition was that of a liquid peridotite. Following the model of a lava lake, such as the Kilauea lava lake in Hawaii (Figure 2.6 a–c), the very early crust would have been a chilled upper margin formed by rapid cooling of the surface of the magma ocean before any fractional crystallization could occur. In this case, the first crust had the same composition as the parental magma (i.e. a peridotite composition). Then, fractional crystallization proceeded with accumulation of olivine crystals at depth and formation of a residual basaltic magma layer below the chilled crust (Figure 2.6d).

Another important difference between the Earth and the Moon is the presence of abundant water on Earth. Liquid water existed on Earth in the early Hadean eon, as we'll see in the next chapter. The primitive Earth's peridotite crust must have turned into **serpentinite** when it came into contact with water, as occurs today when the Earth's mantle interacts with seawater, as it does for example off Portugal. Serpentines are hydrated minerals formed at less than 500 °C by hydrothermal alteration of olivine. The surface of Earth's early crust thus cooled and became hydrated before the entire magma ocean finished crystallizing (Figure 2.7). Reynard et al. (2022) calculated that rafts of a 50 km-thick fully serpentinized crust may have survived for the entire Hadean, owing to the low density of serpentine.

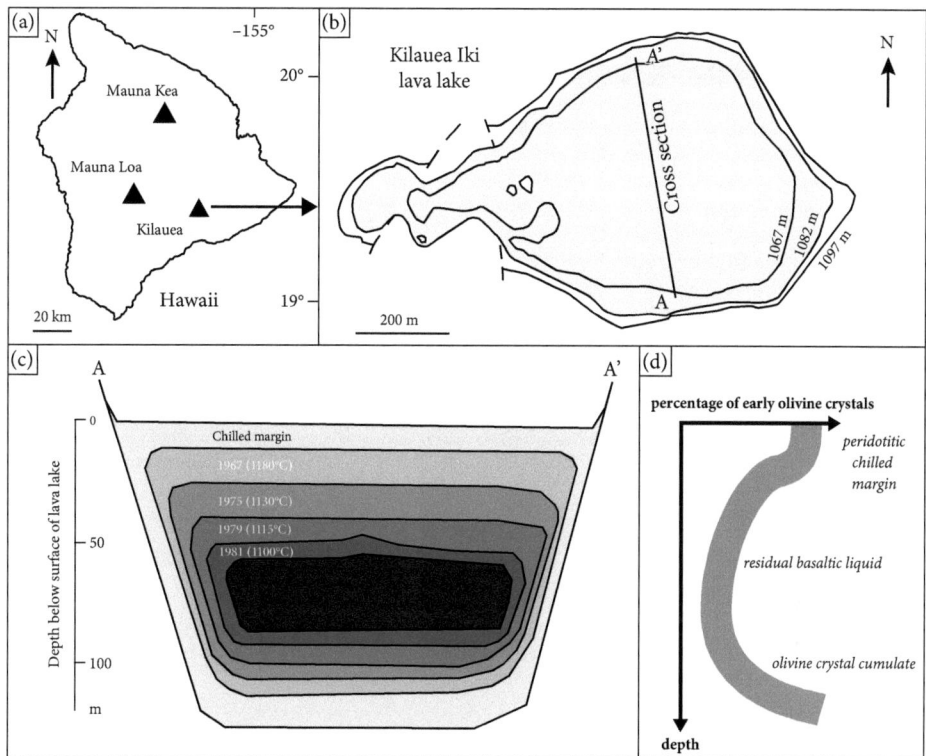

Figure 2.6 **The Kilauea Iki lava lake (Hawaii).** (a) Location of Kilauea volcano. (b) Map of Kilauea Iki lava lake. (c) Vertical section of the crystallizing lava lake. (d) Compositional variation of drilled samples versus depth in the centre of the crystallized lava lake.
Source: Adapted from Helz (2009).

The last liquids remaining from progressive crystallization of the Earth's magma ocean were probably basaltic in composition as a result of ongoing fractional crystallization of the magma beneath the chilled crust. These basaltic magmas could have intruded into the serpentinized primitive crust and reacted with it. Anastasia Borisova experimentally reproduced these postulated interactions under Hadean conditions in the GET laboratory at Toulouse University. She found that, at a depth of around 5 kilometres and a temperature of around 1250 °C, basaltic magmas can induce partial melting of a serpentinized peridotite producing a small volume of tonalitic magma. A **tonalite** is a rock of so-called intermediate composition (namely intermediate between basalt and granite), that resembles a plagioclase-rich granite. Borisova and colleagues further modelled the evolution of their experimental liquids, and they were able to crystallize model zircons whose trace-element and oxygen isotope signatures resemble the Jack Hills zircons. This new model provides the best hypothesis so far to explain the formation of the Hadean zircons and the origin of the crust in which they crystallized (Borisova and Nédélec,

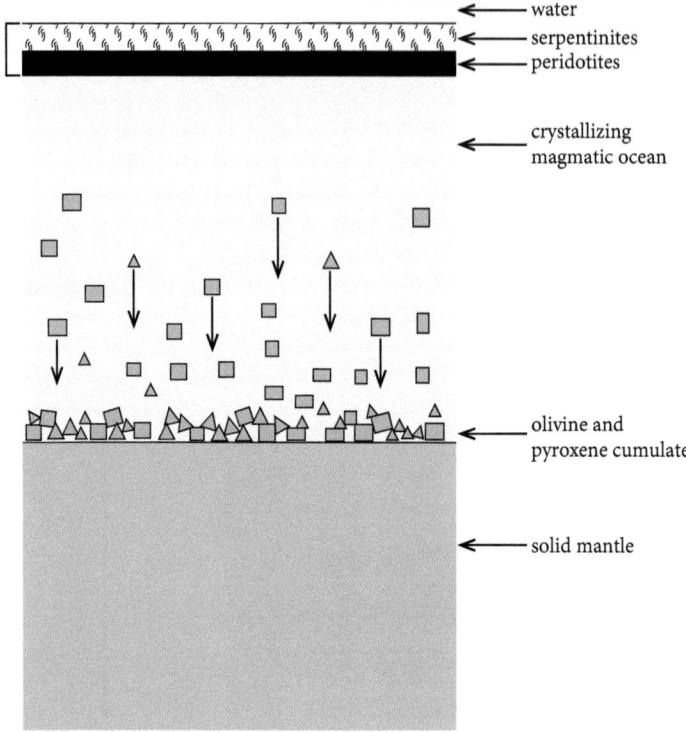

Figure 2.7 Formation of the early Earth's crust. The crust represents the upper part of the magmatic ocean, rapidly cooled and modified by contact with water.
Source: After Hervé Martin, adapted from Albarède and Blichert-Toft (2007).

2021). Recent advances calculate that this process could account for a mass of Hadean crust up to 50% of the current continental mass (Bernadet et al., 2025).

What happened to this primitive crust during the Hadean? We know little about how primitive tectonics operated on the Hadean Earth. However, an efficient recycling process occurred at least episodically through asteroid impacts, which were more numerous at that time than in more recent periods.

Asteroid impacts

A recent database of terrestrial impacts identified 190 confirmed impacts to date.[2] Of course, evidence of many, if not most, impacts over Earth's history has disappeared, since, unlike the Moon's surface which has essentially been frozen since the formation of the basaltic mare, the Earth's surface is constantly being reworked. The largest known meteorite impact crater on Earth is the Vredefort structure in South Africa.

---

[2] See www.passc.net/EarthImpactDatabase/.

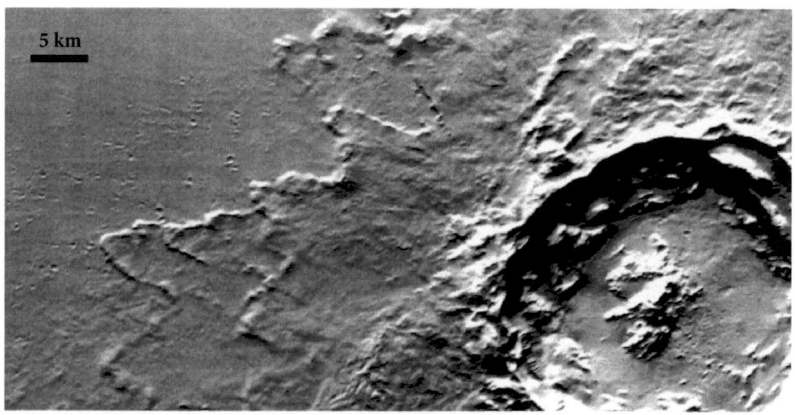

Figure 2.8 The complex impact crater Tooting, preserved on the surface of Mars.
This Mars Reconnaissance Orbiter (2006–2008) composite image shows the central peak, the concentric rings, and the ejecta (the debris ejected from the crater) blanket.
Source: NASA.

Although it has been partially eroded, the impact structure still forms a recognizable arc in the landscape and on maps.

The diameter of the Vredefort structure is about 130 km at the present level of erosion. The structure may have reached 300 km in diameter at the surface. It is a complex structure that resulted from the evolution of the transitional crater created by the impact. Indeed, in the minutes following an impact, the base of the hole excavated by the impacting meteorite sometimes forms a central peak, while the edges collapse along concentric faults (Figures 2.8 and 2.9). The diameter of the final structure is thus much greater than that of the starting crater, and much larger than that of the meteorite that caused it (15 km diameter in the case of Vredefort).

The age of the Vredefort impact has been determined from zircons crystallized in the magmas produced by the impact as a result of localized but very intense heating. It is 2,023 Ma old (Wielicki et al., 2012). The only known large impact structure on Earth older than Vredefort is a 3 Ga structure in Greenland with a diameter of 100 km (Garde et al., 2012). Beds of spherules (bead-like structures that form from impact-generated melts and vapours) in southern Africa and western Australia may represent the ejecta of unpreserved Archean impact structures. The reason for such a large observation gap is that Earth's surface is constantly rejuvenated by erosion, which erases or attenuates landforms. In addition, recent sediments may have covered old impact structures. Finally, plate tectonics deforms the Earth's crust and causes any oceanic crust older than 200 million years to disappear. This means that the chances of identifying very old impact structures on Earth are very slim. However, we can confidently

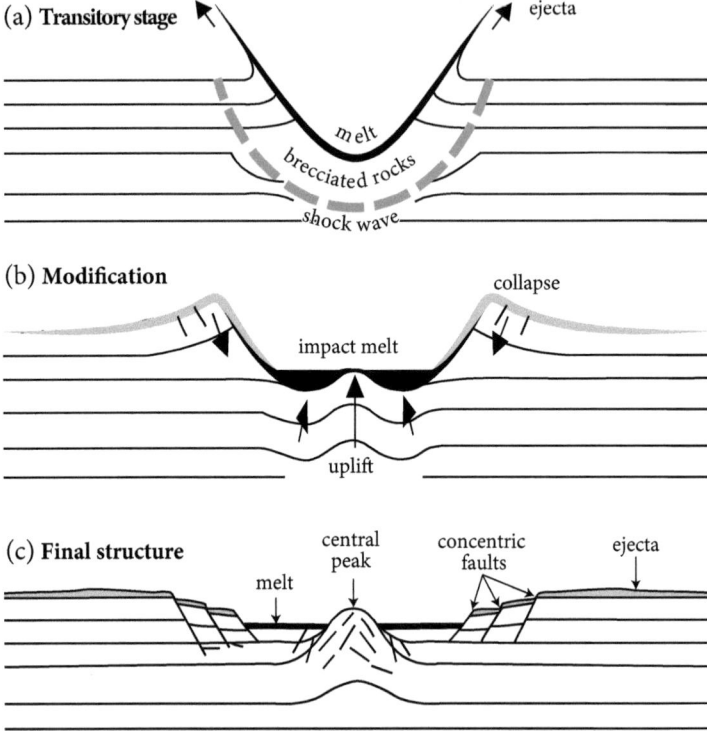

Figure 2.9 Formation of a complex impact crater. (a) The transitional crater after the initial collision. The initial bowl shape cannot be preserved if its diameter is larger than 3 kilometres. (b) Uplift of the core of the structure and collapse of the rim; impact magmas are shown in black. (c) Peripheral collapse continues, producing a final structure with a larger diameter than the transitional crater.

infer that large impacts affected the Hadean Earth surface, just as we see on the Moon (Figure 2.10).

The Moon has 1,700 impact craters with a diameter greater than 20 km, and several hundred thousand smaller craters. Fifteen craters are between 300 and 1,200 km in diameter, making them as large as or larger than the Vredefort structure. The young Earth must have been subjected to a bombardment that was at least as intense, but of which we no longer have any trace.

Many lunar samples were intensely fragmented (**breccias**) or partially melted by impacts. Indeed, under the effect of intense heating at the point of impact, the rocks can melt, producing a new magma. Because magmatic rocks can be dated, these impact melts provide an opportunity to date the impact events. Identification of the successive superimpositions and overlaps of impact structures, along with dates on some of these impacts, provide a chronology of the cratering of the lunar surface (Figure 2.10).

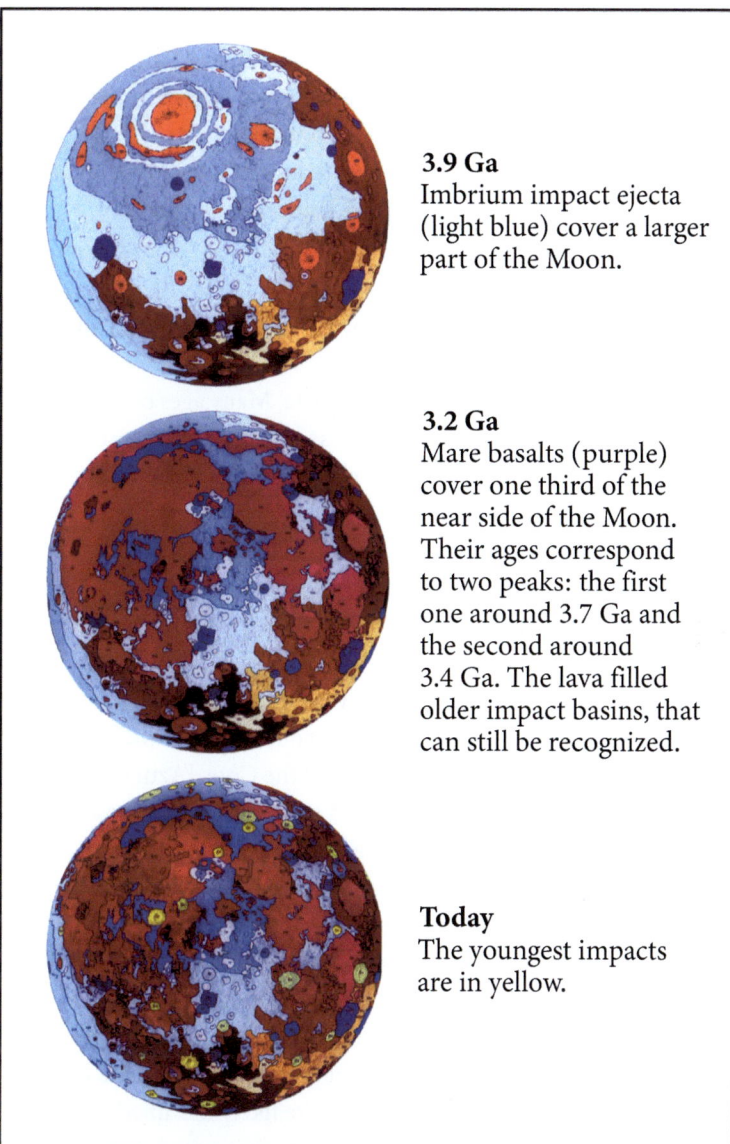

Figure 2.10 Geological maps of the Moon based on ages of samples collected by the Apollo missions and overlain by observations of impact structures. The three maps were obtained by successive removal of the younger impact craters and related ejecta deposits. The youngest impact structures are the smallest (bottom). We note that the ejecta from the Imbrium impact (top) covered a large part of the visible side of the Moon, including the regions sampled during the Apollo missions (compare with Figure 2.3). The Imbrium impact is dated at 3.9 Ga.
Source: Adapted from Wilhems (1987).

In the region of the *Mare Imbrium* ('Sea of Rains'), there is a huge structure called the 'Imbrium Impact'. This structure has an age of 3.92 Ga, according to recent radiometric dating of brecciated rocks attributed to this impact (Liu et al., 2012). The impact basin measures 1200 km in diameter, corresponding to an area that would encompass the whole of metropolitan France. The ejecta would have fallen more than 800 km beyond the edge of the structure. Today, the structure appears to be filled with mare basalts due to a major volcanic episode that occurred around 3.5 Ga, but its circular shape, annular at its edge, clearly identifies the impact structure (Figure 2.3). It is so large that it can easily be seen from Earth with good binoculars.

Evolution of the number of impacts over time

Dating of lunar impact breccias generated a debate. Many ages of these samples cluster around 3.9 Ga. This led researchers to propose a period of intense bombardment at this date. As this episode would have taken place more than 500 million years after the formation of the Moon, it is referred to as the Late Heavy Bombardment (LHB). This hypothesis is counterintuitive because we would expect meteorite impacts and protoplanet collisions to have been most numerous at the very beginning of the formation of the solar system, after which the number and intensity would have steadily decreased. Astronomers have attempted to account for the LHB with a scenario of migration of the giant planets in the solar system, which would have induced major perturbations of the orbits of the smaller inner planets and favoured collisions. This is the so-called Nice model, initially proposed by astronomers at the Côte d'Azur Observatory in Nice, France (Gomes et al., 2005). This model had some success because it accounted for the lunar chronology, but it is now challenged by new considerations and additional dating of lunar samples. A first criticism comes from the interpretation of ages concentrated around 3.9 Ga. It has been suggested that there is a sampling bias leading to an over-representation of Imbrium impact breccias. Indeed, the ejecta of this exceptionally large impact covered a vast area. If we compare Figures 2.3 and 2.10, we see that the objection must be considered seriously. Furthermore, as they had no drilling equipment, the astronauts picked the easiest samples to collect from the lunar soil (Figure 2.4), most of which were isolated blocks that were likely modified by impacts, including the Imbrium impact. Also, new radiometric ages generated with high-performance instruments have confirmed the existence of major impacts with ages greater than 4.0 Ga (Norman and Nemchin, 2014). At present researchers increasingly believe that the whole Hadean was a time of intense impacts and that the most intense bombardment was not confined to the end of the Hadean. Thus, the LHB hypothesis and Nice model have fallen out of favour and been replaced by the hypothesis that several successive major impacts occurred during the Hadean (Hopkins and Mojzsis, 2015).

This discussion has consequences for the interpretation of the end of the Hadean on Earth. It should be remembered that, with the sole exception of the Acasta gneisses which formed at the very end of the Hadean (see Chapter 4), no terrestrial Hadean rocks are known. On the other hand, the hafnium isotopic signature of Hadean zircons shows a change in the origin of magmas during the Hadean–Archean transition (Figure. 2.2). Several researchers proposed that the disappearance of the Hadean crust was a consequence of the LHB. However, this hypothesis clashes with the fact that the oldest lunar crust was preserved despite the LHB, contributing to the view held by many that the latter is now an obsolete hypothesis. Consequently, the Archean–Hadean transition and the corresponding change in magma sources must be interpreted differently. In 2017, I proposed an alternative hypothesis involving specific processes that took place in the Earth's mantle at that time, which will be discussed in Chapter 5.

The contribution of late impacts

The intense asteroid bombardment of Earth during the Hadean must have contributed significant material to Earth. This point is easily demonstrated by considering the abundance of the most siderophile chemical elements (i.e. those with a very strong affinity for iron). Specifically, platinum and its neighbouring heavy elements known as the platinum group elements (commonly abbreviated as PGE),[3] with which we can include gold. These elements, present when Earth initially formed, would have been concentrated in the iron liquid that migrated to form the core. However, although these elements are extremely rare (which accounts for their high value), they are still found in Earth's crust. Therefore, they must have been added to the Earth after core formation. The only possible contribution is from large meteorite impacts during the Hadean eon. This addition of chondritic material following core formation is referred to as 'the late veneer'.

> Earth does not preserve rocks of Hadean age in outcrop, unlike the Moon. Nevertheless, zircons dating from the Hadean have been found in younger detrital sediments. These zircons are the result from destruction of unknown Hadean magmatic rocks, whose nature and mode of formation are heavily debated. Recent experiments suggest that these Hadean protoliths formed by partial melting of a primitive crust with a serpentinized peridotite composition. The Hadean crust seems to have virtually disappeared at the Hadean–Archean transition.

---

[3] For example: palladium (Pd), iridium (Ir), and osmium (Os).

3

Origin of the atmosphere and ocean

The habitability of a planet—its capacity to host living beings—mainly depends on the presence of liquid water. In the solar system today, Earth is the only planet that meets this requirement, although Mars also did in the distant past. The presence of liquid water is possible because the distance from the Sun is neither too great, which is the case of icy giant planets, nor too small, as in the hot surface conditions on Venus and Mercury. However, based on the distance of the Earth from the Sun alone, the average temperature at the Earth's surface should be around −18°C. The average positive temperatures over much of the planet are due to the special composition of the Earth's atmosphere.

Composition of the current atmosphere

Viewed from space, Earth's present atmosphere appears as a thin blue line that envelops our planet. It consists mainly of nitrogen (N_2) and oxygen (O_2).[1] It also contains variable amounts of water in the form of vapour, as well as carbon dioxide (CO_2) and **rare gases** (e.g. **argon**, the most abundant of these) (Figure 3.1). At ground level, the pressure of the atmosphere is around 1 bar (1,000 millibars). All readers who pay attention to weather forecast are familiar with the 1,015 millibar limit that separates anticyclonic conditions from depressions! The lowest layer of the atmosphere is called the **troposphere** (Figure 3.2); this is where aircraft fly and where most weather events take place. The density of the Earth's atmosphere decreases upwards. At the top of the troposphere, at about ten kilometres above sea level, the pressure is slightly less than a quarter that at the surface. Above, the atmosphere becomes increasingly thin.

The amount of water in the atmosphere depends on the temperature. As the temperature decreases with altitude in the troposphere, water vapour condenses into fine droplets that form clouds and give rise to rain or snow. Above the troposphere is the **stratosphere**, where the ozone layer is found. Ozone forms from the dissociation of oxygen molecules (O_2) and rapid recombination of the released oxygen atoms (O) to form O_3. Ozone absorbs much of the Sun's ultraviolet (UV) radiation and therefore shields living beings that might otherwise suffer from possibly mutagenic or carcinogenic doses of UV radiation (Figure 3.2). The ozone layer was less critical in the Precambrian, since life presumably inhabited exclusively aqueous habitats. In fact,

---

[1] To conform to common usage, we refer to both the element (O) and the molecule (O_2) simply as 'oxygen'.

Earth and Life. Anne Nédélecc, Oxford University Press. © Originally published in French La terre et la vie - une histoire de 4 milliards d'années by Anne Nédélec © Odile Jacob, MMXXII; English translation © Oxford University Press (2025).
DOI: 10.1093/9780198945451.003.0004

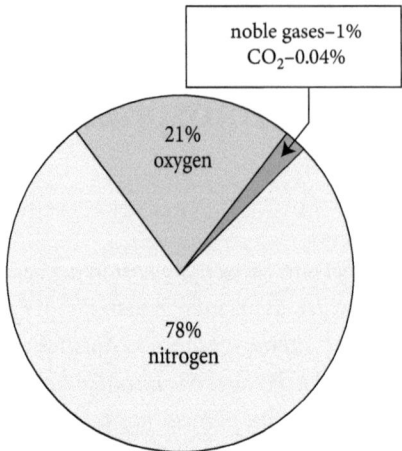

Figure 3.1 **Composition of the Earth's current atmosphere.**

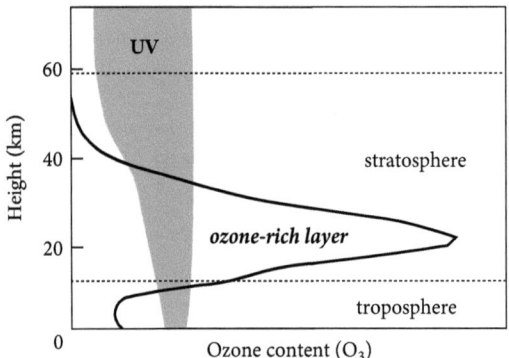

Figure 3.2 **Location of the current ozone (O_3) layer** at the base of the stratosphere.

the Archean atmosphere contained no oxygen, hence no ozone. The current stratospheric ozone layer is in a dynamic state, meaning it is constantly being destroyed and reconstituted. The hole in the ozone layer that appears during the southern hemisphere spring has a natural origin, and this ozone hole recedes during the southern winter. Chlorofluorocarbons (CFCs), such as freon,[2] which was used in refrigerators, prevent the recovery of the ozone layer and were therefore banned in 1987 by the Montreal Protocol. As a result of this international agreement, the ozone layer above Antarctica is progressively recovering.

Conditions at the Earth's surface allow for the presence of water in three states: solid, liquid, or gas (Figure 3.3). Liquid water is a prerequisite for the development of life. Liquid water is abundant on Earth where it makes up the oceans that cover more than two-thirds of the surface, hence the moniker 'blue planet'. If the volume of the world's oceans were evenly distributed over the surface of the globe, this would correspond to a layer of liquid water nearly three kilometres thick. We noted in Chapter 2 that

---

[2] An example is Freon-12 or dichlorodifluoromethane CCl_2F_2, a halogenated methane with no hydrogen. Photolysis of this molecule yields a long-lived reactive chlorine atom in the upper atmosphere, which catalyses the conversion of ozone (O_3) to oxygen (O_2).

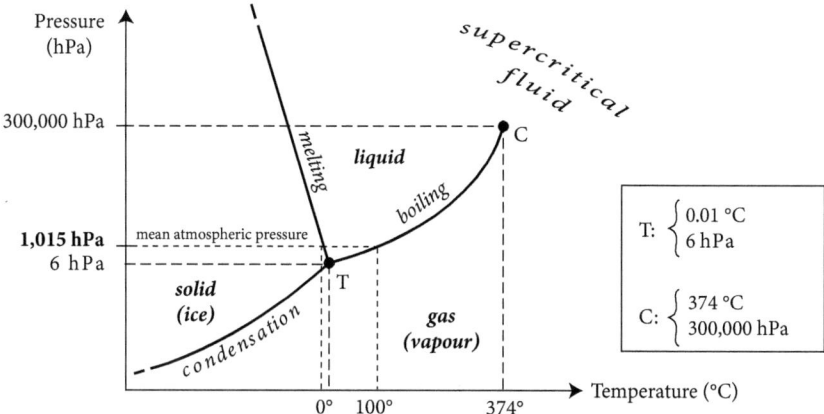

Figure 3.3 Diagram of the physical states of water as a function of pressure and temperature. At the Earth's surface, water can be observed in three states: solid (ice), liquid, or gas (water vapour), depending on temperature. However, at pressures below the triple point T (i.e. about 200 times lower than atmospheric pressure) there are only two possible states of water (solid or gas), as on the surface of comets. Weather forecasts give atmospheric pressure in millibars: 1 millibar = 1 hPa (hectopascal).

oxygen isotopes in the most ancient zircons indicate the presence of liquid water as early as the Hadean. However, the Hadean atmosphere must have been very different from that of today. So, what is the origin of the Earth's atmosphere? And how did it change over time?

Volatile components in the protosolar nebula

Let's go back to the origin of the solar system. The main constituents of the protosolar nebula are hydrogen (H) and helium (He). These elements together account for 98% of the mass of the universe, with hydrogen alone making up 78%. Next in order of abundance are oxygen (O), carbon (C), and nitrogen (N). Our star, the Sun, is made up mainly of hydrogen and helium. However, only the giant planets, such as Jupiter and Saturn, have an atmosphere rich in hydrogen and helium. These very light gases escaped the gravitational attraction of the small terrestrial planets.[3] Earth, therefore, lost its primordial hydrogen-rich atmosphere, inherited from the protosolar nebula.

In the interstellar medium, water is easily formed by reaction between hydrogen and oxygen atoms. At the low pressures of the protosolar nebula, which consisted of gas and dust, water could only exist as gas or ice (Figure 3.3). Other molecules that could have formed from the main elements in the protosolar nebula include methane (CH_4) and ammonia (NH_3). These two molecules are formed preferentially at very

---

[3] In addition to the influence of gravity, the escape of gases from the atmosphere depends on the speed of the particles (atoms or molecules). In uppermost atmosphere, the speed of particles is influenced by collisions between them, and particles that attain the **escape velocity** are lost to space. The solar wind also strips some particles from the upper atmosphere.

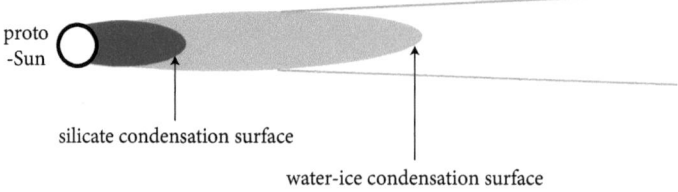

Figure 3.4 Condensation of the protosolar nebula.

low temperatures; at higher temperatures, the oxidized forms CO_2 and N_2 are more abundant (Encrenaz, 2010).

Under the effect of gravity, the protosolar nebula gradually contracted into a disc, while its centre simultaneously heated up, establishing a temperature gradient between the proto-Sun and the periphery of the disc. After the formation of the Sun, the protoplanetary disc gradually cooled. The constituents of the disc successively condensed during this cooling, meaning they transitioned from a gaseous to a solid state. The most refractory constituents (i.e. those that condense at the highest temperatures, over 1,300 °C) appeared first and were close to the Sun. They constitute the refractory mineral grains found in some chondrites and from which the age of the solar system has been obtained. Silicate minerals such as olivine condense at somewhat lower temperatures. Volatile constituents, those that are mostly found in a gaseous state on our planet, can only condense at the low temperatures encountered far from the Sun (Figure 3.4).

The condensation of water vapour to ice delineates a thermal boundary conveniently known as the 'snow line' or 'ice line'. At the very low pressures of the protoplanetary disc, this boundary corresponds to a temperature of around −100 to −120 °C; it was located beyond the planet Mars (Lecar et al., 2006). Water ice is followed shortly afterwards, or further out in the nebula, by condensation of methane, ammonia, and carbon dioxide. This is the realm of the giant gaseous planets, and beyond, of the icy comets.

Origin of terrestrial water

Earth formed by the accretion of chondritic material. Only carbonaceous chondrites contain water and carbon in significant quantities, but these represent only a minority (4%) of the meteorites. It has therefore been proposed that terrestrial water may have come from comets, whose highly elliptical orbits periodically bring them from the distant solar system into the range of Earth's orbit. Near the Sun, a comet begins to sublimate, forming the characteristic tail, sometimes visible to the naked eye. Analysis of the tail of Halley's comet[4] during its last passage in 1986 showed the presence of abundant water.

---

[4] Halley's comet comes back regularly every 75 or 76 years. It is certainly the comet represented on the Bayeux tapestry which tells of the conquest of England by William, Duke of Normandy, in 1066.

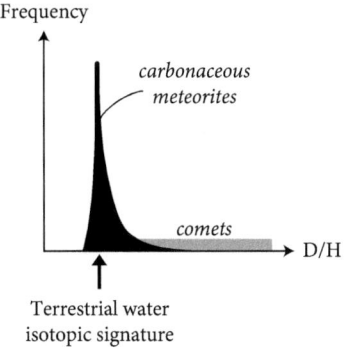

Figure 3.5 Hydrogen isotope data. The D/H ratio of comets is clearly higher than that of terrestrial water, whereas the latter is superimposed on that of carbonaceous chondrites.
Source: After Bernard Marty, in Gargaud et al. (2005).

While not discounting the possibility that comet impacts on Earth may have been important, in particular in the origin of life, hydrogen isotope data rule out a mainly cometary origin for terrestrial water. Hydrogen is the simplest of all chemical elements since its nucleus contains only one proton. However, there is an isotope of hydrogen whose nucleus also contains a neutron, giving it an atomic mass of 2. It is called **deuterium**. Although it should be written as ^{2}H, it is commonly denoted by the letter D.

The D/H ratio of water in the oceans has been measured and compared with that of carbonaceous chondrites and comets (Figure 3.5). The result is clear: terrestrial water does not come from comets, but rather from carbonaceous chondrites. Some contribution from comets cannot be ruled out, but it must be modest. Nitrogen isotope data yield comparable results. Thus, both the water in the oceans and the nitrogen in the atmosphere are derived from chondrites, and more specifically carbonaceous chondrites, which contain mainly H_2O and CO_2 in their volatile inventory.

Where is the water in carbonaceous chondrites found? Some 'free water' can be found in their pores, but most of the water is bound in hydrated minerals. The hydrated minerals of carbonaceous chondrites are mainly serpentines, which we have already learned form when olivine is altered by water. Serpentine is stable at temperatures up to a few hundred degrees. In addition to serpentines, other secondary minerals, such as oxides and hydroxides, formed at the expense of metallic iron, and even some carbonate minerals occur. The degree to which carbonaceous chondrites were altered before they eventually collided with Earth depends on the conditions they experienced in the protoplanetary nebula: pressure, temperature, amount of water vapour present. These conditions depend on the place of formation of these meteorites in the nebula, as well as on the size of the asteroids from which they were fragmented.

The largest asteroids Ceres and Vesta display different mineral compositions. Vesta seems to have a dry basaltic crust. Ceres, located further away from the sun, shows hydroxylated minerals, carbonates, and even water ice on its surface. Hence the composition of Ceres is more akin the composition of carbonaceous chondrites (McCord et al., 2022). Dedicated astronomical missions explored smaller hydrated rocky asteroids that could have brought water and organic matter to Earth. For instance, the NASA OSIRIS-Rex[5] mission reached the asteroid Bennu at the end of year 2018 and fully characterized its mineralogical and chemical composition by spectroscopy. Bennu is among the most aqueously altered asteroids studied so far. It contains abundant unheated clay minerals (Hamilton et al., 2019). Water supply to Bennu initially occurred when the asteroid's precursor came in contact with water ice contained in the outer protoplanetary disc. During the accretion process, internal heating by shocks and decay of short-lived isotopes melted the ice and generated aqueous fluids that interacted with the primary mineral and metallic phases to form secondary minerals.

Degassing of the magma ocean

The formation of our planet by the accretion of chondrites and other larger rocky objects led to heating, which induced melting and segregation. The water and carbon dioxide contained in some objects—either in free form in pores or hidden in hydrated silicates, hydroxides, and carbonates—then dissolved in the magma ocean. A magma can contain a significant amount of dissolved water, especially at high pressures or depths (Figure 3.6). At low pressures near Earth's surface, a hydrated magma will reach the water saturation point at which the excess water cannot remain in solution and escapes, either as water vapour at the surface or as a supercritical fluid at shallow depth. In the same way, a magma may also contain some dissolved carbon dioxide. As carbon dioxide is less soluble than water in a silicate liquid, it is expected to degas earlier and more completely than water—the process by which a magma releases its dissolved gases is referred to as **degassing**.

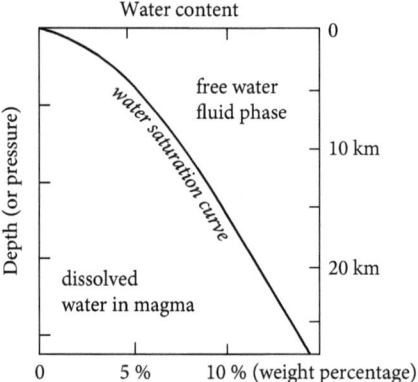

Figure 3.6 Water saturation curve of a magma as a function of pressure. When the pressure decreases close to the surface, the amount of water dissolved in the magma drops very sharply. The excess water is then degassed from the magma, usually as steam.

---

[5] Origins, Spectral Identification, Resource Identification, and Security—Regolith Explorer.

The sudden outgassing of water-rich magma can be highly explosive, as is the case for many volcanoes from the Pacific Ring of Fire, and many in Indonesia, like Krakatoa. These explosive eruptions illustrate the tremendous volume of water vapour that can be degassed from a magma. Volcanoes emit other gases as well, including sulphur dioxide (SO_2) and carbon dioxide. Sulphur dioxide has a short residence time in the terrestrial atmosphere as it combines with liquid water in clouds to form dissolved sulphuric acid that rapidly falls to the ground as acidic rains.

In short, the magma ocean that covered the young Earth must have released large quantities of gases, especially carbon dioxide and water vapour. In 2008, Lindy Elkins-Tanton of the University of Arizona calculated the volume of gas released, assuming an initial dissolved water content of 0.05% and dissolved carbon dioxide of 0.01% in a magma corresponding to complete melting of the mantle. The result far exceeds the combined volumes of the present atmosphere and ocean. Indeed, if we consider a 2,000 km-thick magma ocean with the volatile contents stated above, it would degas almost ten times the volume of the current oceans! This simple calculation suggests that the magma ocean did not degas completely. Alternatively, the magma ocean may have been much thinner than 2,000 km, or the content of dissolved volatile constituents was lower, or some part of the atmosphere escaped to space. Note that CO_2, which is slightly less soluble than water in magmas, would have degassed a little earlier and more completely than water, leading to a CO_2-rich early atmosphere.

The Earth's atmosphere formed by the degassing of the magma ocean is called 'primitive', to differentiate it from the 'primordial' atmosphere of the protosolar nebula, which was rich in hydrogen and has not been preserved. The primitive atmosphere is therefore a secondary atmosphere of volcanic origin. But as we will see later in the book, this atmosphere evolved as life originated and proliferated.

Information provided by noble gases

The six noble gases, from helium (He) to radon (Rn), are found in the last column on the right of Mendeleyev's period table (Figure 1.8). All these gases are chemically inert, which means they do not participate in any chemical reaction, making them potentially excellent tracers.

Helium is the second most abundant element in the universe. Primordial helium, consisting mainly of helium-3 (^{3}He), was largely lost through atmospheric escape. However, some helium remains in the Earth's atmosphere. This helium is mainly ^{4}He, which is radiogenic, produced by the decay of radioactive elements such as uranium.[6] The outgassing of the relatively uranium-rich continental crust thus gradually releases ^{4}He into the atmosphere. The stable isotope ^{3}He, which is not radiogenic, is much rarer: it is one million times less abundant than ^{4}He in the atmosphere and only comes from the outgassing of the mantle. Despite its extremely low levels, the ratio

---

[6] This is alpha radioactivity, a form of nuclear fission in which the parent nucleus splits into two daughter nuclei, one of which is a highly ionized nucleus ^{4}He, called 'alpha particle'.

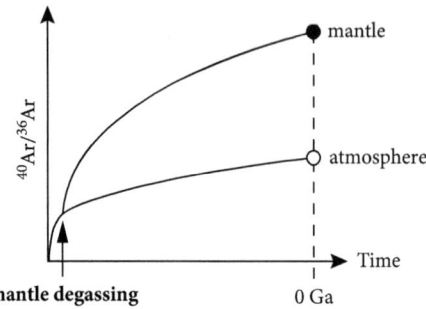

Figure 3.7 Comparison of argon isotope ratios in the mantle and in the atmosphere. If mantle degassing had occurred yesterday, mantle and atmosphere ratios would be the same. If massive degassing happened early, most ^{40}Ar produced at that time would have been transferred to the atmosphere and the atmospheric ratio would reflect that of ancient mantle, although the atmospheric ratio still evolved because of subsequent minor mantle degassing. The current mantle ratio is much higher than the atmospheric ratio because ^{40}Ar progressively increased due to the radioactive decay of ^{40}K.
Source: Adapted from Marty (2020).

of $^3He/^4He$ can be measured in basalts, which are rocks formed by partial melting of the Earth's mantle. The basalts of oceanic islands, corresponding to 'hot spots' where plumes rise from the deep mantle, have a higher ratio of $^3He:^4He$ than basalts formed at oceanic ridges. Consequently, they originate in a part of the mantle that has retained its primitive characteristics (i.e. it is less degassed than the upper mantle).

Argon, the most abundant noble gas in the Earth's atmosphere, provides additional information. The common argon isotope is ^{36}Ar. Another isotope, ^{40}Ar, is radiogenic and is only produced by the decay of radioactive potassium ^{40}K. No ^{40}Ar existed when the Earth was formed, and $^{40}Ar/^{36}Ar$ has increased progressively in the mantle through the decay of ^{40}K. Measurements of this argon isotope ratio in the atmosphere yield information on the time of degassing. Today's $^{40}Ar/^{36}Ar$ isotope ratio in the mantle (measured from recent basalts) is much higher than the $^{40}Ar/^{36}Ar$ in the atmosphere, which is strong evidence that degassing mainly occurred at the beginning of Earth's history (Figure 3.7). After this early degassing, $^{40}Ar/^{36}Ar$ in the mantle increased sharply because most initial Ar (mainly as ^{36}Ar) was lost to degassing. Ongoing decay of ^{40}K is responsible for the progressive increase of this ratio at a slower pace.

Formation of the ocean

The outgassing of the terrestrial magma ocean produced a warm and dense atmosphere, rich in water and carbon dioxide. Other constituents were present in the atmosphere immediately after the largest impacts, namely a mist of molten silicate droplets that fell rapidly, gradually brightening the horizon.

At the beginning of the Hadean eon, when the surface temperature fell below the boiling point of water, it began to rain—a deluge that delivered the water of the oceans.

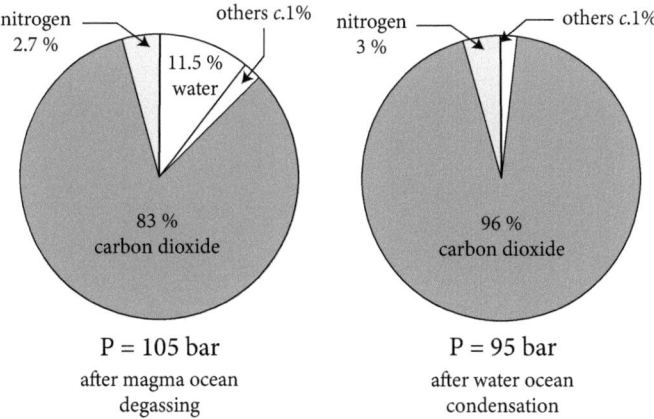

Figure 3.8 Composition of the early Hadean atmosphere before and after ocean formation.

It has been calculated that it took 1 to 2 million years for liquid water to appear and form the primordial ocean, with an average thickness of 2,500 m (Zahnle et al., 2010). Under higher pressures than today, water vapour became liquid above 100 °C (Figure 3.3), and the ocean was therefore initially hot. Chemical reactions between ocean water and the solid crust trapped some water in the form of secondary hydrated minerals, such as serpentines (Figure 2.6), that result from the alteration of olivine. At the same time, the atmosphere and ocean were still cooling. After a few tens of millions of years, the situation would have 'normalized', although the Hadean 'normal' was much different to that of today. The atmosphere had lost most of its water, but it was still rich in CO_2 and denser than the current atmosphere (Figure 3.8). No oxygen was yet present.

The atmosphere of other terrestrial planets

Now, let's compare the Earth and the nearby planets. From what we have just discussed, we expect to find mainly carbon dioxide, water, and nitrogen in the atmosphere of the terrestrial planets. However, none of other terrestrial planet atmospheres have this composition. With a radius of 2,440 km, Mercury is barely larger than the Moon. It has therefore not been able to retain a substantial atmosphere.

Venus, on the other hand, is almost the Earth's twin in terms of size, but its surface conditions are very different. Analysis of the current Venusian atmosphere identifies carbon dioxide as the main constituent. The Venusian atmosphere is dense (93 bars, instead of 1 bar at Earth's surface) and surface temperatures reach almost 500 °C. The atmosphere of Mars is similar in composition to that of Venus, but with a vastly lower pressure (Figure 3.9). Carbon dioxide and nitrogen contents are consistent with the magma ocean degassing hypothesis, but water is apparently lacking in both cases.

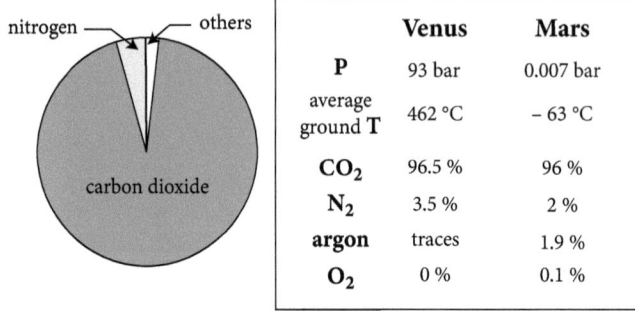

Figure 3.9 Comparison of the atmospheres of Venus and Mars.

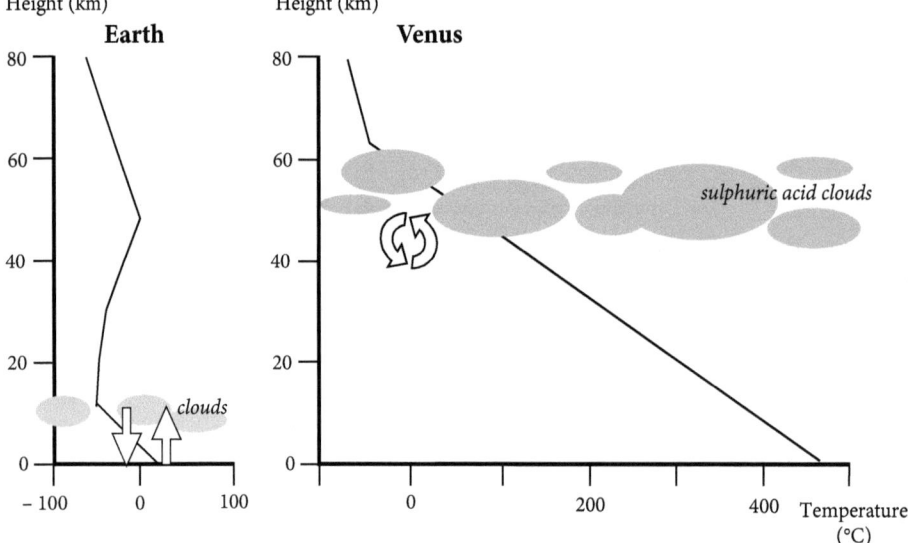

Figure 3.10 Comparison of atmospheric temperatures for Earth and Venus. White arrows show the water cycle with phase transitions from vapour to liquid and vice versa. Colder conditions on Earth induced early formation of the ocean and limited water loss in the upper atmosphere.

Source: Adapted from Wordsworth and Pierrehumbert (2013).

Owing to Venus's proximity to the Sun, it has high surface temperatures. Water vapour has combined with sulphur dioxide (SO_2) emitted by volcanoes to form thick clouds of sulphuric acid (H_2SO_4), contributing to the inhospitable nature of the planet. At the top of the thick Venusian troposphere (~45 km), temperatures are low enough for water to condense. Nevertheless, liquid water likely never reaches the ground, because falling liquid water (rain) rapidly vaporizes again in the upper troposphere. Water vapour is also lost through photolytic reactions in the upper atmosphere, resulting in dry surface conditions (Figure 3.10).

The Earth escaped this fate because its distance from the Sun allowed the condensation of liquid water from the atmosphere, because temperatures rapidly decrease at the top of the troposphere (around 10 km). Evaporated water falls to the surface as rain or snow. Thus, the Earth's upper troposphere is an efficient cold trap for water (Figure 3.10).

The main component of the Martian current atmosphere is carbon dioxide (96%), followed by nitrogen. Oxygen is only present in trace amounts and is thought to result from photolytic reactions in the upper atmosphere.[7] The current atmosphere of Mars is thin: its pressure is only 6 millibars. A large portion (50 to 88%) of the early atmosphere was loss to space because of the low gravity of the planet, as Mars has only one tenth of Earth's mass. This is supported by the larger isotopic D:H ratio of the Martian atmosphere compared to the Earth's ratio, as isotopically lighter H_2 and H_2O molecules escaped more easily (Jakosky, 2021). The early Martian atmosphere was likely much thicker and would have resulted in a warmer climate than today. This early atmosphere would have precipitated liquid water similarly to Earth's early atmosphere, since the older Martian southern highlands display networks of valleys and canyons demonstrating the presence of liquid water in the past (Forget et al., 2007). However, the existence and duration of a martian liquid ocean are debated. The northern lowlands of Mars are the best candidates to host such an ocean. Indeed, several deltaic features have been identified at the same elevation along these northern plains. Thus, geological evidence suggests the existence of an early ocean equivalent to about a 100-metre-thick global layer (Carr and Head, 2015). Nevertheless, the duration of a liquid water ocean must have been short. Because of the distant faint young Sun and a progressively thinning atmosphere, the Martian climate became colder and the water ocean quickly froze. Evaporation and sublimation contributed to the loss of its water or to its redistribution on elevated terrains and in the icy polar caps (Turbet and Forget, 2019). Today, the surface of Mars is on average too cold to allow the presence of liquid water, except at depth.

The greenhouse effect

Let's return to the early Earth's atmosphere and its evolution after the formation of the ocean. At that time, the CO_2 content of the atmosphere was still high. With progressive cooling of the ocean and atmosphere, carbon dioxide dissolved in water, yielding bicarbonate (HCO_3^-) and carbonate (CO_3^{2-}) ions, which can participate in the precipitation of calcium carbonate (e.g. limestone) at the bottom of the ocean due to hydrothermal alteration of the seafloor (see Chapter 6). This process progressively lowered the carbon dioxide content of the ancient Earth's atmosphere. The current terrestrial atmospheric CO_2 content is very small (compared to the ancient Earth) as most carbon is now contained in carbonate minerals and organic matter (Figure 3.11).

[7] The reaction is: $2 CO_2 = 2 CO$ (carbon monoxide) $+ O_2$.

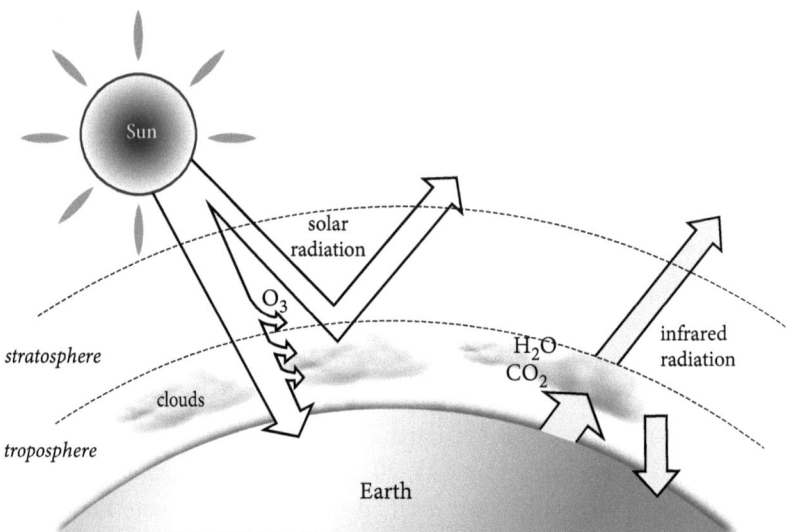

Figure 3.11 Earth's energy budget under current conditions. The greenhouse effect warms the lower layers of the atmosphere because certain gases, such as carbon dioxide and water vapour, absorb the infrared radiation emitted by the Earth's surface. The internal heat flux (i.e. the heat emitted because Earth is cooling off) is negligible in this energy balance that regulates Earth's climate system.

The N_2 content, however, has remained mostly unchanged. Therefore, the current low CO_2:N_2 ratio reflects this progressive decrease of atmospheric CO_2 through time. Mars still preserves a high atmospheric CO_2:N_2 ratio, pointing to a different evolution consistent with an apparent lack of carbonates. This hypothesis remains to be tested in the light of the most recent observations by the rover missions.

Carbon dioxide plays an important role in climate because it is a greenhouse gas (as is H_2O vapour). Today, the Earth receives an average of 1,368 W/m² of energy from the Sun. This value is called the **solar constant**. Only a fraction of this energy passes through the atmosphere and reaches the ground. The rest (30%) is reflected into space. This reflected fraction, which does not contribute to Earth's energy balance, factors into the **albedo** or reflectivity of the surface. The albedo is the ratio of the reflected light energy to the total incident light energy. Since it is the ratio of two energies, it is a dimensionless quantity between 0 and 1, with higher values indicating greater reflectivity. The Earth's surface, warmed by the sun, also gives off energy into space in the form of infrared black body radiation.[8] Some of this radiation is absorbed by greenhouse gases that are transparent to incident visible light but not to

---

[8] In physics, a black body is an ideal object that absorbs all the energy it receives, hence the term 'black', since all visible light is absorbed. Under the effect of thermal agitation, the black body emits radiation whose wavelength depends only on its temperature. At room temperature, the radiation emitted is in the infrared range, which is invisible to the human eye, and the body appears effectively black. At higher temperatures, the body will appear red, then white.

infrared light.[9] The lower layers of the atmosphere are therefore heated by this process (Figure 3.11).

Without the greenhouse effect, the average temperature at the Earth's surface would be −18 °C, a situation possibly detrimental to the development of life since the entire surface would be frozen. Thanks to the greenhouse effect, the Earth's average temperature today is +15 °C.[10] What about the ancient Earth, whose atmosphere was rich in CO_2?

The faint young Sun paradox

As we have learned, liquid water has been present on Earth's surface since early in the Hadean. We also know that the Hadean atmosphere was very rich in carbon dioxide, even if some of this carbon dioxide was progressively sequestered into the cooling ocean.

We must also consider that the Sun was then in its infancy, and therefore less active than today. That is, the Hadean Sun shone less brightly than it does now. Specifically, the solar constant 4 billion years ago was about a quarter of what it is today. If we calculate the surface temperature with an atmosphere identical to the present one, the energy balance becomes incompatible with the presence of liquid water on the surface during the Hadean and even throughout the Archean, and average surface temperatures above freezing are not reached until two billion years ago. This result flagrantly contradicts geological data: it is the paradox of the faint young Sun.

The reader has probably found the solution: the Hadean or Archean atmosphere is obviously not identical to the present atmosphere. It must contain a higher proportion of greenhouse gases to compensate for the low luminosity of the Sun. The calculations are then repeated, testing increasing proportions of CO_2 in the atmosphere.

Calculations show that a minimum of 10% of CO_2 is needed in the atmosphere to obtain a positive average temperature at the Earth's surface at the beginning of the Archean period. This is a much higher value than the current content, which is around 0.04%, or 400 ppm![11] The composition of the Earth's atmosphere in the past was therefore very different from its current composition, particularly because of its high carbon dioxide content and low oxygen content. Finally, it is worth noting that carbon dioxide is not a pollutant, as is sometimes claimed in the media. It is a normal constituent of the atmosphere of terrestrial planets. Life appeared on Earth when the carbon dioxide content of the atmosphere was much higher than it is today.

---

[9] A more accurate term than 'greenhouse effect' is 'radiative forcing', because a greenhouse, whose windows prevent the exchange of air between the outside and inside, is not a precise analogy. Nevertheless, throughout this book, we will use the term 'greenhouse effect' because it is widely used and understood by the general public.

[10] This is an average. Temperatures measured at the surface of the globe range from −80 °C to +55 °C.

[11] The so-called pre-industrial (before 1750) CO_2 content of the atmosphere was slightly below 0.03%. Atmospheric CO_2 levels subsequently increased because of the burning of fossil carbon (coal and oil), which is driving current anthropogenic global warming.

The Earth's atmosphere originated from the outgassing of the magma ocean. This primitive atmosphere contained water vapour and carbon dioxide. Water vapour condensed rapidly to form the oceans, possibly after a few million years. The greenhouse effect of the high atmospheric CO_2 content compensated for a cooler Sun throughout the Archean, accounting for ample geological evidence for a liquid ocean. Atmospheric CO_2 progressively decreased through time due to dissolution in the ocean and carbonate precipitation in relation with sea-floor alteration.

4
The oldest rocks on Earth

In contrast to the Hadean, Archean rocks outcrop in many places around the world (Figure 4.1). In Europe, large Archean terranes are found in Finland and Ukraine. Canada, Greenland, southern Africa, southern India, and western Australia are also reference regions for the Archean eon. The Archean lasted for one and a half billion years, between 4 and 2.5 Ga. However, there are few terranes that date to between 4 and 3.6 Ga compared to younger Archean terranes. In this chapter, we will explore information from the oldest rocks on Earth, found in Canada, Greenland, and South Africa.

Acasta gneisses (Canada)

The Acasta gneisses in north-western Canada are the oldest rocks known to date. Gneisses are **metamorphic** rocks, that is, rocks that have been modified at depth by elevated temperatures and pressures. These rocks were deformed after their crystallization and acquired a schistose or laminated aspect, typical of gneisses (Figure 4.2).

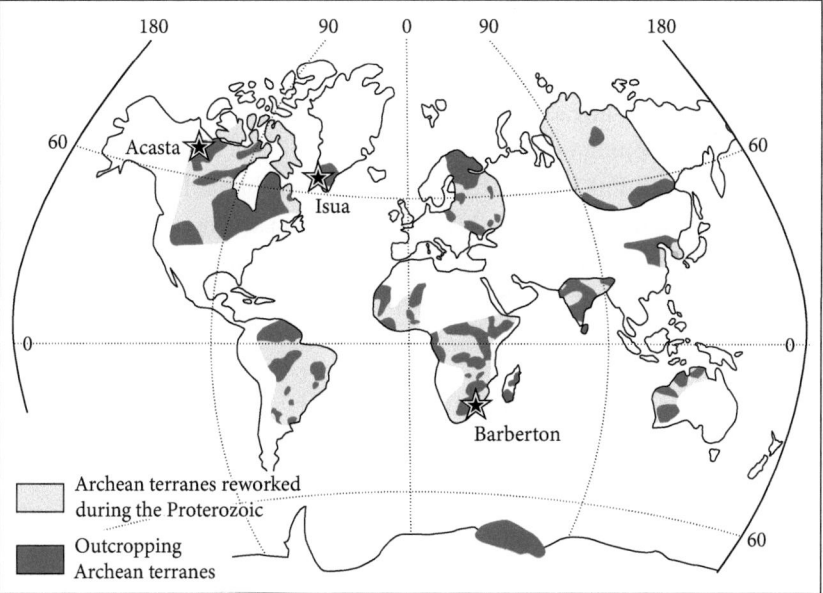

Figure 4.1 Map of Archean terranes in the world.

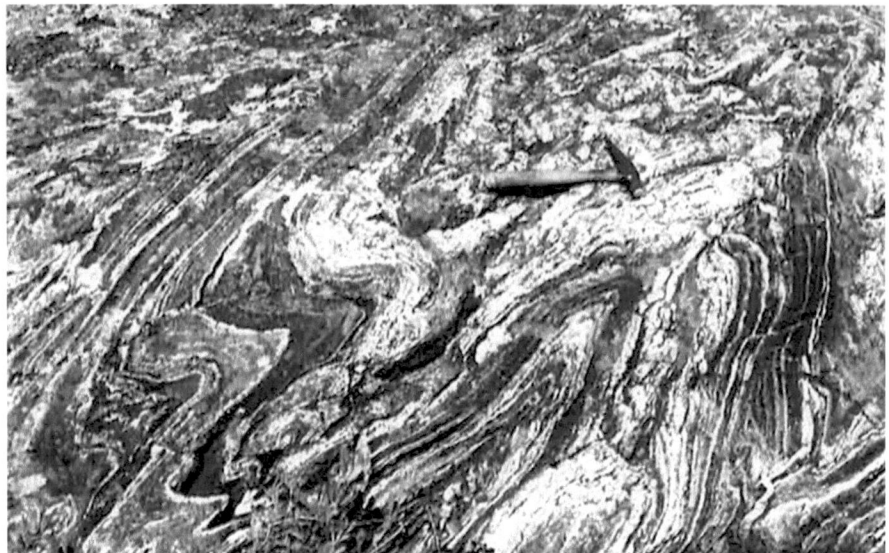

Figure 4.2 The Acasta gneisses: deformed rocks of basaltic (dark) and granitic (light) compositions; hammer for scale.
Source: Paul Hoffman.

The Acasta gneisses comprise dark, abundant, metamorphosed volcanics of basaltic composition, and lighter gneisses resembling granites in composition. Only the light gneisses were dated, because they contain zircon. Their ages range from 3.9 to 3.6 Ga. One sample, the Idiwhaa **tonalite**, yielded an age of 4.02 Ga, corresponding to the very end of the Hadean eon, close to the Hadean–Archean transition (Bowring and Williams, 1999). A tonalite is a magmatic rock, containing slightly less silica than a granite.

We know that granitic rocks are emblematic of the continental crust. But can we deduce that there was already a continent in northern Canada at the end of the Hadean? The outcropping extent of the Idiwhaa tonalite is very small (less than one square kilometre: Reimink et al., 2016), and it is not possible to draw any conclusion in the field about the geological setting before deformation and metamorphism. In addition, the Idiwhaa tonalite has particular characteristics showing that it cannot have been formed in the same way as the Archean granitic rocks.

Isua (Greenland)

The age of the Isua terrain on the south-west coast of Greenland is between 3.9 and 3.7 Ga. Several teams have studied the Isua area, despite the difficult conditions of snow and ice—not to mention the cost of such remote fieldwork. The rocks include a very old succession of volcanic examples, which include pillow lavas recognized to be the result of submarine lava flows. These basalts are associated with siliceous and ferruginous sediments formed by chemical precipitation at the bottom of the ocean. The whole group of rocks (lavas and sediments) is referred to as a **'supracrustal'** series, meaning they are a combination of volcanic and sedimentary rocks that accumulated on the surface of the crust.

In detail, two types of volcanic rock have been identified: basalts, the iconic volcanic rock (this is still true today!), and more magnesium-rich rocks (Szilas et al., 2015). Basalts are known to be the product of partial melting of mantle peridotites. However, the geological context of the formation of Isua's volcanic rocks is debated. In several places, these rocks have been altered or deformed. The bedrock on which the basaltic lavas were deposited is unknown.

Gneissic rocks outcrop widely in the vicinity but are slightly younger in age (3.8–3.6 Ga). The contact between the supracrustal series and the gneisses is deformed, which complicates the study of their relationships.

Barberton (South Africa)

Given the challenging climatic conditions in these icy northern places, neither Acasta nor Isua can be the promised land of Archean geology. We will therefore travel to the southern hemisphere and a more hospitable climate, namely the Barberton area of South Africa (Figure 4.3). Barberton is a small, mining town, founded in the late 19th century. Although the gold rush of that time is long over, Barberton remains relatively active. The town is located 360 km east of Johannesburg, not far from the border with Eswatini (formerly Swaziland). I visited Barberton with colleagues. We left Johannesburg and drove due east for almost three hours on the

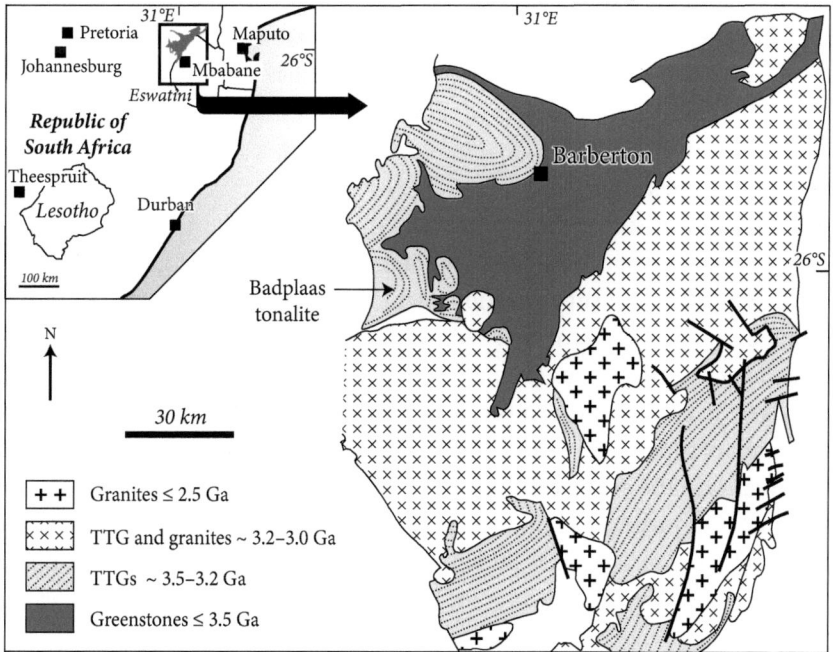

Figure 4.3 **Geological map of the Barberton area (South Africa).** Two main rock types are described: greenstones and TTGs (tonalite, trondhjemite, and granodiorite).

High Veld Plateau, whose monotony is only broken by waste hills from coal mines. The High Veld Plateau is underlain by 300-million-year-old (Carboniferous) Karoo sediments and has an average elevation of 1,500 m. The end of the journey was a pleasant surprise. From the eastern edge of the plateau, the road drops quickly (more than 1,000 metres of descent), with a spectacular scenery. This descent represents a real plunge into the past, as the oldest rocks in the Barberton area date back to 3.5 Ga.

Barberton is a place known to all geologists around the world. The geology of the area has been studied in detail and is a reference for the knowledge of the ancient Archean. Its importance is such that the area was designated a UNESCO World Heritage Site in 2018.[1] Two types of rocks have been described: greenstones and granitic rocks in the broad sense, the so-called **TTGs**.[2] This duality is found in all the Archean terrains of the world.

Greenstones

The Barberton Greenstone Belt, as it is known to geologists, is an elongated area of about 100 km (Figure 4.3). Its scalloped outline is bordered or cross-cut by granitic intrusions. The belt consists mainly of basaltic volcanic rocks with a light metamorphic imprint, hence their greening due to the development of a secondary mineral, **chlorite**. The rocks are in an exceptional state of preservation considering their age. Usually, basaltic rocks are intensely weathered in tropical latitudes, so they are often only detected by the red colour of their weathering products on tracks and by the topographic lows they leave behind in the landscape. This is not the case in Barberton: the greenstone belt forms the high hills above the town. The rocks here have been well preserved because they were impregnated with silica shortly after their formation, which hardened them and preserved all the details of their structure. This process of silicification is discussed in Chapter 6. In addition, a few sedimentary rocks are observed in the greenstone belt.

Before getting to know the Barberton Greenstone Belt, the reader should be reminded of the basic principle of stratigraphy (i.e. the study of geological **strata**, or layers). The principle of superposition—which allows the relative dating of strata—states, quite obviously, that younger sediments are deposited on top of older sediments and are then covered by even younger sediments. This principle also applies to volcanic rocks: the lava from which they solidified flowed over rocks that must be older. The principle of superposition applies even if all the strata have been deformed, folded, or fractured, as in the example in Figure 4.4.

Unlike basaltic magmas, granitic magmas, which are very viscous, do not reach the surface and therefore crystallize at depth. The granites are called **plutonic** magmatic rocks, whereas basalts are volcanic magmatic rocks. The principle of superposition

---

[1] See https://whc.unesco.org/fr/list/1575.
[2] Standing for tonalite, trondhjemite, and granodiorite, as explained later in the chapter.

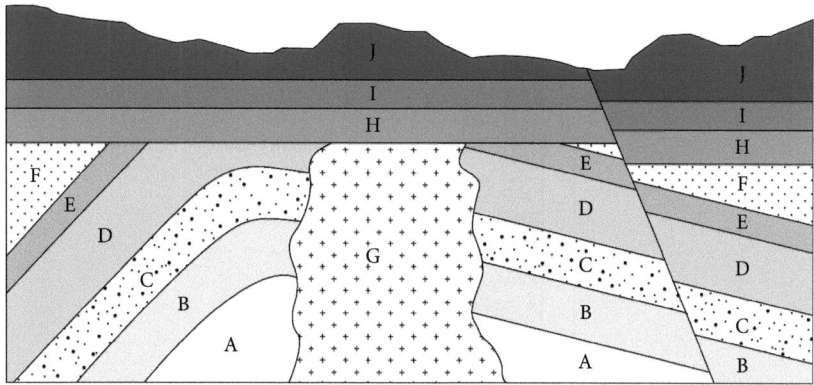

Figure 4.4 Schematic section illustrating the relative dating of rocks and tectonic events. The strata A to F were deposited successively in this order, then deformed by folding. Folding was followed by the intrusion of granite G, and then erosion was responsible for the development of a surface (called 'unconformity') on which strata H, I, and J were deposited. The final event was the formation of a fault cutting through all strata on the right-hand side of the figure.

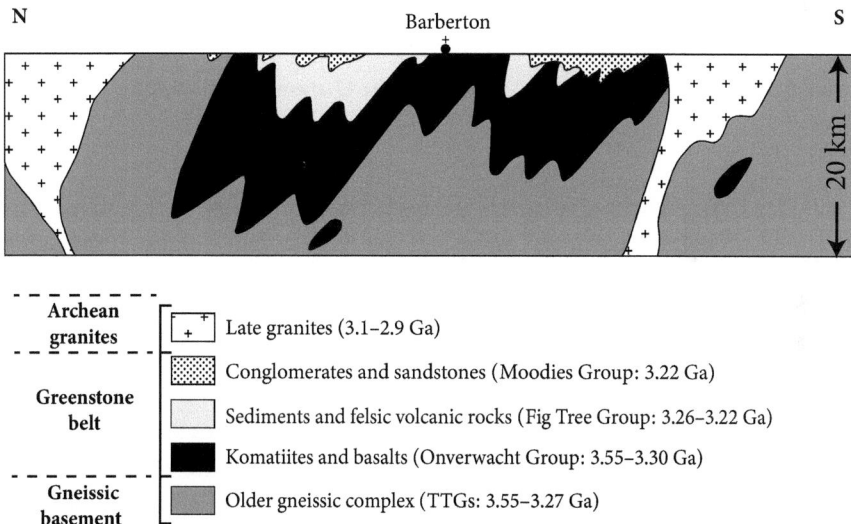

Figure 4.5 Geological cross-section a few kilometres east of Barberton. The section shows the synclinally folded greenstone belt and the late cutting granites.

cannot be applied to granitic rocks. However, the principle of cross-cutting relationships establishes that granites must be younger than the strata they cut. Volcanic and plutonic rocks share the enormous advantage that they can be dated precisely using the U/Pb method on zircons. Therefore, they provide temporal benchmarks that complement the relative dating of sedimentary strata and faults using the principles of superposition and cross-cutting relationships.

Figure 4.4 is an imaginary geological cross-section, whereas the following figure (Figure 4.5) is a realistic cross-section of the Barberton area. The Barberton greenstone

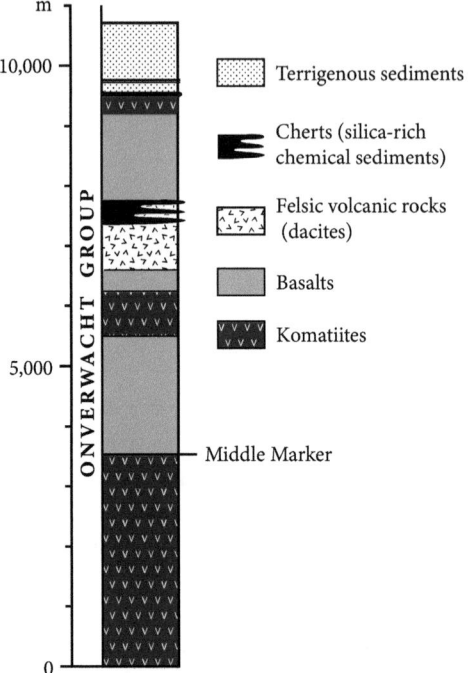

Figure 4.6 Simplified stratigraphic column of the Onverwacht Group.

belt is folded into a syncline, resembling a folded basin. Because of this structure and after erosion, all of the rocks in the series are observed, from the oldest in the centre to the youngest at the periphery. A total thickness of 17 km for the Barberton series can be estimated by reconstructing or flatting out the synclinal structure. This thickness must correspond to a very long duration of volcanism and sediment deposition—in this case almost 300 million years. The lower part of the Barberton Greenstone, the thickest and oldest part, is called the Onverwacht Group and comprises 90% volcanic rocks (Figure 4.6). The remaining 10% of the Onverwacht Group are sediments: mainly siliceous chemical sediments known as **cherts**, as well as sandstones. The sandstones were deposited only at the top of the series, whereas cherts occur at different levels of the stratigraphic column. The first (and oldest) cherty deposit is located in the so-called Middle Marker. It is only a few metres thick.

About half of the volcanic rocks in Barberton are basalts, sometimes showing the pillow-like appearance typical of an underwater flow (Figure 4.7). Basalts are magmas produced by partial melting of mantle peridotites at depths of around 100 km. Laboratory experiments have shown that the degree of peridotite melting required to generate basaltic magmas is 15 to 25%. Basalts are common rocks and today constitute the bulk of the lavas extruded at oceanic ridges as well as in continental rifts. They also represent the majority of lavas released at hot spots where basaltic flow piles can reach several kilometres in thickness. These giant outpourings occur as **oceanic plateaus** and large continental magma provinces.

Figure 4.7 Barberton greenstones (a) Archean pillow basalt outcrop (3.5 Ga) in the Barberton area. (b) Close-up view of basalt pillows (diameter: 30–40 cm). The coin for scale shows the edge of a pillow, where its margin was rapidly cooled by the water into which it was extruded.

Komatiites

Other volcanic rocks that occur in abundance in the lower part of the Onverwacht Group are the **komatiites** (Figure 4.6). These rocks were first described in 1969 by two South African geologists, the Viljoen brothers (Viljoen and Viljoen, 1969). The rock name derives from the Komati River near which they were first observed. Komatiites often display curious elongate skeletal olivine crystals, measuring several centimetres long and a few millimetres wide (Figure 4.8). These elongate crystals are evidence of very rapid cooling at the surface of a magma that must have been uncommonly hot.

Komatiites are almost exclusively Archean in age. They are characterized by their high magnesium content, more than double that commonly found in basalts. Komatiites are therefore rocks whose chemical composition resembles that of mantle peridotites. To obtain such magnesium-rich magmas, the mantle must have melted by almost 40%, whereas basalts result from a more moderate degree of partial melting of the mantle. The mantle is made up of minerals such as olivine and **pyroxenes**, which are so-called ferromagnesian silicates, but which actually contain much more magnesium than iron. The iron component of these minerals melts more readily than the magnesium part. At the beginning of the peridotite melting process, the resulting magma therefore has a composition richer in iron than the mantle from which it originated, as is the case of basaltic magmas. If the degree of partial melting increases, the magnesian part of the minerals begins to melt and the resulting magmas become more

Figure 4.8 Typical komatiite texture. Elongate greenish grey multi-centimetric crystals characteristic of komatiite and indicating rapid crystal growth from a magma that was very hot when extruded. The coin gives the scale.

magnesian than the basalts. This is the case for komatiitic magmas. Thus, the ratio of Mg:Fe in the volcanic rock is an indicator of the percentage of melting.

The formation of komatiitic magmas required higher temperatures than occur today in the mantle, where most magmas are formed. However, high-temperature melting was possible, and even expected, in the Archean. This is the reason why komatiitic volcanism has now completely ceased. In other words, Archean komatiites are witness to the conditions that once prevailed in the Earth's interior.

TTGs

In addition to basalts and komatiites, the Archean terranes contain rocks that were initially described as granitic. However, to mimic a well-known commercial advertisement, they look like granites, they have the structure of granites, they contain quartz and feldspars like granites, but they are not real granites! The difference lies in the chemical composition of their feldspars. Archean granitic rocks contain sodium plagioclase feldspar but no potassium feldspar. This may seem a rather subtle difference, but it is puzzling for the specialist. These Archean rocks are tonalites and **trondhjemites**. We already mentioned tonalites in connection with the Acasta gneisses, the oldest dated rocks on Earth. Trondhjemites, named after the town of

Trondheim in Norway, are light-coloured tonalites that are richer in quartz than tonalites. There is also a third type of rock that contains a little potassium feldspar, and therefore looks more like granite: granodiorite. Let's forget all these rock names and keep only the acronym that is usually invoked by geologists: the TTGs from the first letter of their respective names (tonalites, trondhjemites, granodiorites).

The magmas that produced the TTGs were rich in silica and therefore viscous, like all granitic magmas. This explains their slow ascent in the crust. As a result, these magmas began to cool and crystallize as they rose, making them even less mobile, until they eventually ceased to move at depth below the surface. If TTGs (or granites) are observed on surface outcrops today, it is because millions of years of erosion have removed the thickness of the original crust that concealed them.

The Barberton TTGs are between 3.5 and 3.2 billion years old. The oldest TTGs are the same age as the oldest greenstones (3.5 Ga in both cases). They correspond therefore to coeval magmatism, of obviously different origins. Basalts and komatiites were derived from partial melting of the mantle. What was the source of the magmas that gave the TTGs? The difference in mineralogical composition between TTGs and granites is minor but reflects a difference in their chemical composition and therefore in the initial nature of the magmas. Therefore, the TTGs cannot have formed from the same source rocks as younger granites.

It is known that granitic rocks are typical of the continental crust. Granites can have two different origins. First, they can be derived from magmas produced by melting of pre-existing continental crust; these are known as crustal granites (implying both crustal and continental origin). These magmas represent crustal recycling. Other granitic rocks form at subduction zones and derive from magmas produced by melting of mantle peridotites in the presence of water, a process to which we shall return in the next chapter.

Neither of the two processes mentioned above for recent granites can generate the sodic magmas from which the TTGs are derived. In 1985, in his PhD thesis on the TTGs of Finland, Hervé Martin proposed that the TTGs originated from partial melting of hydrothermally altered basalts transformed into **amphibolites**. Amphibolites are metamorphic rocks rich in amphibole, as indicated by their name. They result from the metamorphism of hydrated basalts that were transported to depth where they were subjected to high pressures and temperatures. Hydration of the basalts is required, otherwise the amphiboles, which are hydrated minerals, could not be formed. This process occurred early due to the action of seawater on submarine basalts.

During metamorphism, amphibolites form at temperatures above 500 °C and they begin to melt at 750 °C. Hervé Martin was able to specify that TTG magmas formed from amphibolites containing **garnet**, a metamorphic mineral that forms at medium to high pressures (or depths). In this case, garnet indicates a pressure of about 10 **kilobars** (10 kbar), or 10,000 times atmospheric pressure. This is the pressure prevailing at a depth of about 30 km in the crust, and it is therefore the depth that must be reached to obtain garnet-bearing amphibolites. The conditions for the formation of

TTG magmas are thus relatively well defined: temperatures above 750 °C and pressures above 10 kbar (Martin, 1986). This model, which quickly achieved consensus, was then supported by experimental melting studies.

This hypothesis implied that source rocks for TTG magmas—namely partially melted Archean garnet amphibolite—should exist. But at the time, no examples of such rocks were found. With Jean-François Moyen, then lecturer at Stellenbosch University in South Africa, I showed that an enclave of garnet amphibolite collected from a 3.2 Ga TTG south of Barberton, in the Badplaas tonalite (Figure 4.3), could be one of these missing samples. This rock contains localized clear zones, rich in quartz and feldspars, that indicate the onset of partial melting. In addition, mineral analysis in this sample made it possible to determine the conditions of this incipient melting: 760 °C and 11.5 kbar—that is, corresponding to a **geothermal gradient** (increase of temperature with depth) of about 20 °C/km (Nédélec et al., 2012).

In his seminal 1986 paper, Hervé Martin proposed that the geodynamic process that brought basalts that had already been hydrated at the surface to depth was **subduction**, a well-known manifestation of **plate tectonics** that we will return to in the next chapter. Subduction is the process that drives oceanic crust deep into the mantle, when this crust reaches an age greater than 200 million years. This process explains why virtually no ancient oceanic crust is observed at the Earth's surface. However, modern subduction does not produce TTG magmas. This is because hydrated basalts in the oceanic crust cannot start to melt when they are driven to depth, as current geothermal gradients in subduction zones are generally less than 15 °C/km. Under these conditions, basalts dehydrate but do not melt. In contrast, in the warmer Archean subduction zones, the basalts entrained at depth could begin to melt to form the TTG magmas (Martin, 1986).

In all Archean regions worldwide, the same rocks, TTG and komatiites, are observed and the same conclusions are obviously drawn from both of these unusual rock types, namely that the Archean mantle was hotter.

The terrestrial heat flux

But where does the internal heat of the Earth come from? Is this heat measurable at the surface? Let us start with the last question. The heat removed from the surface is the Earth's heat flux. The heat flux is expressed in watts (W) per unit area. This is a loss of energy, because the Earth's interior is warmer than its surface. It should be noted in passing that this internal heat is much less than the heat received at the surface from the Sun. The current heat flux can be estimated from temperature measurements in deep boreholes. The temperature of ocean sediments can also be measured with a probe. These temperature measurements must be combined with measurements of the thermal conductivity of the rocks to derive the heat flux. It is estimated that the current total heat flux for the entire globe surface is 46 TW (terawatts, or one thousand billion watts, or 10^{12} W). This may sound like a large number, but it translates into only about

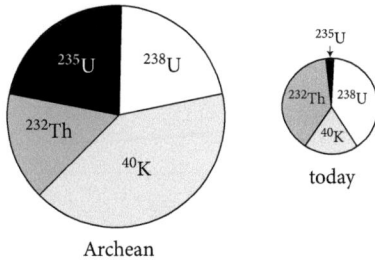

Figure 4.9 Participation of radioactive isotopes in the Earth's heat flux. The early Archean heat flux was larger than the present one and was primarily due to the decay of potassium-40. Today, the isotopes ^{238}U and ^{232}Th play the main role, whereas ^{40}K and especially ^{235}U play only a minor role.

75 milliwatts per square metre on average, a low value indeed compared to the solar constant, which approaches 1,400 W/m².

About half of the Earth's heat flux comes from the decay of radioactive elements, which always produces heat. Today, the main isotopes involved are uranium-238, thorium-232, and potassium-40 (Figure 4.9). These elements are much more abundant in the crust (especially the continental crust) than in the mantle. Nevertheless, the mantle represents such a huge volume in our planet that the radioactive elements it contains play an essential role despite their very small quantities in percentage terms.

Knowing the half-life of these elements, we can calculate the quantity of radioactive elements that was present at the beginning of the Archean eon, and therefore the resulting radiogenic heat production. The result would be a radiogenic heat flux in the early Archean about four times greater than now (Figure 4.9). We note in passing that uranium-235 played an important role in heat generation in the Archean but is less important today. This discrepancy is due to its much shorter half-life than that of uranium-238, and therefore its faster decay (i.e. most of the original uranium-235 has been exhausted). Similarly, other more rapidly decaying isotopes contributed heat early in Earth's history, including isotopes that have now gone extinct due to their rapid radioactive decay, such as aluminium-26 and iron-60.

In addition, during the accretion phase, Earth received large amounts of heat from the dissipation of the kinetic energy of impacts. Indeed, some of the heat deposited by large impacts remained stored in the deep Earth. Finally, a third contribution to the heat of Earth's interior is the latent heat release produced by the crystallization of the solid inner core from the surrounding iron liquid. Estimation of the current core–mantle boundary heat flow ranges from 5 to 15 TW (Lay et al., 2008). Estimation of this heat flow in the past is still more difficult and is related to the debate about the age of the inner core.

There is no doubt that the Earth generated more heat in the Hadean and Archean than it does today. The evolution of surface heat flow proposed in Figure 4.10 reflects the major role of radioactive decay in internal heat production.

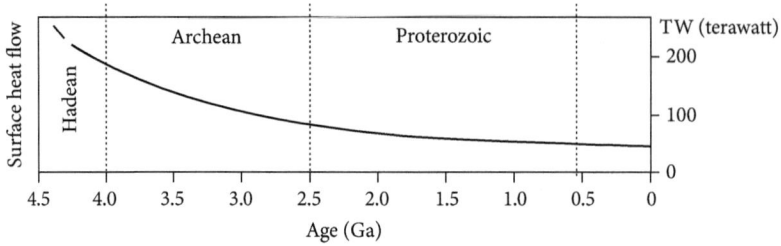

Figure 4.10 Decrease of internal heat production over the history of the Earth. There are large uncertainties on the suggested figures for the ancient Earth.
Source: Adapted from Nimmo (2022).

The modes of heat transfer within our planet and their magmatic and tectonic consequences are the subject of the next chapter.

The ancient Earth's interior was hotter than it is today, allowing the formation of unique magmatic rocks that do not form today: komatiites and TTGs. Komatiites are volcanic rocks that result from extensive melting of the mantle. TTGs are granite-like rocks that are typical of the Archean continental crust worldwide.

5
Earth's internal cooling through time

The interior of our planet is hot as a result of the decay of radioactive isotopes of uranium, thorium, and potassium in the solid mantle, and also because some of the primordial heat from the formation of the planet remains in its core. In this chapter, we will see how the Earth dissipates its internal heat and the manifestations of this heat transfer, namely magmatism and tectonics. These processes are still active today, as evidenced by volcanic eruptions and earthquakes. They were likely more intense in the warmer ancient Earth. Before reconstructing the past internal activity of our planet, we must understand the forces at work.

Mode of internal heat transfer

The transfer of terrestrial heat from the interior to the exterior of the planet takes place in two different ways. Near the surface (i.e. in the crust and the uppermost mantle), heat is transferred by **conduction**, which does not involve movement of rocky material. As rocks are poor conductors of heat, this method of heat transfer is not very efficient. The consequence is a significant temperature increase with depth near Earth's surface. The mean temperature rise, or **geothermal gradient**, is 20 to 30 °C per kilometre in the crust: workers in deep mines experience this steep geothermal gradient every day.

Beyond a depth of 100 to 200 kilometres in the mantle, rocks are more easily deformed than at the surface due to the effect of temperature. They soften to some extent and may even slowly flow, even though, as seismic waves tell us, they are solid. Under these conditions, heat is transferred by mantle **convection**. The motion of the solid mantle arises from the small density difference between the hot—and therefore less dense—material at the base of the mantle, and the colder—hence denser—material at the top of the convecting mantle. The warm material tends to rise and the cold material to fall. Convection is easy to visualize in a container heated from the bottom, such as a pan on a hotplate. Upward currents transfer hot material from the lower mantle to the surface, where it cools, becomes denser, and eventually sinks back into the warmer, less-dense areas. Convection is a very efficient way of transferring heat and tends to homogenize temperatures throughout the convecting layer, resulting in a much smaller geothermal gradient than in the case of conduction.

Note that convection is only possible in the solid mantle when the temperature is above about 1,300 °C. At lower temperatures, mantle peridotites cannot convect

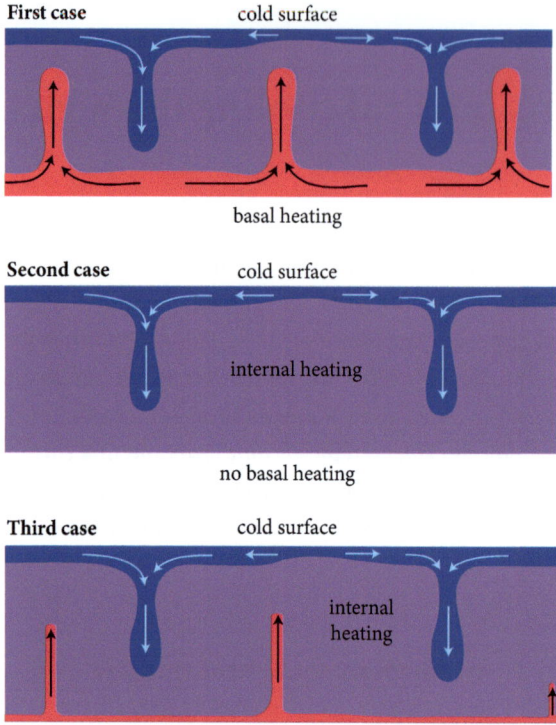

Figure 5.1 Convection patterns according to the origin of the heat. The first case corresponds to basal heating and produces both upwelling and downwelling plumes. The second case corresponds to internal heating and does not show any rising plumes. The Earth's mantle corresponds to the third case, combining internal and basal heating, and resulting in the formation of both upwelling and downwelling plumes, although the former are currently less important.
Source: Adapted from Yanick Ricard/Planet-Terre—ENS-Lyon.

because they are too rigid. The crust and the upper non-convective part of the mantle form the **lithosphere**, the rigid envelope of our planet. The boundary between the convecting mantle (the asthenosphere) and the non-convecting mantle (the lithosphere) is an **isotherm**, a virtual surface corresponding to a temperature of about 1,300 °C. It does not define a compositional change, as does the surface separating the crust from the upper mantle.

The situation is more complex in the mantle than in the common example of a pot heated from below. Indeed, part of the heat is produced in the mantle itself by the decay of its radioactive isotopes. In the case of internal heating without basal heating, convection produces descending cold plumes but no rising hot plumes from the base (Figure 5.1). The Earth's mantle is an intermediate case combining internal heating and basal heating from the core. This results in sinking cold material, as observed in oceanic subduction zones, but also in hot ascending plumes as suggested by intraplate hot-spot volcanoes such as Reunion or Hawaii.

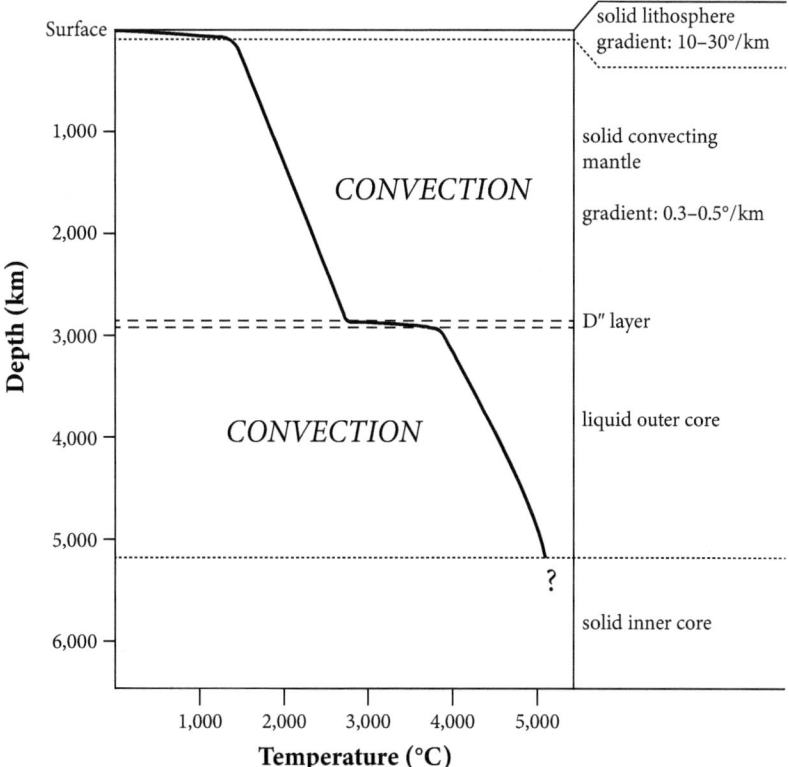

Figure 5.2 The Earth's geotherm. The temperature profile inside the globe displays steep slopes in regions where heat is transferred efficiently by convection. Shallow slopes define boundary layers where heat is transferred by conduction, namely the D" layer and the lithosphere. The geotherm is not shown in the solid inner core where temperatures are highly uncertain.

The convective mantle is limited by two layers called 'thermal boundary layers', one at the surface—the lithosphere—the other at depth—the D" layer at the core–mantle boundary. Inside thermal boundary layers, there is no convection, but only heat transfer by conduction. Due to the low efficiency of conduction compared to convection, the thermal boundary layers coincide with large temperature variations. Below the D" layer, convection is the mode of heat transfer in the liquid outer core. Sharp changes in slope of the Earth's temperature profile or **geotherm** highlight these different modes of internal heat transfer (Figure 5.2).

Rising plumes and hot spots

Reunion Island is an isolated hot spot in the Indian Ocean. Its Piton de la Fournaise is one of the most active volcanoes in the world: it erupts almost every year, fortunately without much risk to the local population. Hawaii is a hot spot in the middle of the Pacific Ocean: its Big Island has several active volcanoes including Kilauea (Figure 2.6), also one of the most active in the world. It is located at the south-eastern end of a chain of 80 mostly undersea volcanoes: the Hawaiian–Emperor seamount chain. The study of seismic wave velocities shows that the Big Island of Hawaii is located above a column of material that is slightly hotter than the surrounding mantle. The speed of seismic waves depends directly on the density of the material, with higher densities corresponding to higher velocities. As there are no compositional differences in the mantle, a zone of lower seismic velocities in the mantle is conventionally interpreted in terms of higher temperatures. However, these differences are quite small and precise geophysical images of ascending hot **plumes** in the mantle have only been made in the last few decades (French and Romanowicz, 2015). Plumes such as the one below Hawaii originate at the core–mantle boundary. These plumes are solid, which might seem counter-intuitive since rising hot plumes are often drawn in a beautiful red colour reminiscent of lava or magma. Hot plumes rise because their density is slightly lower than the surrounding mantle. Their rising is very slow, at most 20 cm per year (Steinberger and Anstretter, 2006). Thus it would take a few tens to more than one hundred million years for a plume to rise from the base of the mantle to the surface. At shallow depth, the plume intersects the solidus of mantle peridotites and melting begins (Figure 5.3).

Plumes occur beneath both the oceans and the continents. They are responsible for **large igneous provinces** (LIP) where they reach the surface. Where the magmas generated by plumes ascend through the continents, they form piles of basaltic flows that can be up to a few kilometres thick. These thick piles of basalts are known as 'traps', from the Swedish word *trappa*, which means 'stairs', because the partially eroded basalt flows resemble a staircase, as in the Deccan region of India (Figure 5.4).

Human beings have never witnessed the intense volcanism responsible for traps, which is fortunate, when you consider that single flows within the Deccan Traps can be up to 1,000 km in extent (Self et al., 2008). In the oceanic domain, hot spots

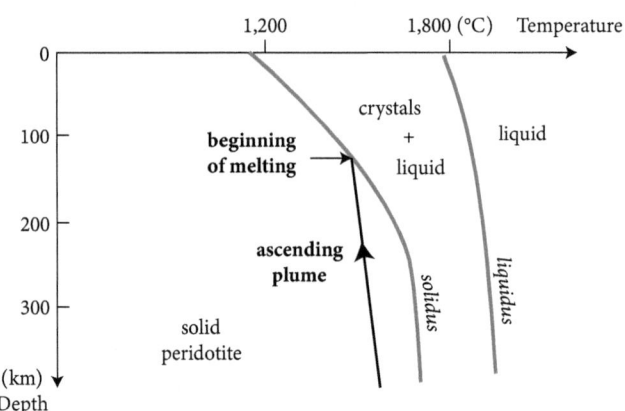

Figure 5.3 **Conditions for melting of an ascending plume in the mantle.**

Figure 5.4 The Deccan Traps (India) as seen near Mahabaleshwar, south of Mumbai. The pile of dark-coloured basaltic flows dating from 66 Ma is over 1,000 m thick.
Source: Thierry Adatte.

generate oceanic plateaus, which are also huge accumulations of basaltic flows. Examples include the partially emergent Kerguelen Plateau in the Indian Ocean and the entirely submarine Ontong Java Plateau, east of New Guinea in the Pacific Ocean. These plateaus date back to 110 and 120 Ma, respectively.

Plate tectonics

The Earth is an active planet, as we are regularly reminded whenever a destructive volcanic eruption or earthquake takes place. Earthquakes, like active volcanoes, are not randomly distributed across the globe. After the Second World War, systematic mapping of the seafloor revealed a series of submarine mountain chains stretching over 60,000 kilometres in length. Along these oceanic **ridges**, basaltic oceanic crust is regularly created, resulting in few centimetres of **oceanic accretion** per year. Because Earth's volume does not change, this expansion of oceanic lithosphere at ridges must be balanced by loss of oceanic lithosphere elsewhere. Indeed, the oceanic lithosphere disappears in the deep oceanic trenches, which are also outlined by volcanic arcs. This phenomenon, known as oceanic **subduction**, occurs widely around the edges of the Pacific Ocean, for example along the Andes and Japan. On a world map, oceanic ridges and subduction zones make it possible to draw the boundaries of the **plates** that make up the lithosphere (Figure 5.5).

Lithospheric plates are mobile with respect to each other. Therefore, their boundaries, most of which are zones of either convergence or separation, are the site of tectonic deformation and therefore of earthquakes. Many plates have both an oceanic part and a continental part, each with very different characteristics (Figure 5.6), but the base of the plate always corresponds to the 1,300° isotherm in the mantle. This is the case for the Eurasian Plate, which includes part of Atlantic and Arctic oceans. In fact, the Atlantic Ocean spans four different plates: Eurasia, North America, South America, and Africa. Only a few plates, such as the Pacific and Nazca plates, comprise only oceanic lithosphere.

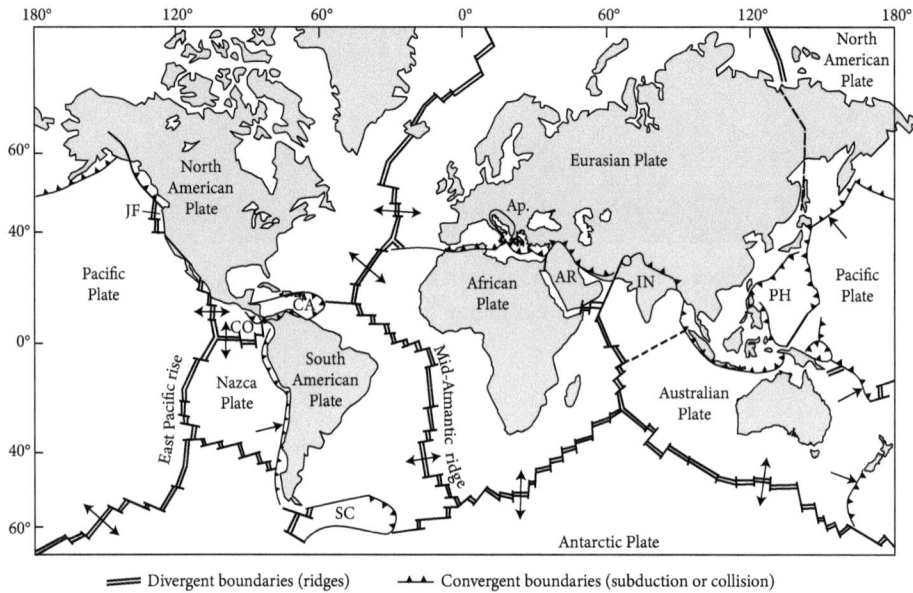

Figure 5.5 Map of lithospheric plates. Arrows indicate the relative motion of the plates. The smallest plates are abbreviated: Arabia (AR), Caribbean (CA), Cocos (CO), India (IN), Juan de Fuca (JF), Philippines (PH), and Scotia (SC). The Apulian promontory (Ap.) is a northward extension of the African plate, responsible for the formation of the Alps.

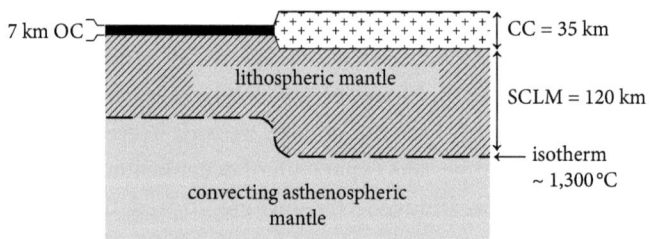

Figure 5.6 The two types of lithosphere (oceanic and continental). CC: continental crust, OC: oceanic crust; SCLM: sub-continental lithospheric mantle.

The oceanic part of a plate is always basaltic in composition. The continental part is thicker, but less dense, because it contains granitic rocks, which are less dense than basaltic rocks. This is the reason why the continents emerge above sea level. When two continental plates converge, neither can really sink below the other because there is not sufficient contrast in density. The result is a continental collision that generates a large mountain range like the Himalayas between the Indian and Eurasian plates. In contrast, an oceanic plate will sooner or later disappear into the mantle by subduction after it has cooled and grown sufficiently dense. This is the reason why no ocean floor today is older than 200 million years, whereas some continental rocks are more than 3 Ga old. The subduction of oceanic plates can be regarded as the manifestation of the sinking of cold plumes predicted by convection models and required to balance the rising hot plumes. Recent advances in seismology have made it possible to identify ancient subducted oceanic plates deep in the mantle, such as the Farallon plate beneath America (Figure 5.7).

Plate tectonics, which accounts for the mobility of lithospheric plates and their peripheral deformation, is the surface manifestation of mantle convection. It has been

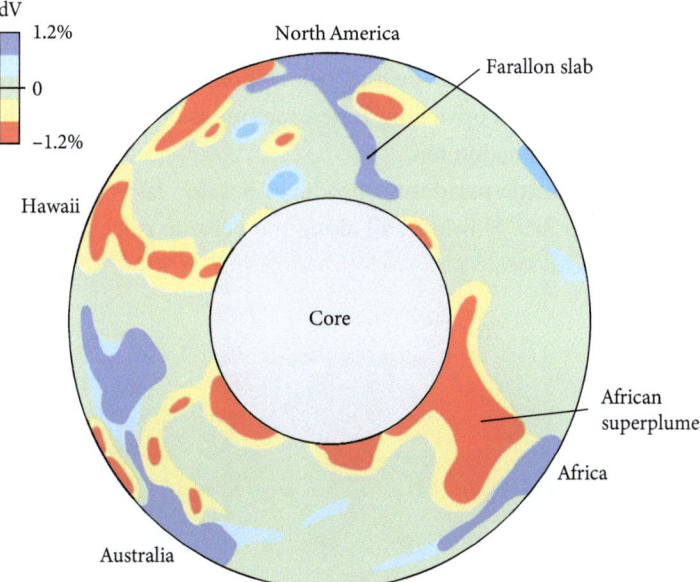

Figure 5.7 Mantle seismic tomography. Image of the small differences in seismic wave velocities (dV) with respect to a radially uniform model of the Earth's interior seismic velocity structure. A lower speed (negative dV in red) corresponds to a less dense mantle material, which is likely due to a slightly higher temperature. Conversely, higher seismic velocities (positive dV in blue) correspond to denser and thus colder mantle. Subducted plates can be traced as far as the core–mantle boundary in some cases. Hot mantle plumes (red) can also be recognized, although they appear to be of variable width and are not perfectly vertical.
Source: Adapted from Zhao et al. (2013)

accepted by geologists for about 50 years and now provides the frame of reference for virtually all geological studies. The mantle convection that drives plate tectonics is different from the convecting plumes that give rise to hot spots. Some twenty mantle plumes have been identified, and most occur away from plate boundaries. The spatial independence of plumes and plates is nicely illustrated by the Hawaiian–Emperor seamount chain, which is the track generated by motion of the Pacific plate to the north and then to the west-north-west above a fixed hot spot.

Magma genesis due to plate tectonics

The world's volcanoes occur in only three geodynamic contexts: at the apex of rising mantle plumes, along oceanic ridges or other zones of divergence, and along subduction zones (Figure 5.8). This means that outside these geodynamic sites, the Earth's mantle does not melt. Indeed, it must be emphasized that the mantle is solid almost everywhere, despite the fact that it very slowly convects.

We have already explained how melting occurs when ascending plumes intersect the solidus, generally at around 150–200 km depth (Figure 5.3). Melting also occurs at shallow depth under oceanic ridges, where the upper convecting mantle upwells. Here, melting starts at around 100 km depth, and unlike plumes, where temperature is an additional major factor, partial melting below spreading ridges is the consequence of decompression only. In all upwelling zones, melting is incomplete, with only 10 to 25% of the mantle becoming molten.

Partial melting of mantle peridotites produces basaltic lavas that can sometimes be observed at the surface. However, all along the oceanic ridges, lavas flow quietly

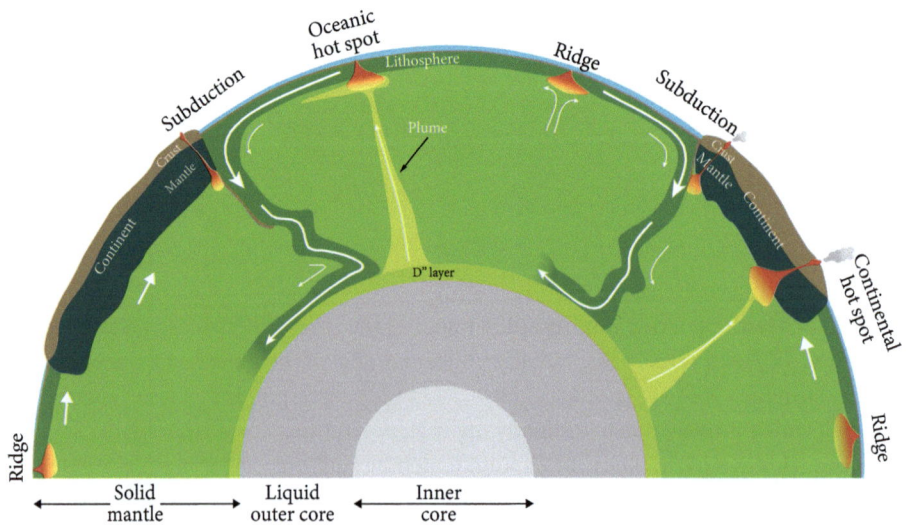

Figure 5.8 Convection and partial melting sites in the Earth's mantle. Thicknesses of oceanic and continental lithosphere and D″ layer have been exaggerated. In red: magma production zones.

and undetected into the deep ocean. Mid-oceanic volcanism was observed for the first time *in situ* as fresh basalt pillows forming at the Atlantic ridge south of the Azores. This observation occurred during underwater diving expeditions using manned submersibles. These were part of the Franco-American FAMOUS[1] project in the early 1970s to explore the Atlantic ridge. Three underwater vessels were used: one American, the *Alvin*, and two French ones, the *Cyana* saucer and the *Archimède* bathyscaphe (Juteau and Maury, 1999). As basaltic lavas cool, cracks appear and seawater can seep into the newly formed fractures. This is the origin of submarine hydrothermalism, which alters the basaltic rocks: secondary hydrated minerals, such as chlorite and serpentine, grow at the expense of pyroxene and olivine.

Volcanism along subduction zones has a different explanation, as there is no upwelling of warmer mantle in this context. As the cold oceanic plate sinks, it warms up gradually and dehydrates, releasing water that seeps into the overlying mantle wedge. Water percolating through mantle peridotites lowers their melting point by a few hundred degrees. Thus, magmatism associated with subduction zones is due to the influence of the water added to basalt during hydrothermal alteration after it formed at the ridges (Figure 5.9). Subduction-related magmas, such as **andesites**, are slightly

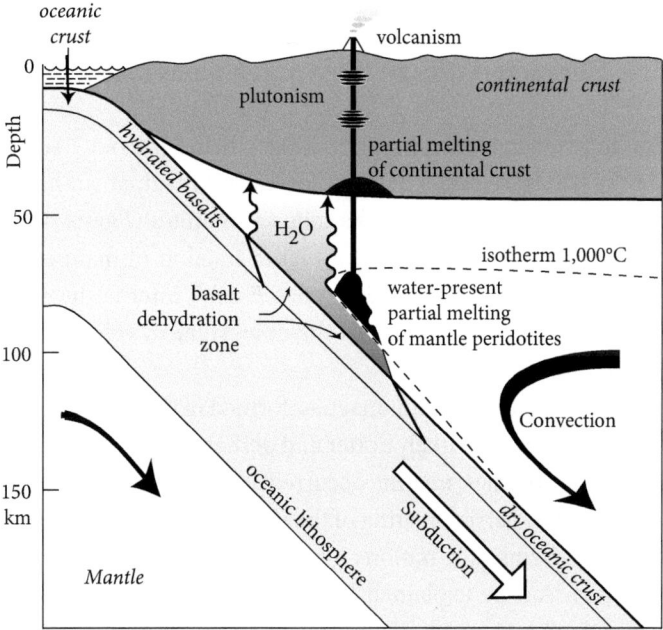

Figure 5.9 Vertical cross-section of a current subduction zone under the Andes.
Magmas originate in the sufficiently hot mantle (at least 1,000 °C) due to water release by the dehydration of the subducted oceanic crust. At the base of the continental crust, mantle magmas can evolve from an andesitic to a granitic composition by assimilation of continental crust and fractional crystallization.
Source: Adapted from Wyllie (1984).

[1] French American Mid Ocean Undersea Survey.

richer in silica than ordinary basalts. Andesites take their name from the Andes Mountains, where they are abundant. Water that facilitated the melting of the mantle ends up in solution in the newly formed magmas. These water-rich andesitic magmas are often responsible for violent and deadly explosive eruptions when degassing at shallow depth.

Volcanism can be aerial or submarine, and it can be spectacularly explosive. Yet only a fraction of magma generated in the mantle is extruded by volcanoes—about one tenth of magma reaches the surface. The remaining magma crystallizes within the crust or at the base of the crust. The solidification of magmas at depth generates **plutonic** rocks, which crystallized slowly due to the higher temperatures. There is no difference in the chemical composition of plutonic rocks with respect to their volcanic counterparts, but there is a difference in the size of their minerals, because slow cooling allows for prolonged crystal growth. The plutonic equivalent of a basalt is called a **gabbro**, and the equivalent of an andesite is a **diorite**. The oceanic crust, resulting from the crystallization of magmas produced at the level of the ridges, is mainly formed of gabbros. In subduction zones, andesitic magmas that do not reach the surface produce rocks that are dioritic to granitic in composition. These plutonic rocks are the main constituents of the continental crust.

Characteristics of the Archean magmatism

Plate tectonics and magma genesis as described in the previous section occur on the present-day Earth. How might plate tectonics and magmatism have operated on the ancient Earth? Unique volcanic rocks called komatiites (Figure 4.6) formed only in the Archean. They were produced by a higher fraction of mantle partial melting than occurs today. This is not surprising since Earth's internal heat was greater in the Archean. However, komatiites are not the only Archean volcanic rocks; Archean basalts are also abundant.

Komatiites are thought to represent magmas formed in the hot mantle plumes of the Archean. Archean plumes were likely hotter and perhaps more numerous than current plumes. In these hotter plumes, melting occurred at depths of 200 to 300 km below the surface and reached 40% partial melting of the mantle, thus yielding hot, magnesium-rich magmas. Small chemical variations have been identified between komatiites of different ages, which can be explained by different depths of melt initiation in the mantle. For example, the Munro komatiites in Canada (2.7 Ga) resulted from slightly shallower melting than the Barberton komatiites (3.5 Ga). This pattern reflects the thermal evolution of the mantle, which cools progressively over time. Logically, late Archean mantle plumes were cooler than those of the early Archean, but warmer than present-day plumes (Figure 5.10).

Warmer Archean conditions also explain how TTG magmas were formed. At that time, melting of subducted hydrated basaltic oceanic crust was possible (Figure 5.11), whereas in younger subduction zones, subducted oceanic crust dehydrates without

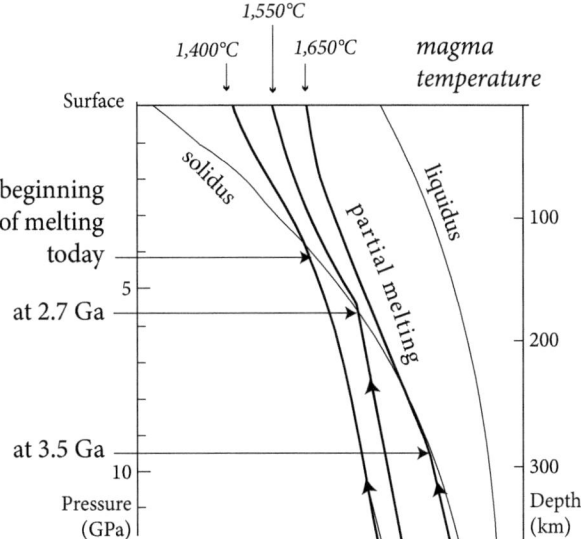

Figure 5.10 Conditions for the genesis of hot-spot magmas in the mantle. Melting begins when the rising plume (arrow) intersects the peridotite solidus. The hotter the ascending plume (corresponding to a right shift in the figure), the greater the depth at which it encounters the solidus, allowing a greater percentage of melt to form and reach the surface.

Source: Adapted from Herzberg (1992).

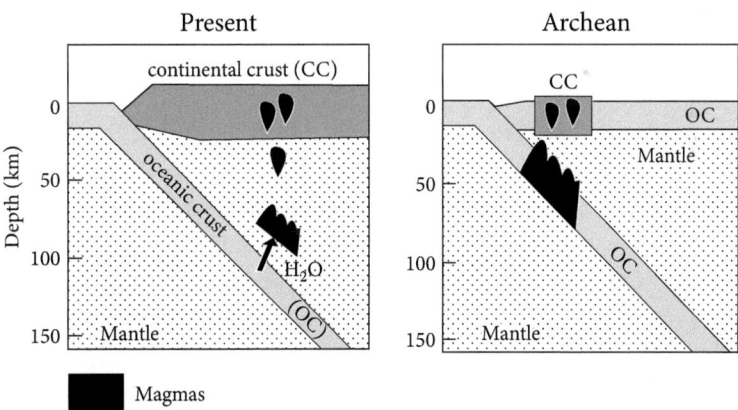

Figure 5.11 Vertical sections comparing subduction through time. Today magmas are produced in the mantle wedge due to water-driven melting, whereas, as hypothesized by Hervé Martin, during the Archean, they were formed at shallower depths within the subducted oceanic crust as a result of the higher mantle temperatures.

melting. The source of subduction magmas and the places in which they formed in the Archean were therefore different from the present day. However, this hypothesis assumes the existence of subduction in the Archean, which is not obvious, and is widely debated.

When did plate tectonics begin?

A priori, Precambrian **geodynamics** obeys the same laws as current geodynamics. However, a comparison with the planets closest to the Earth raises questions about these ancient periods. Venus, so like the Earth in size, shows no large-scale linear structures, which are a hallmark of plate tectonics on Earth. Rather, volcanoes are more or less uniformly distributed on Venus's surface. Thus, plate boundaries appear to be absent, and volcanoes on Venus appear to be the product of hot spots. Mars also has huge hot-spot volcanoes, such as Olympus Mons, which reaches 22 km in elevation. The possibility of embryonic plate tectonics in the Martian past is debated, but even if some form of plate tectonics occurred early in Martian history, it never reached the scale of plate tectonics on Earth. Therefore, it seems that the Earth is a special case. But how did plate tectonics initiate in the first place?

Geological evidence for ancient plate tectonics is mostly associated with convergent plate boundaries where large mountain ranges originate, such as the Alps and the Himalayas. Ancient continental collision chains may reach lengths of 1,000 km or more, and are readily recognized by geologists by their intensely deformed and metamorphosed rocks, even if the mountains themselves have long since been eroded. Ancient oceans, on the other hand, have largely been lost to subduction. Nevertheless, remnants of ancient oceanic crust that escaped subduction are found here and there, along ancient convergent plate boundaries. These oceanic relics are known as **ophiolites**,[2] because of their greenish colour and shiny appearance resembling a snakeskin, which is due to the abundance of serpentine. This is how, eventually, the location of the extinct oceans can be found.

There is no doubt that plate tectonics was operating during the Proterozoic eon in a manner that must have been very similar to the present. But what about during the Archean? Fortunately, there are numerous Archean terranes across the globe that provide clues to this question. The Superior province in Canada (named after Lake Superior) is an example. This province contains a series of plutonic rocks (TTGs and granites) aligned with greenstone belts and their associated sedimentary successions over hundreds of kilometres in length (Figure 5.12).

Detailed study of the Abitibi region of Québec, well known for its gold mines, has identified and dated three successive volcanic and sedimentary cycles between 2,735 and 2,697 Ma, representing intense geological activity spanning some 38 million years. These structures are interpreted to record rapid subduction along an elongated protocontinent and correspond with a global phase of continental growth around 2.7 Ga (Figure 5.13).

It is such a context that Hervé Martin imagined to explain the origin of Archean TTGs. Archean subduction then only differed from more recent subduction in that the plates would be driven faster, linked to faster convection in the mantle. Instead of

---

[2] From the Greek *ophis* (snake) and *lithos* (rock).

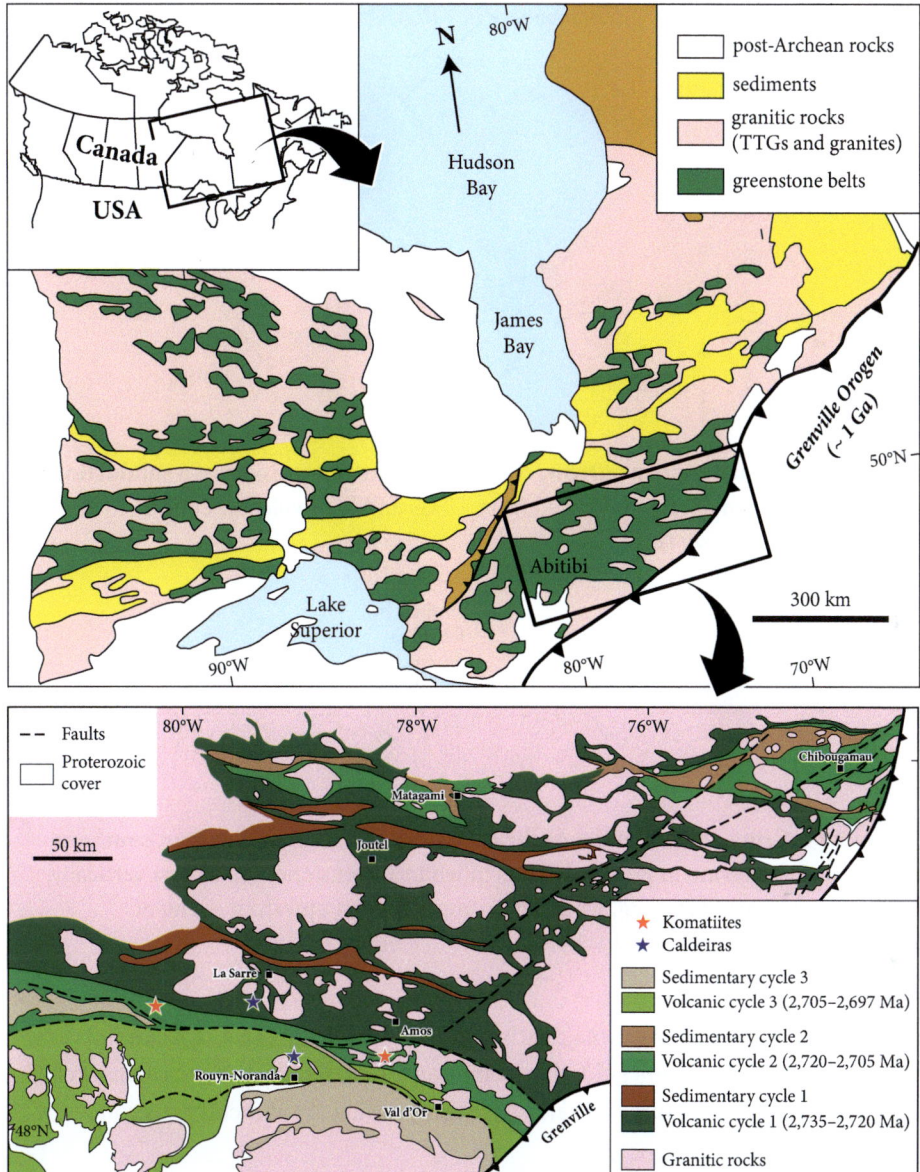

Figure 5.12 The Archean Superior Province and detailed geological map of the Abitibi region of Québec. Three cycles consistent with subduction have been recognized in the Abitibi greenstone rocks between 2,735 and 2,697 Ma; they correspond to three east–west elongated microcontinents successively juxtaposed to form a larger continent.
Source: Adapted from Chown et al. (1992).

dehydrating by heating at depth as it does today, the oceanic crust itself would partially melt and produce the TTG magmas and their volcanic equivalents that are typical of Archean crust. Transition to the thermal conditions of modern subduction happened at some point during the Archean, accounting for the progressive replacement of TTG

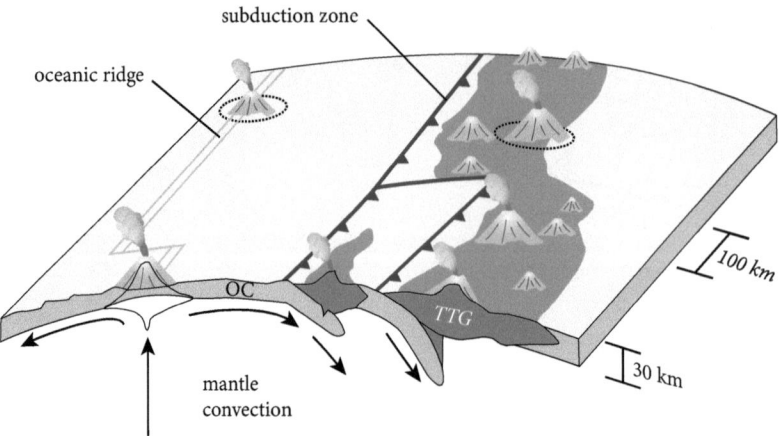

Figure 5.13 Formation of a narrow, elongated Archean continent at a subduction zone. The continental crust is formed from TTGs and their volcanic equivalents.
Source: Adapted from Condie (1994).

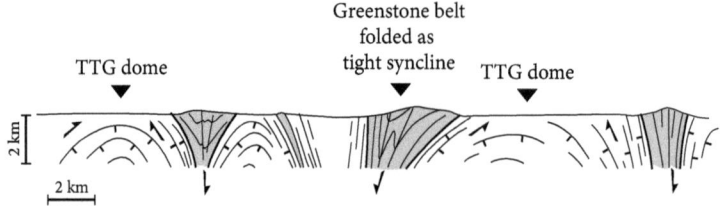

Figure 5.14 Cross-section of the Archean basement near Holenarsipur, southern India. Greenstones are in grey and TTGs in white. Greenstones dip almost vertically, suggesting downward sinking, while TTG domes suggest upward movement.
Source: Adapted from Bouhallier et al. (1995).

magmatism by 'true' granitic magmatism in the continental crust. We can thus envisage a mode of plate tectonics closely resembling the present by the end of the Archean. However, certain structures in early Archean terrains are difficult to reconcile with plate tectonics.

Before plate tectonics

In the Holenarsipur area of western Dharwar, southern India, some 150 km west of Bangalore, TTG domes are sandwiched by greenstones terrains that appear to have sunken relative to the domes, based on vertical deformation fabrics (Figure 5.14). The basalts and komatiites making up the greenstone belts are 3.3 Ga old, the same age as the surrounding TTGs. The deformation of the TTGs to form domes took place after they had crystallized in the crust. Today, after billions of years of erosion, both the TTGs and the greenstone belt rocks outcrop at the same topographic level.

It looks as if the greenstones were dragged down due to their much higher density than the TTGs, whereas the latter tended to rise. The deformation features demonstrate roughly vertical movements within the Archean crust. The driving force behind this style of deformation appears to be the difference in density between the greenstones and the TTGs. This density-driven deformation is called **gravity tectonics**. The same relationships between TTGs and greenstone belts is also seen in the Pilbara region in Australia, where the rocks are 3.4 Ga old.

The development of gravity tectonics in the Early Archean (i.e. more than 3.2 Ga ago) seems incompatible with plate tectonics. Indeed, plates must be sufficiently rigid to be passively dragged along the surface of the globe before plunging obliquely into subduction. The buoyancy-driven movements recognized in these early Archean terrains, on the contrary, suggest a soft crust without rigidity, or at least a crust incapable of supporting the enormous weight of the stacked lava of the greenstone belts. These observations (along with others not mentioned here) strongly suggest that modern-style plate tectonics did not begin until after 3.2 Ga.

What was the geodynamic context of the formation of the TTGs at the beginning of the Archean? A process referred to as **sagduction** has been proposed, whereby the cold, dense greenstones descend like cold drops under the effect of gravity (Figure 5.15). This mechanism, which is somewhat different from subduction, is incompatible with plate tectonics, but accounts for the temperature and pressure conditions required for the formation of TTG magmas to be achieved. In other words, sagducted drops of hydrated greenstone basalts were the protoliths from which the TTG magmas were generated. The residual greenstone belts were metamorphosed into garnet amphibolites during their entrainment at depth.

If this hypothesis for the formation of TTGs and greenstone belts is correct, why did the style of plate tectonics change at 3.2 Ga? This question remains a topic of lively debate and study. One possibility is that the as the Earth cooled, it crossed a **rheological** threshold that resulted in a rigid lithosphere that no longer accommodated sag-like deformation.

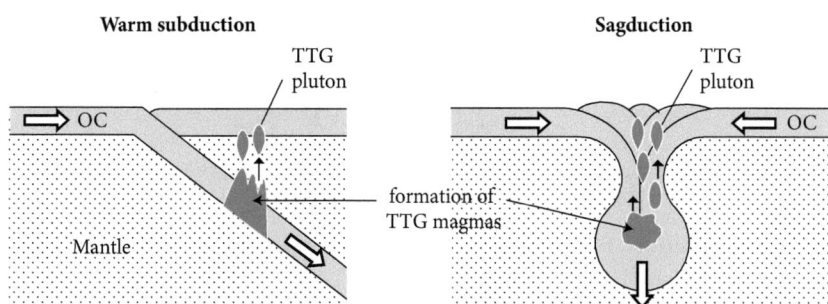

Figure 5.15 Two geodynamic contexts explaining the genesis of TTG magmas.
Source: Adapted from Moyen and Martin (2012).

The Hadean–Archean transition

If we step back in time to the end of the Hadean, 4 Ga, many more questions arise about how early crust formed and deformed. Logically, we expect the interior of the Earth to have been much warmer at that time, resulting in more voluminous production of mantle magmas. Under the higher potential temperatures in the Hadean mantle, magma genesis would have initiated at even greater depths than the 300 km estimated for the genesis of the oldest Archean komatiites. However, magmas generated by partial melting of the mantle between 330 and 410 km depth are richer in iron than their mantle protoliths and possess an unexpected characteristic: they are denser than the mantle from which they formed (Lee et al., 2010). Therefore, magmas remained trapped at this level in the mantle instead of rising towards the surface (Figure 5.16).

This hypothesis is not yet unanimously accepted, and the Hadean times still retain their share of mystery. Nevertheless, it explains why the isotopic data from the Hadean zircons indicate a lack of mantle-derived magmatism (Figure 2.2). Importantly, unlike the magmas, the unmelted part of the ascending plumes did not become stuck in the density trap. Thus, the residual hot plume continued its ascent, as shown in Figure 5.17. But the mantle of the rising plume had changed composition because it had left iron-rich magma in the density trap. The modified plume was richer in magnesium than the surrounding mantle, and it was **refractory**, meaning it was harder to melt. This refractory mantle also had a lower density than the ordinary mantle because

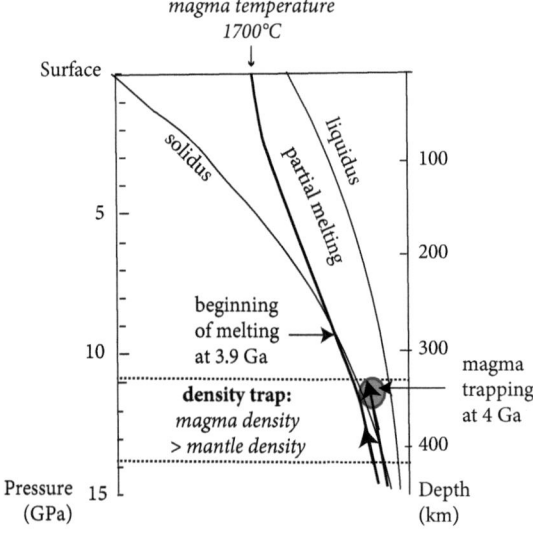

Figure 5.16 Conditions for the genesis of hot-spot magmas at the Hadean–Archean transition. The hotter Hadean plumes (right arrow) begin to melt in the density inversion zone, so that the resulting magmas remain locked at 330 km depth, while early Archean magmas generated higher in the mantle reached the surface.
Source: Adapted from Nédélec et al. (2017).

it had lost iron, a heavier element than magnesium. Because of its buoyancy, it could not easily be returned to the deep mantle by the cold downdrafts of convection. Thus, this buoyant, Mg-rich refractory mantle remained and became integrated with the thickening lithosphere as it cooled.

At the Hadean–Archean transition, or more precisely around 3.9 Ga, the onset of mantle melting (i.e. the depth at which mantle plumes intersected the solidus) was just above the density trap due to the progressive cooling of the planet. The production of mantle magmas at the surface was then at its maximum, covering the Earth's surface and engulfing the Hadean crust, of which no significant trace has yet been found (Figures 5.16 and 5.17).

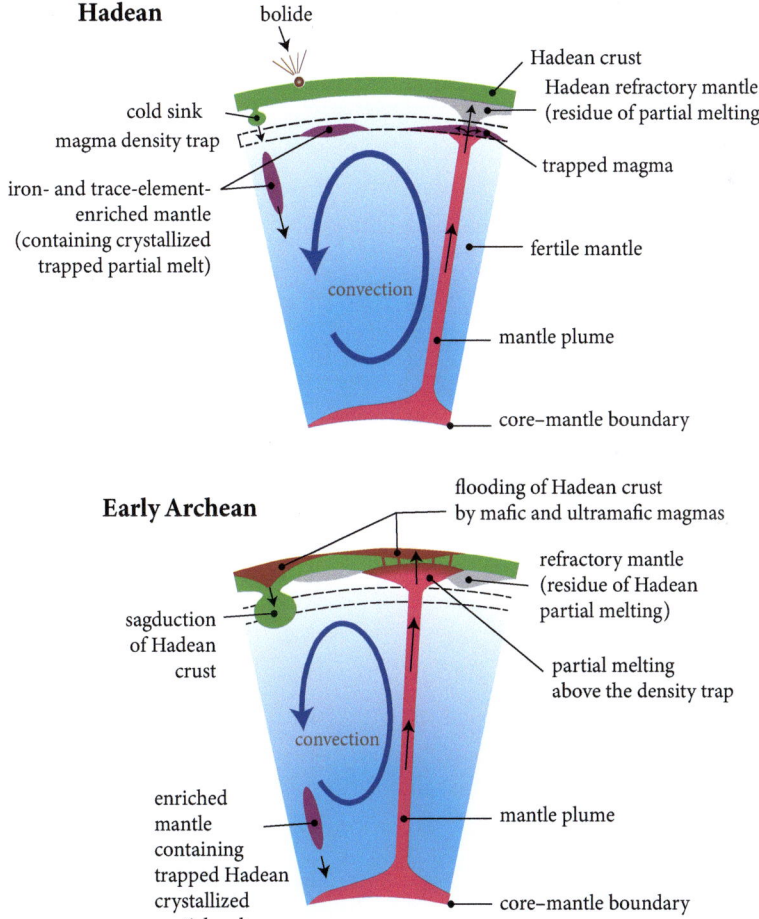

Figure 5.17 Comparison of the fate of mantle magmas in the Hadean and early Archean. Hadean magmas remained trapped in the mantle and were then driven to depth by convection, while refractory mantle accumulated near the surface in ascending plumes. Around 3.9 Ga, magmas reached the surface in large quantities and covered the entire Hadean crust.

Source: Adapted from Nédélec et al. (2017).

These events would have resulted in a complete resurfacing of our planet. Such a scenario could explain both the disappearance of the Hadean crust and the renewal of mantle magmatism that characterizes the early Archean. Nevertheless, it is still debated. Of course, the disappearance of the Hadean crust was not instantaneous: this 'catastrophic' scenario probably played out over the first few hundred million years of the Archean.

> Earth's internal heat causes melting and deformation of rocks, whose surface manifestations are volcanism and tectonics. Plate tectonics in its present form has existed for about three billion years. In the early Archean eon, the lithosphere was hotter and less rigid, and it easily bent under the weight of thickly stacked lavas that resulted from the intense magmatism at that time. In contrast, mantle magmatism seems to have been rare during the Hadean, because dense, iron-rich magmas were trapped at depth. The Hadean crust disappeared after the Hadean–Archean transition.

6

The Archean ocean

Having studied the inner engine of the ancient Earth, let us now discover its environmental landscapes. We already know that the Archean atmosphere lacked oxygen but was rich in carbon dioxide, hence there was a significant greenhouse effect. Surface conditions were therefore compatible with the presence of liquid water. Sediments deposited between 4 and 2.5 Ga provide clues to reconstructing the Archean environment, generally in the oceans, but in some instances on land. It is now possible to imagine the primitive world in which life emerged.

An Archean beach in the moonlight ...

In the present time, 70% of the Earth's surface is covered by oceans. Thus, the Earth is truly a 'blue planet' when seen from space. During the Archean eon, the extent of Earth's surface with emergent land was much smaller than today (only a few per cent of the Earth's surface, compared to 30% today); it was virtually a water world. Furthermore, the continents likely did not stand high above sea level, with the notable exception of volcanic islands. Indeed, few Archean sediments show evidence for subaerial deposition. A particularly informative type of Archean sediment is found in c.3.2 Ga Moodies Group, in the upper, youngest part of the Barberton Greenstone Belt in South Africa (Figure 4.4) (Simpson et al., 2012). The Moodies Group sediments are quartz-rich sandstones, whose characteristics indicate fluvial to near-shore deposition in an arid climate. No terrestrial plant or animal life existed in the Precambrian. Land surfaces were always barren. It was not until the Paleozoic era that the continents were finally colonized by living plants and animals. The sandstones of the Moodies Group show some particularly interesting features. These strata are relatively easy to study on the outcrop, as they have been turned upright by tectonics. One type of deposit is finely bedded and made up of alternating layers of sandstones and clay. Each sandstone layer is 1 to 2.5 mm thick, whereas the clay layers are a little less than 1 mm thick (Figure 6.1). These sediments show features familiar to sedimentologists, allowing the identification of rhythmic foreshore deposits, called **tidalites**—the oldest ever observed on Earth.

Sands alternate with finer-grained sediments deposited as the tide ebbs, as can be observed today in the Bay of the Somme or in the Bay of Mont Saint-Michel, both of which open into the English Channel. The Moodies Group tidalites also display regular variations, interpreted as alternating spring tides (at full moon or new moon)

Figure 6.1 Moodies Group tidalite as seen in outcrop. In this section, bedding is almost vertical. The thicker, lighter beds were deposited during stronger spring tides, whereas the darker, finer and more clayey beds were laid down during weaker neap tides. Two beds contain layers that are oblique to overall bedding and seem to intersect the others. These are cross beds, which form in moderate-strength flows in shallow water, as might be expected in a foreshore setting.
Source: Pierre Thomas/Planet-Terre—ENS Lyon.

and neap tides (when the moon is at its first or last quarter). Therefore, at first glance, these 3.2 Ga tidalites are strongly reminiscent of modern tidal sediments. However, a closer look shows subtle but important differences. A systematic counting of the beds and their thicknesses reveals that the lunar month lasted 20 days, not the 28 days of today (Eriksson and Simpson, 2000). This difference is because the Moon was closer to the Earth during the Archean than it is today. This means it appeared wider in the sky and revolved faster around the Earth. Earth's rotation itself was also faster, making one complete rotation in about 20 hours as compared to 24 hours today.

Little land mass emerged

The sediments of the Fig Tree Group, just below the Moodies Group, and those of the Onverwacht Group, even lower in the Barberton Greenstone succession, are quite different from the Moodies Group sediments. These rocks have ages of between 3.5 and 3.3 Ga. They are predominantly composed of shaley sediments deposited in the ocean, but in places they contain beds of siliceous chemical sediments called **cherts**. Cherts, which are made of a microcrystalline version of quartz, are produced by the precipitation of silica directly from water. This was a common sediment type during the Archean eon, but it is relatively rare today. Conversely, Archean sedimentary successions contain few carbonate rocks, though these rock types are abundant in more recent strata. Moreover, the Archean carbonates are mostly **dolostones**—that is, rocks

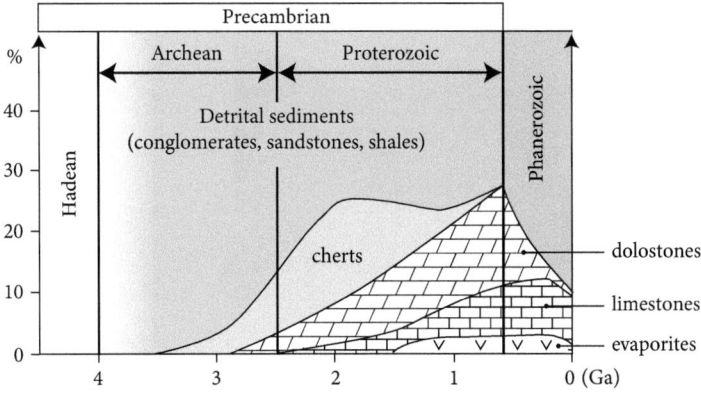

Figure 6.2 Distribution of sedimentary rock types over geological time. The figure provides percentages of the different sediments and cannot account for the relatively low volume of sediments in the Archean eon with respect to younger times.

containing both magnesium and calcium, in addition to carbonate[1]—as opposed to limestones which have only a small amount of magnesium (Figure 6.2). Nevertheless, abundant carbonate minerals precipitated in the hydrothermally altered marine volcanic rocks. Archean strata also lack **evaporites**—rocks made of salt minerals that commonly precipitate in lagoons or other shallow, restricted basins in arid settings.

The composition of Archean detrital rocks indicates that they were mostly derived from the destruction of basaltic crust. This observation is especially true for the oldest sedimentary rocks. However, towards the end of the Archean, the upper crust becomes increasingly rich in TTGs and granites such that, compositionally, the late Archean crust resembles the current continental crust. Many early Archean basalts and komatiites are found in areas considered to be the embryos around which early continents grew, such as the Barberton area, meaning these early crustal remnants are denser than younger, more granitic crust (Figure 6.3).

The composition of the crust has consequences for Earth's topography. Today, granitic continental domains, whose average altitude is about 800 metres above sea level, are different from oceanic domains, whose basaltic crust is on average 5,000 metres below the ocean surface. This contrasting topography is due to the different mineralogical composition and therefore the different density of the two types of crust (Figure 6.4). A simple calculation can be used to determine whether the continents are emerging or not. Whatever the type of crust, above a certain reference depth, or compensation surface below which density is uniform, such as the base of the crust, the rocks are considered to be in gravitational equilibrium. This application of Archimedes principle to the continental crust is known as **isostasy**. A 35 km-thick

---

[1] A dolostone is a rock consisting mainly of dolomite, a mineral which is a calcium and magnesium carbonate with the formula $CaMg(CO_3)_2$, whereas a limestone is formed of calcium carbonate (without magnesium), such as calcite $CaCO_3$.

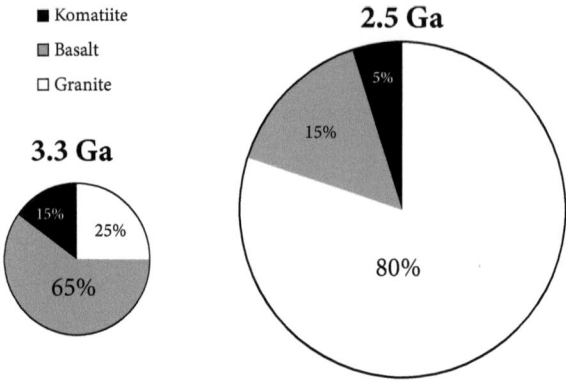

Figure 6.3 Composition and relative extent (circle areas) of the 'continental' crust at 3.3 and 2.5 Ga.

Source: Adapted from Tang et al. (2016).

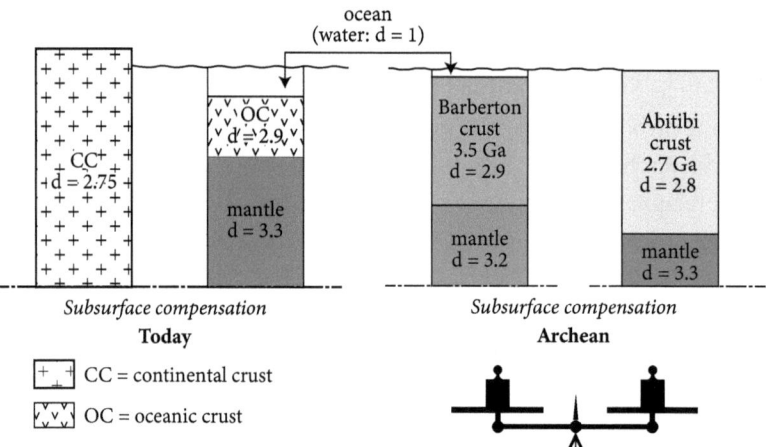

Figure 6.4 Application of the isostasy principle. Columns above the compensation surface are weighed for comparison. Today, the weight of the column consisting of 35 km of continental crust (with density 2.75 g/cm³), of which 1 km is above the ocean surface, is equivalent to that of the column consisting of 5 km of water, 10 km of oceanic crust (density 2.9) and 19 km of mantle (density 3.3). This value is the usual density of the mantle beneath the crust. However, the very ancient Archean crust is underlain by a somewhat more magnesian mantle with a density of only 3.2. During the Archean, a dense Barberton-type crust (at 3.5 Ga) was in isostatic equilibrium under 1 km of water, while a more granitic (and therefore less dense) crust, such as the crust of the Abitibi at 2.7 Ga, was at approximately sea level and possibly slightly emergent.

column of present-day continental crust, which stands about 1 km above sea level and has a density of 2.75 g/cm³, exerts the same pressure on the compensation surface in the mantle as a 10 km-thick column of oceanic crust with a density of 2.9 g/cm³, overlain by 5 km of water (with a density 1.0 g/cm³) and resting on a 19 km-thick column of mantle with a density of 3.3 g/cm³.

Now, let's compare two types of Archean continental crust: an older, Barberton-type (3.5 Ga) crust and a younger crust, such as the 2.7 Ga Abitibi crust. They do not have the same density, as they do not contain the same proportion of granitic rocks (TTGs and granites) and basaltic rocks *sensu lato*. Our isostatic calculation suggests that the early Archean crust was submerged under 1 km of water (Figure 6.4). Therefore, only a few volcanoes would have been above the water surface at that time. From 3.2 Ga onwards, however, the continental crust became less dense as granitic rocks became more abundant. The surface of the continental crust rose therefore closer to sea level. In some places, it might even have formed a low topography that was subjected to aerial weathering and erosion just as mountains and hills are today.

Therefore, we need to revise the idea of 'continent' for such ancient times. The Archean continents did not have the same composition as those of today, and they did not provide large areas of emergent land until the latter Archean. Therefore, if we are to look for evidence of early life, we must focus on the marine settings, such as basaltic ocean floor.

Submarine hydrothermalism and the origin of cherts

Archean volcanic rocks are often associated with the chemically precipitated siliceous rocks, or cherts, which we described previously. The impressive state of preservation of the Barberton greenstones is due to widespread silica impregnation. Where did all this silica come from? Evidently this silica was extracted by seawater that circulated through fractures in the newly formed volcanic crust. Seawater seeped in, and as it descended through the fractures, it heated up (because of the geothermal gradient) and reacted with basaltic and komatiitic rocks. The heated seawater was then buoyant and rose again, finally emerging at hydrothermal vents with a composition strongly modified by the reactions that took place at depth. This style of hydrothermal circulation is well known in the present-day ocean in the vicinity of the mid-ocean ridges, where newly formed basaltic oceanic crust is still warm. In 1977, in the Pacific Ocean not far from the Galapagos Islands, American scientists in the small submersible *Alvin* observed for the first time active chimneys rising above the seafloor from which hot water gushes out at more than 300 °C. Pressure at the depth of 2,000 m allows water to remain in the liquid phase at these high temperatures. The plumes of hot water resemble black smoke, because they are loaded with iron sulphide particles, such as **pyrite**, which is why these submarine vents are known as 'black smokers' (Figure 6.5). Previously unknown ecosystems teem with life in the vicinity of these oceanic hydrothermal vents, where the first link in the food chain is made up of microorganisms that exploit the energy emerging from the vents, unlike the photosynthesizing plants and phytoplankton that form the base of the food web in most continental and oceanic settings today. Together with the rich microbiota that have colonized the black smokers, crustaceans, molluscs, and giant tubeworms thrive in the dark abyss.

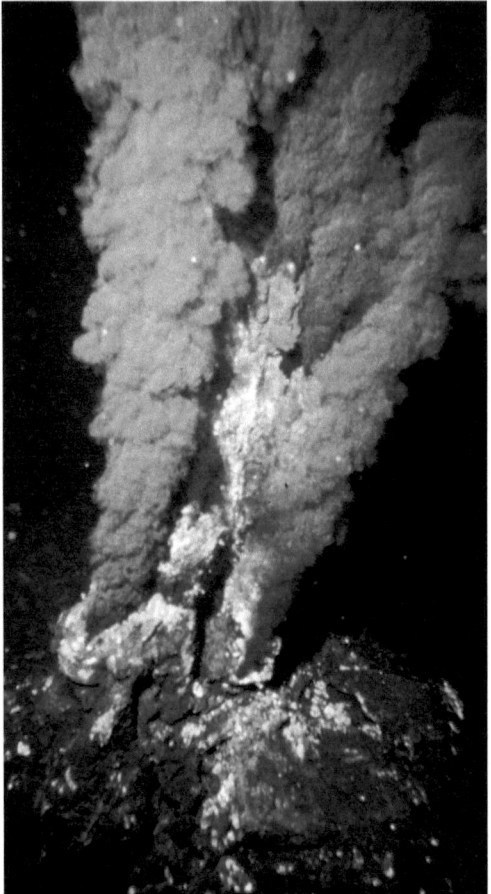

Figure 6.5 A black smoker near the Mid-Atlantic Ridge. The temperature of the fluid emitted from this vent is around 300 °C, far above the boiling point of water at the Earth's surface, because of the high pressure prevailing at a depth of around 2,500 m. Hydrothermal chimneys like this can reach a height of several tens of metres, with a diameter of a few metres at the base.

Circulation of seawater in the oceanic crust induces water–rock reactions. Seawater is normally loaded with dissolved salts. In present-day seawater, the main ions are Na^+, Cl^-, SO_4^{2-} (the sulphate ion), and Mg^{2+} (Figure 6.6). This composition results from the weathering of the continental surfaces by runoff and from the hydrothermal alteration of the oceanic crust.

Heated to more than 300 °C during its circulation in the fractures of the oceanic crust, seawater becomes chemically very aggressive and reacts with the oceanic crust. Calcium and iron, which are abundant in basalt minerals, are carried away as dissolved elements in the hydrothermal fluid. They are replaced by sodium and magnesium from seawater, which form secondary alteration minerals, such as chlorite (a greenish hydrated mineral), that replace the original magmatic minerals such as olivine and

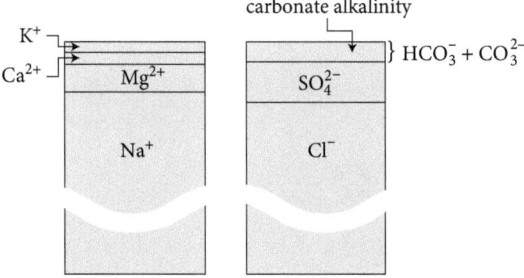

Figure 6.6 Composition of current seawater. NaCl is the most important dissolved salt. Left: cations (positively charged), right: anions (negatively charged). Carbonate (CO_3^{2-}) and bicarbonate (HCO_3^-) ions are grouped together because they are linked by equilibrium reactions to CO_2 dissolved in seawater. These inorganic carbon species contribute most of the alkalinity of seawater.

pyroxene. Upon emerging at the seafloor, the hydrothermal fluids react with the cold seawater. The dissolved elements Ca^{2+} and Fe^{2+} react with the sulphate ion (SO_4^{2-}), which is abundant in present seawater. Thus, anhydrite, a calcium sulphate, precipitates and builds the chimney at the exit point of the hot fluid. Dissolved iron (Fe^{2+})[2] contributes to small grains of pyrite (FeS_2) (Figure 6.7), which, as we already learned, make the vent fluids black. These pyrite grains settle rapidly around the hydrothermal vents.

Marine hydrothermal fluids and reactions must have been different in the Archean ocean. Basalts and komatiites were obviously strongly altered hydrothermally, but Archean seawater differed from present-day seawater because of a different atmosphere composition. The Archean atmosphere was devoid of oxygen, but rich in carbon dioxide, such that seawater was anoxic, and by extension contained little or no dissolved sulphate. Conversely, it was rich in dissolved CO_2 and likely in HCO_3^- (the bicarbonate ion) and CO_3^{2-} (the carbonate ion), which participate in the following reactions in seawater:

$$CO_2 + H_2O = H^+ + HCO_3^- = 2H^+ + CO_3^{2-}.$$

A high dissolved CO_2 content (left) shifts the equilibria to the right, suggesting a rather acidic seawater because of its H^+ content. Today, seawater **pH** is around 8; neutral pH is at 7; lower values correspond to an acidic solution, higher values to a basic solution. Archean seawater is assumed to have been only slightly more acidic, with a pH around 7. Indeed, the H^+ ions of the above reactions are rapidly neutralized by hydrothermal reactions with the minerals in the oceanic crust. Alteration of the seafloor therefore likely buffered the seawater pH.

---

[2] Strictly speaking, the notation Fe(II) should be used instead of Fe^{2+}.

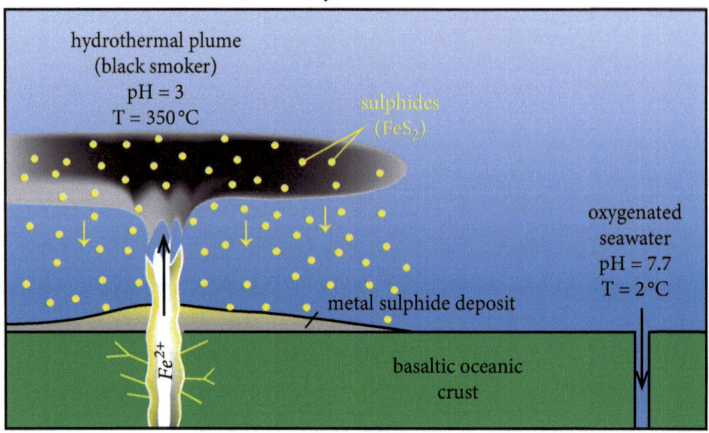

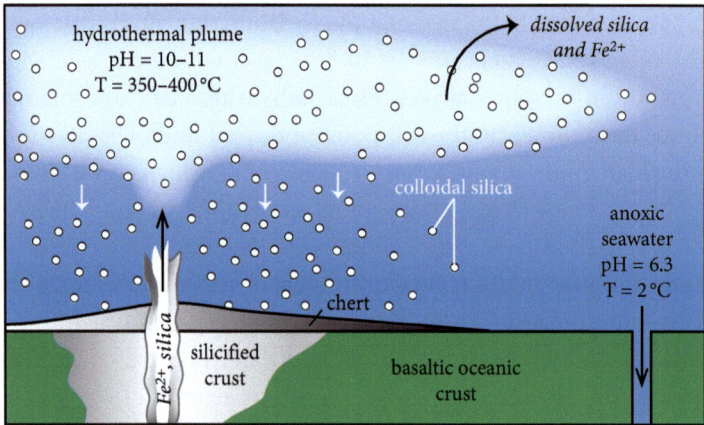

Figure 6.7 Comparison of present-day and Archean submarine hydrothermal smokers. Dissolved ferrous iron is shown as Fe^{2+} in both cases.
Source: Adapted from Shibuya et al. (2010).

However, the circulating fluid in the seafloor fractures may have become basic, that is, its pH rose as a consequence of mineral reactions with this hydrothermal fluid. At pH > 9, hydrothermal alteration of basalts releases dissolved silica. At a lower pH, the dissolution of silica is impossible. The differing pH values of present-day and Archean submarine hydrothermal fluids explain why Archean submarine volcanic rocks are impregnated with silica in the vicinity of ancient hydrothermal vents (Figure 6.7). Silica was dissolved from the Archean seafloor but precipitated rapidly in contact with the more acidic seawater. Note that the exact pH of the Archean seawater is unknown and is a matter of debate. Maarten de Wit, professor at Nelson Mandela University in Port Elizabeth, South Africa, recently discovered an ancient hydrothermal field with three silica vents still standing in a chert layer sandwiched between two pillow basalt flows of the Onverwacht Group (de Wit and Furnes, 2016). These pipes, some 5 m in diameter and 14 m high, are a few dozen metres apart. They are now enclosed in a thick pillow basalt flow, itself sealed by the overlying flow. This underwater hydrothermal field was active a little over 3.4 billion years ago.

The rest of the dissolved silica in the hydrothermal fluids was released into the Archean seawater where it then precipitated as chert, which is so common in the vicinity of Archean greenstone belts such as Isua and Barberton (Figure 6.8a). Morgane Ledevin, who studied the Barberton cherts in detail for her PhD research, showed that the dissolved silica from the hydrothermal vents first forms a colloidal suspension—a whitish, opalescent water, loaded with nanometric to micrometric silica particles—as soon as the pH of the hydrothermal fluid drops a little (becomes less basic) on contact with seawater (Ledevin et al., 2014). These particles settle to the seafloor where they first form a ductile siliceous mud (Figure 6.8b), which when compacted forms chert. Hydrothermal reactions also explain the dissolution of iron. In the oxygen-free Archean ocean, iron was stable in a dissolved state, which is impossible today.

Finally, the high concentration of dissolved bicarbonate and carbonate also drove the precipitation of hydrothermal calcite ($CaCO_3$) in Archean seafloor basalts, whereas calcite does not precipitate readily in the current oceanic crust. The amount of precipitated calcite changed during the Archean. Early Archean greenstone belts display the traces of active hydrothermal alteration, such as in Isua (3.8 Ga) and Barberton (3.5 Ga). The 3.2 Ga altered oceanic crust exposed in the Cleaverville area of Western Australia was also pervasively carbonatized, reflecting a higher dissolved CO_2 content of seawater than in recent times (Shibuya et al., 2007). By contrast, carbonation of late Archean to Paleoproterozoic volcanic rocks is at a lower level, suggesting a significant decrease of seawater CO_2 concentration and therefore a lower CO_2 content of the atmosphere at that time (Shibuya et al., 2013).

Figure 6.8 Archean cherts within the Barberton Greenstone. (a) Thin chert layers (coin for scale), which are normally white, unless they contain impurities; black chert seen here contains carbonaceous material. (b) Black chert within a sandstone. The siliceous mud that formed the chert deformed sand deposited on top of it before consolidation, hence the flame-like structures near the tip of the pen.
Source: (a) the author; (b) Marianne Ledevin.

Temperature of the Archean oceans

The temperature of the Archean oceans is a matter of debate. It has long been thought that Archean seawater was much warmer than today. This hypothesis was based originally on oxygen isotope data from ancient cherts, supplemented by silicon isotope data after this technique became available. The ratios of these stable isotopes are partially controlled by the temperature at the time of the rocks' formation. The first isotopic analyses of the Barberton cherts suggested temperatures around 75 °C. But how should this result be interpreted?

Cherts have been shown to have a hydrothermal origin. Isotopic analysis of the siliceous hydrothermal vents recently discovered at Barberton provides temperatures over 100 °C. As we learned, the hydrothermal fluids emitted at current black smokers are about 300 °C (Figure 6.7). However, they cool very rapidly as they mix with cold seawater. In the same way, the temperatures obtained by the isotopic analyses of the Archean cherts probably only have a local significance corresponding to a stage of mixing between the hydrothermal fluid and the seawater. It would be hazardous to draw any general information on the temperature of the whole of the Archean ocean from these isotopic results. Nevertheless, let's remember that there were high temperatures in the vicinity of the Archean hydrothermal vents. We will see in Chapter 7 that such settings may have been the cradle of the first life on Earth.

Sulphates or sulphides? A consequence of the composition of the atmosphere

In the present-day ocean, hydrothermal fluids react immediately with sulphate ions in seawater to form pyrite, as described in the context of black smokers. The sulphate ions in modern seawater come mainly from the oxidative weathering of pyritic rocks exposed to air. Indeed, the instability of pyrite on the surface of the continents is illustrated by the oxidation of pyrite crystals in cheap slate roofs, which became oxidized in less than thirty years. Iron in pyrite is then transformed into iron hydroxide (rust) and sulphur forms sulphate ions that are carried away by run-off and ultimately washed into the ocean.

In the Archean, on the few emergent land masses, pyrite remained stable when exposed to the atmosphere, as demonstrated by the Witwatersrand **conglomerate** in South Africa (Figure 6.9). This coarse detrital rock contains rounded quartz pebbles, as well as pyrite pebbles. These pebbles were transported by a stream and deposited on the coast of a continent at around 2.9 Ga. They were then covered by other detrital sediments and have remained preserved since. The pyrite pebbles were exposed to air during weathering and erosion, but they show no sign of oxidation. Such a scenario is unimaginable today—detrital pyrite is never found in clastic sediments. Hence, this pyritic conglomerate is clear evidence that the Archean atmosphere lacked free oxygen.

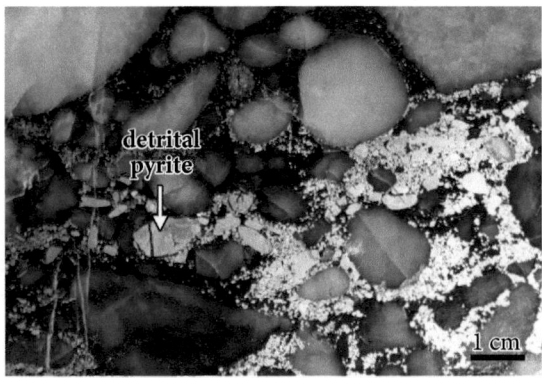

Figure 6.9 A polished slab from the Witwatersrand Conglomerate. The slab is exhibited in the Geosciences Building Hall at Stellenbosch University (South Africa). Among the rounded white quartz pebbles, the smaller pyrite pebbles are recognized by their bright golden appearance, showing no signs of oxidation.
Source: The author.

What happened to dissolved iron in the ocean?

We already saw that ferrous iron could remain in solution in the **anoxic** (oxygen-free) Archean ocean. Only a small amount of dissolved oxygen in water is required to oxidize ferrous iron (Fe^{2+}) to ferric iron (Fe^{3+}), which precipitates as insoluble iron hydroxides and oxides such as hematite. The oxidation of the iron dissolved in the ocean led to the deposition of very special rocks: *banded iron formations*, commonly referred to by geologists using the acronym **BIF**. Banded iron formations are chemically precipitated sediments characterized by alternating siliceous and ferruginous layers, which give the appearance of banding (Figure 6.10). Coexistence of silica and iron oxides in BIFs provides clues as to the nature of the waters in which they precipitated. These were deep-sourced waters, strongly influenced by submarine hydrothermalism.

The very fine layers of BIFs, millimetre to more than one centimetre thick, can be followed over long distances. They were therefore deposited below the wave base in a quiet environment, far from emerging land, based on the typical absence of detrital sediments. Calculation of the deposition rate of Hamersely BIFs in Australia suggests a maximum rate of 1 mm/year (Trendall et al., 2004), Hence, the finest banding may correspond to an annual process, possibly with a seasonal control.

BIF were described through the whole Archean and even as early as in the 3.8 Ga old Isua rocks. Nevertheless, BIFs became very abundant between 2.7 Ga and 2.4 Ga, during a peak in deposition at the end of the Archean (Figure 6.11) corresponding to observed current quantities reaching several billions of tonnes. BIFs disappear from the geological record after 1.8 Ga, but reappear briefly at the end of the Precambrian at about 700 Ma, as will be discussed in Chapter 13. This temporal distribution implies that after 1.8 Ga, the world ocean was oxygenated, at least near its surface.

Figure 6.10 A 2.7 Ga BIF sample from Carajás, Brazil. Note the double centimetric and millimetric banding, suggesting a depositional process influenced by climatic and/or seasonal rhythms. The siliceous chert beds are red due to tiny dispersed particles of hematite (iron oxide), while the massive hematite beds are grey.
Source: The author.

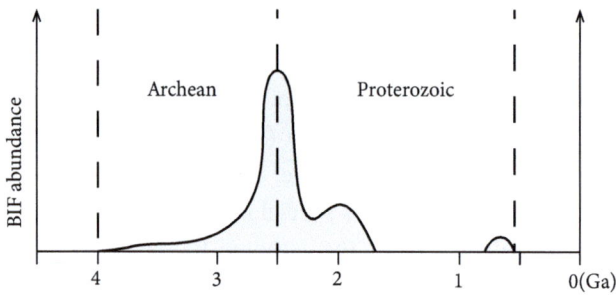

Figure 6.11 Temporal distribution of BIFs. BIFs were very abundant during the Archean–Proterozoic transition, then disappeared c.1.8 Ga—only to reappear briefly at the end of the Precambrian.
Source: Adapted from Bekker et al. (2010).

What was the source of the oxygen that precipitated the iron oxides and gradually emptied the ocean of all its dissolved iron? The common answer is that it was a biological waste product, produced by **photosynthesis.** Indeed, photosynthesis enables cells to produce their own organic (carbonaceous) matter from CO_2 thanks to the energy of sunlight. In the current marine environment, photosynthetic cells use CO_2 dissolved in the water and form oxygen as a byproduct:

$$CO_2 + H_2O \ (+ \text{sunlight}) = CHOH \ (\text{organic matter}) + O_2$$

Photosynthesis is limited to the upper 200 metres of the ocean, which defines the photic zone where sunlight penetrates. Deep waters containing the dissolved iron and silica released by hydrothermal vents must have come into contact with the surface waters where photosynthetic organisms live to precipitate insoluble iron oxides. In the late Archean, the photosynthesizers that produced the oxygen in the surface oceans were probably cyanobacteria (a group of bacteria of very ancient origin).

Nevertheless, minor BIFs formed as early as 3.8 Ga, as observed in Isua. Oxygenic photosynthesis is unlikely to have occurred at that time. Other microbial metabolisms may have contributed to the oxidation of dissolved ferrous iron. An iron-dependent anoxygenic photosynthesis called 'photoferrotrophy' is the likely predecessor of oxygenic photosynthesis. Photoferrotrophic bacteria grow using light and ferrous iron to produce their own biomass from dissolved CO_2. Ferric iron is a byproduct of this reaction and is readily precipitated as hydroxides and oxides. Other bacteria may directly generate ferric iron without light using only the energy derived from chemical reactions—a process called 'chemolithotrophy'. Photoferrotroph and chemolithotroph bacteria still exist today in special environments. They have been studied in detail and cultured in order to estimate the bacterial mass required to deposit the huge amounts of Archean BIFs. The required bacterial densities are in the range 10^4 to 10^5 cells/cm^3, one or two orders of magnitude less than the typical bacterial density in current seawater (Konhauser et al., 2002).

The productivity of planktonic photosynthetic organisms, such as cyanobacteria, follows a seasonal cycle, compatible with the rhythmic banding of BIFs. Importantly, it is only the iron oxide layers that require a rhythmic pacing. Precipitation of the silica layers would have been solely dependent on pH conditions, and could theoretically have occurred as soon as the deep water containing dissolved silica became less basic than the hydrothermal fluid, which occurs rapidly away from the hydrothermal vents, as we saw in the formation of cherts.

Giant iron ore mines

Most iron used around the world comes from BIFs deposited about two and a half billion years ago at different depths in the ocean. Normally, the crust of ancient oceans disappears through subduction into the mantle. A recent detailed study of BIFs from the Carajás mining area in Brazil (Rossignol et al., 2023) shows that BIFs were deposited in sharp and conformable contact immediately above the volcanic rocks of the Parauapebas large igneous province aged 2.7 Ga, evidencing a link between the formation of a deep oceanic plateau and BIF formation. Thus, remnants of ancient oceanic crust may have been preserved, because the crust was too thick to sink or to be subducted easily, as in the case of the Caribbean oceanic plateau formed about one hundred million years ago. Other BIFs were deposited on shallower platforms along continental nuclei and were also preserved. The largest deposits in the world are the late Archean (c.2.5 Ga) Hamersley deposits on the southern edge of the Pilbara craton in Australia.

Australia, the world's largest producer of iron ore, produced 945 million tonnes in 2023. Brazil is the world's second-largest producer, with two important mining regions: the Iron Quadrangle (*Quadrilátero Ferrifero* in Portuguese) in the state of

Figure 6.12 View of part of the N4 iron mine (Carajás, Brazil).
Source: Vale (CVRD) company.

Minas Gerais, and the Carajás region in the state of Pará. The iron from the BIFs is mined there in huge quarries (which geologists still call 'mines'). Thanks to Afonso Nogueira, a colleague from the University of Belem, and to the managers of the Vale company (formerly CVRD)[3] in Carajás, I had the opportunity to visit the huge N4 iron mine, which extends several kilometres in length (Figure 6.12).

The iron ore consists mainly of hematite and contains up to 67% iron by weight. Mining is relatively simple: extraction and transport, without on-site processing. Every evening, explosives are placed at the working face. They break up the rock into huge blocks, which are then picked up by mechanical shovels and placed in giant trucks, then unloaded onto a conveyor belt. The ballet of trucks is never-ending. The conveyor belt dumps the blocks into the wagons of an ore train. A purpose-built, 800-kilometre-long railway through the Amazon rainforest carries the ore to the port of San Luis in the Maranhão state, east of Pará. There it is loaded onto ships bound mostly for China, the world's largest iron consumer and Brazil's main customer.

> The continental surfaces were still small during the Archean eon. The composition of the Archean ocean was different from today due to the carbon dioxide-rich and oxygen-free atmosphere. Vigorous oceanic hydrothermalism stripped silica and ferrous iron from the oceanic crust and pumped them into seawater. Precipitation of silica and oxidized iron from seawater produced cherts and banded iron formations (BIFs). The oxidation and precipitation of iron was likely mediated by

---

[3] CVRD, the Rio Doce Valley Company, founded in 1942 in Minas Gerais, has today become the multinational company Vale, whose headquarters are still in Brazil.

microorganisms akin to the oxygenic photosynthesizing bacteria in surface oceanic waters. Life was therefore already present in the Archean and exerting a strong influence on the terrestrial environment.

7
Origin of life

When and how did life appear on Earth? We know that the Earth was probably habitable as early as the Hadean, since its surface temperature, except at times of giant meteorite impacts, allowed the existence of liquid water. This is indeed the fundamental prerequisite for life. But was life already present in the Hadean? How old are the earliest fossils? How did these primitive organisms live? And where did the molecules that early life used to metabolize and reproduce come from? We'll try to answer these questions, but the first question that fascinates biologists and philosophers alike is how do we define life? A living being is recognized as capable of maintaining its physical integrity in its environment and capable of reproducing itself. Another critical feature of life is that the process of reproduction gives rise to evolution.

The molecules of life

Living matter, or more precisely, the matter that constitutes living beings, is made up of carbon, hydrogen, and oxygen, with a little nitrogen, a pinch of phosphorus, and a trace of sulphur. These elements make up organic molecules, which are often of large size; these molecules are divided into carbohydrates, lipids, proteins, and **nucleic acids.**

Carbohydrates have the basic formula $C_x(H_2O)_y$. Some carbohydrates, such as glucose (Figure 7.1), are the essential fuel for **cellular metabolism.** Others serve as reserves, for example starch or glycogen. Other long-chain carbohydrates, such as cellulose, are used to build the wall that protects or stiffens the membrane of bacterial and plant cells. Finally, a simple five-carbon sugar, ribose, and its variant, deoxyribose, are

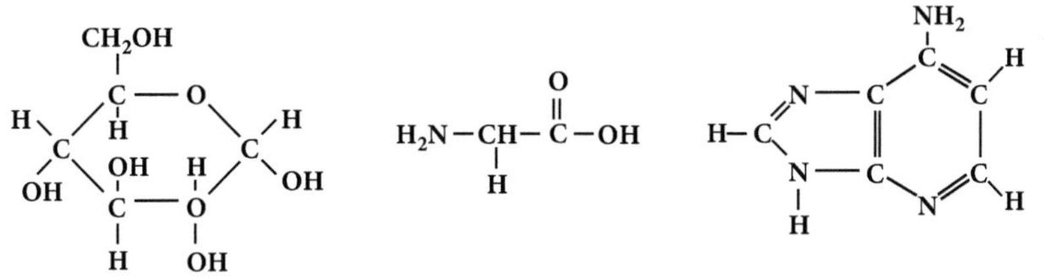

glucose glycine adenine

Figure 7.1 Some basic molecules found in living things. A sugar (glucose), the amino acid glycine (which is a common component of proteins), and a nitrogenous base or nucleobase (adenine, which is a component of DNA).

DOI: 10.1093/9780198945451.003.0008

used in the composition of **RNA** (ribonucleic acid) and **DNA** (deoxyribonucleic acid), which are essential molecules for the functioning and reproduction of cells.

Lipids make up the cell membranes of all living things. They contain mainly carbon and hydrogen. Proteins are complex macromolecules that ensure the functioning of living cells—for example by acting as enzymes, biocatalysts that increase the rate of cellular reactions. They are made up of long chains of **amino acids**, which are folded into complex three-dimensional structures.

The order (biologists say the 'sequence') of the amino acids in the molecular chain of a protein is dictated by the information stored in DNA as a gene, or piece of DNA. DNA carries the genetic information transmitted by a cell to its descendants. In 1953, two researchers, the American James Watson and the Briton Francis Crick, who were working in the physics department of the University of Cambridge in England, published the structure of DNA, in the form of two chains made up of alternating deoxyribose and phosphate linked by nucleobases. The monomeric unit, or **nucleotide**, comprises a deoxyribose, a phosphate, and a nucleobase. There are four nucleobases: adenine (A) (Figure 7.1), cytosine (C), guanine (G), and thymine (T).[1] The nucleobases are associated in complementary pairs, always in the same way: A only with T and C only with G, and thus form the rungs of a ladder. Finally, the ladder turns on itself like a double helix (Figure 7.2). The discovery of the structure of DNA earned Watson and Crick the Nobel Prize in 1962. For the sake of completeness, it should be noted that Watson and Crick used X-ray images of DNA taken by another British researcher, Rosalind Franklin (Franklin and Gosling, 1953). Unfortunately, the latter, who died at the age of 37 in 1958, did not receive the Nobel Prize, which is never awarded posthumously.

The genetic code

The sequence of nucleotides is translated into proteins following a universal code in the living world: the genetic code. A sequence of three nucleotides, or codon, corresponds to a single amino acid, and a sequence of codons to a protein. The information contained in the DNA in the form of codon sequences is transcribed by means of RNA, a single-stranded nucleic acid complementary to one of the two DNA strands. This so-called messenger RNA (mRNA) brings the copy of the genetic information to small organelles present in the cytoplasm of the cells, the **ribosomes**. Ribosomes were discovered by the Romanian–American biologist George Emil Palade, who was awarded the Nobel Prize in Physiology and Medicine in 1975. They are micro-machines able to read the information and to assemble the corresponding amino acids into a protein chain; that is, they translate DNA's genetic messages into proteins. Small molecules of RNA called 'transfer RNA' (tRNA) transport specific amino acids. Each tRNA has a three-nucleotide sequence complementary to a three-nucleotide codon in the mRNA. This sequence is called an 'anticodon'. At one end of the tRNA, the anticodon matches

---

[1] Thymine is replaced by another nucleobase, uracil, in RNA.

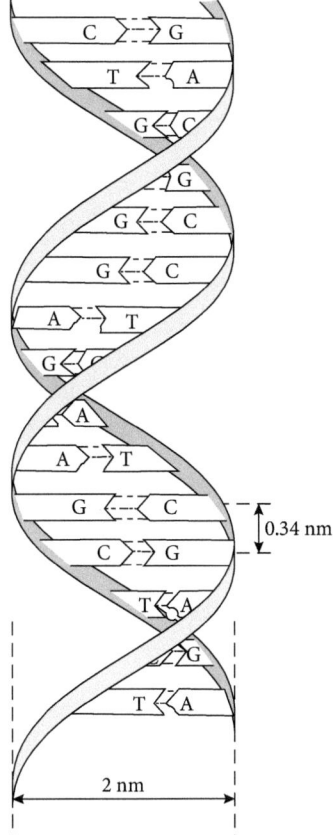

Figure 7.2 Structure of the DNA molecule. Two complementary strands are wound to form a double helix; A, C, G, and T refer to the four nucleobases: adenine, cytosine, guanine, and thymine.
Source: Adapted from Watson and Crick (1953).

the codon, whereas at the other end, the transported amino acid is attached to the growing protein chain (Figure 7.3).

An experimental approach to prebiotic chemistry

The molecules described above show that the chemistry of life is mainly a chemistry of carbon. But where, when, and how did the elementary molecules necessary to make the macromolecules of life come together? In 1871, in a letter to one of his correspondents, Darwin imagined that life could have appeared in a 'warm little pond' on the Earth's surface. In this favourable environment, ammonia and other simple precursors could have reacted to form amino acids and even simple proteins. This idea was the basis of an experiment carried out in 1953 by the young American biologist Stanley Miller, a member of Professor Urey's laboratory at the University of Chicago.

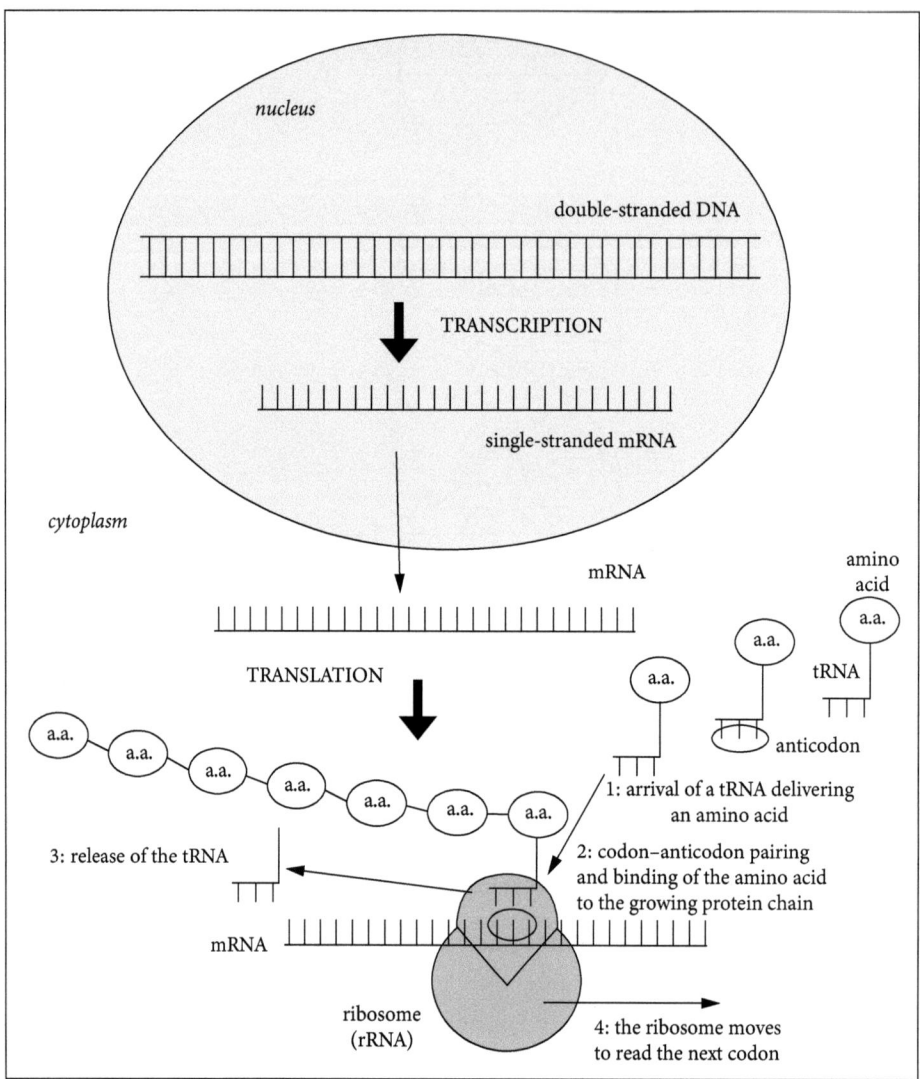

Figure 7.3 Genetic information and protein synthesis. In eukaryotic cells, transcription occurs in the nucleus: it produces an mRNA molecule from the DNA template. During translation, ribosomes synthesize amino-acid chains (future proteins) with the participation of specific tRNAs.

Stanley Miller used a gas mixture of hydrogen (H_2), ammonia (NH_3), and methane (CH_4) to represent what at the time was thought to be the composition of the early atmosphere. This gas mixture was contained in a flask where electrical discharges were produced episodically, mimicking lightning (Figure 7.4). Water vapour from a bubbling bath flowed through the flask. After a week, brown-coloured muck had accumulated at the bottom of the reactor. These organic compounds included amino acids, which are essential building blocks for life.

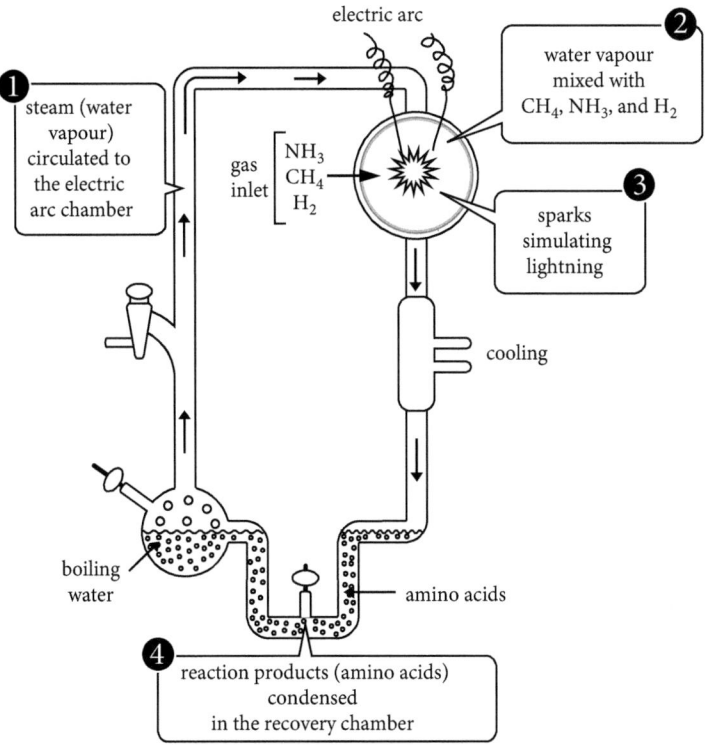

Figure 7.4 The Miller experiment (1953). A device designed to reproduce the genesis of prebiotic molecules in Earth's early atmosphere.

Unfortunately, we now know that Miller's experiment did not reliably reproduce Earth's early atmosphere, which was less reducing and richer in CO_2 than CH_4. Indeed, CO_2, the most oxidized form of carbon, is highly stable. Reduction and hydrogenation of CO_2 to form organic compounds is difficult to achieve abiotically.

Other more recent experiments into the origin of organic molecules are worth mentioning, including the work of Alan Schwartz of Radboud University in Nijmegen, the Netherlands, in 1982. Schwartz and his collaborators showed that adenine, one of the nucleobases of DNA and RNA, can be synthesized at low temperatures (−2 °C) from a frozen solution of hydrocyanic acid (HCN). In solution HCN tends to polymerize spontaneously, and under the effect of ultraviolet radiation, adenine is formed. The main objective of this experiment is to show that nitrogenous bases can be obtained by a cold synthesis route, applicable to the extraterrestrial environment.

Despite these compelling results, we are still a long way from having synthesized a nucleic acid in the laboratory. It is as if we have only made a brick, whereas we are trying to build a house! We still do not know how the macromolecules of life are formed. Nevertheless, the building blocks of life seem to be formed easily and are even present in some meteorites and comets.

Organic matter in meteorites and comets

We have seen that some meteorites contain carbon. These are carbonaceous chondrites, some of which contain amino acids. One of them, the Murchison meteorite, which fell in Australia in 1969, weighed more than 100 kg, and contained about fifteen amino acids and nitrogenous bases, including uracil, which characterizes RNA. The carbon-isotopic signature of these extraterrestrial compounds shows unambiguously that they were produced by an abiogenic process. To determine this isotopic signature, ratios of the stable carbon isotopes (^{12}C and ^{13}C) contained in the carbonaceous matter are measured[2] and compared to a reference standard in order to identify an excess or a deficit of carbon 13. If an excess of ^{13}C is observed with respect to the standard, the carbon isotopic signature is said to be positive. In the opposite case, it is said to be negative.

The isotopic signature of extraterrestrial carbon molecules is strongly positive with an excess of ^{13}C (Martins et al., 2008), whereas carbon molecules produced by living beings on Earth always have a negative signature typical of biological processes. For example, living cells that carry out photosynthesis prefer to use CO_2 molecules containing light carbon (^{12}C) rather than ^{13}C. Their own organic matter, as well as that of the cells that feed on it, will therefore always have a deficit of ^{13}C, and therefore a negative carbon isotopic signature.

Nevertheless, the building elements of life (i.e. the simple carbonaceous molecules) may have been produced by extraterrestrial chemistry. These molecules were probably present already during the accretion of our planet. However, it is unlikely that these molecules would have been preserved under the very high temperatures that led to the formation of a magma ocean at the beginning of the Earth's history. It is therefore necessary to consider renewed contributions from falling carbonaceous meteorites or comets after the surface temperature of our planet cooled enough for liquid water to exist.

Comets originate from the icy far reaches of the solar system. They may contain organic molecules synthesized at low temperatures during the condensation of the protoplanetary nebula. They are commonly described as dirty snowballs, because organic inclusions darken the ice. As they approach the Sun, they begin to sublimate: the ice changes from a solid to a gas, forming the characteristic comet tail. Three remarkable space missions provided more knowledge of these objects and confirmed the presence of organic molecules.

The first of these missions was NASA's *Stardust*, which aimed to return samples from the tail of comet 81P/Wild to Earth. The probe was launched in 1999 and reached its target in 2004. Using a cleverly designed tennis racket-shaped device, the probe captured thousands of particles in an ultralight porous medium called aerogel (Figure 7.5). The capture device then folded up, protecting the aerogel until it returned

---

[2] The results are expressed using the delta notation (already seen in Chapter 2 for oxygen isotopes): $\delta^{13}C$ of the sample (in ‰) = {[($^{13}C/^{12}C$) sample / ($^{13}C/^{12}C$) PDB standard] −1} × 1,000. The international PDB standard is the ratio measured in calcite of a belemnite rostrum from the Pee Dee formation in South Carolina.

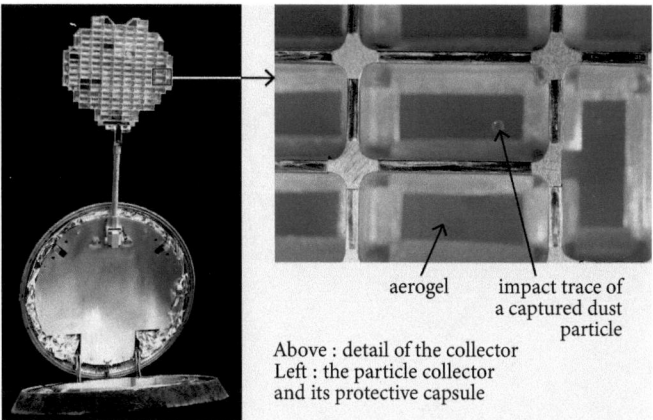

Figure 7.5 The cometary particle-collection device of the US *Stardust* mission. The device is now on display at the Smithsonian Air and Space Museum.
Source: NASA.

to Earth. Among the many particles captured was the amino acid glycine, whose strongly positive $\delta^{13}C$ isotopic composition left no doubt that it was extraterrestrial in origin (Elsila et al., 2009).

In 2005, NASA launched another probe, appropriately named *Deep Impact*, intended to drop an impactor on the nucleus of comet Tempel 1. The probe analysed the fragments thrown up by the impact and determined the presence of dust, water ice, and hydrogen cyanide ice.

In 2004, the European Space Agency (ESA) launched the *Rosetta* spacecraft towards comet 67P/Churyumov–Gerasimenko, better known as Chury. Ten years after its launch, *Rosetta* entered Chury's orbit (Figure 7.6) and sent the lander *Philae* to the comet's surface. The landing was partially successful: *Philae* was badly positioned on the rough surface of the comet and could not communicate with Earth. Nevertheless, the mission enriched the catalogue of cometary organic molecules. Many organic compounds or their precursors (methane CH_4, ammonia NH_3, hydrogen cyanide HCN, thioformaldehyde CH_2S, etc.) have been detected on the surface of the cometary nucleus, some of them (e.g. acetone, C_3H_6O) for the first time on a comet. Despite their diversity, these compounds are always small molecules. For example, propane (C_3H_8) and butane (C_4H_{10}) have been identified, but no compounds with more than four carbon atoms have been found (Altwegg et al., 2007). Furthermore, we now know that organic matter only occurs as grains or even clusters on the cometary surface, rather than being dispersed throughout its ice, as was previously thought.

All these results confirm that extraterrestrial prebiotic chemistry may have provided the buildings elements for life, but no incontrovertible evidence of extraterrestrial life has yet been found in the solar system.[3] However, the precursor molecules of life,

---

[3] The discovery of numerous exoplanets (3,639 in 2,729 extrasolar systems since 2017) is opening new prospects for the search for extraterrestrial life.

Figure 7.6 Comet Chury photographed by the *Rosetta* probe. It is about 6 km long and 2 km wide. The resolution of this image is 70 cm.
Source: ESA.

perhaps brought by comets, may have assembled in favourable places and conditions on the early Earth to give rise to the first cells. The question of how and when life originated remains open.

In search of the oldest fossils

Until recently, the oldest recognized fossils were about three and a half billion years old and were found in Australia and South Africa. The Australian fossils come from the Archean Warrawoona series of rocks in Western Australia, near one of the hottest places on the planet, which, in wry Australian humour, is named North Pole. The Warrawoona series is mostly a pile of lava flows that erupted underwater, perhaps as part of an oceanic plateau. The sedimentary part of the succession includes cherts of hydrothermal origin and, at the very top, some carbonate horizons. The volcanic rocks date the Warrawoona series to about 3.5 Ga. The Apex basalts and cherts from these series became famous when in 1984 geologist Roger Buick identified carbonaceous filaments in these cherts (Figure 7.7) and suggested that these filaments might have a biological origin.

This interpretation has not been unanimously accepted by specialists, which highlights the challenge in identifying fossils of this age. To begin with, Archean fossils are bound to be simple traces of microbial life, and hence very small. Both morphological and geochemical evidence is necessary to advocate for a biological origin of any putative fossils. Morphological arguments can be made by comparing fossil shapes with those of modern microbial communities. As for geochemistry, the presence of

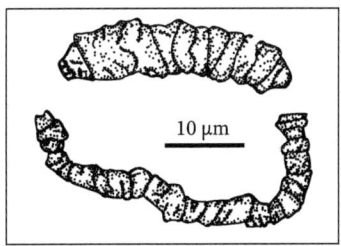

Figure 7.7 Filamentous microfossils from the Apex chert in Western Australia.
Source: Adapted from Schopf and Kudryavtsev (2012).

carbon and sometimes other elements characteristic of living organisms, such as nitrogen, must be demonstrated. Carbon isotope ratios are a valuable tool to discriminate between biological and non-biological origins, because biological reactions, such as photosynthesis, always favour light carbon (^{12}C) at the expense of heavy carbon (^{13}C), meaning they generate strongly negative carbon isotopic signatures. Ideally, these isotopic measurements should be analysed *in situ*.

Let us return to the Apex cherts. Brasier et al. (2002) negated the biogenicity of the recognized filamentous microstructures and proposed that they were secondary artefacts, although they did not dispute that they were formed in a hydrothermal setting. Hickman-Lewis et al. (2016) recently found more convincing evidence for life than the filaments described above. The newer discovery comprises submillimetre fragments of carbonaceous laminae, sometimes rolled up at the edges (Figure 7.8). This sort of shape is typical of fragments of modern **microbial mats** (carpet-like concentrations of microbial colonies) that have been ripped in places but not completely torn apart. Advanced microanalysis techniques have shown that these fragments contain some nitrogen and even traces of sulphur, in addition to carbon. These small fragments of a very old microbial mat from 3.46 Ga ago could be the oldest known fossils of well-established biogenicity. They were protected by the precipitation of the silica mud, which then hardened to form the Apex chert without compaction or deformation of the microbial mats encased within. Oxygen isotopes of the Apex chert indicate that the silica in the chert precipitated at just over 100 °C. This suggests that the microbial mats were constructed by thermophilic bacteria, capable of withstanding high temperatures in the vicinity of hydrothermal vents.

In other silicified rocks from the same region of Australia that are only slightly younger (3.43 Ga), laminated domes and cones are observed, which may correspond to ancient indurated microbial mats called **stromatolites**, formed in shallow marine deposits close to the shore. Stromatolites, which were widespread at times in Earth's history prior to the emergence of animals, are relatively rare today. However, they still exist in a few special environments such as Shark Bay, a UNESCO Word Heritage site in Western Australia. Shark Bay is about 8,000 square kilometres, with an average depth of 9 metres. It is characterized by shallow banks, islands, and peninsulas, providing many protected marine areas, some of which are hypersaline because of a

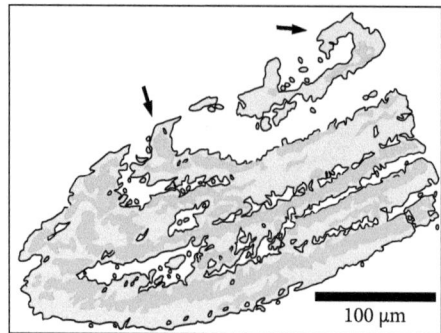

Figure 7.8 Small carbonaceous fragment consisting of leaflets curled up at the edges (arrows), found in the Apex chert (Australia). This fragment is a chip of microbial mat dating back 3.46 Ga.
Source: Adapted from Hickman-Lewis et al. (2016).

high evaporation rate. Stromatolites are irregularly laminated buildups that form due to **cyanobacteria**, which are photosynthesizing bacteria (and not algae, as their former name of 'blue green algae' suggested). Spirulina, which is well known as a food supplement, is an example of a cyanobacterium that is cultivated and marketed in the form of flakes with a beautiful dark blue-green colour after drying. In nature, certain cyanobacteria live in filamentous colonies in settings hostile to most grazing animals, like Shark Bay.

Stromatolites may form microbial mats in the tidal zone or other shallow-water settings (Figures 7.9a and b). Fine sedimentary particles are often found trapped in the felted filament mat, and successive layers of mat and sediment can become cemented (most commonly by calcium carbonate). The result is a construction of indirect biological origin, distinct from more familiar recent bioconstructions, such as corals. Corals precipitate calcium carbonate crystals by the action of their cellular metabolism, whereas microbial mats, including stromatolites, trap existing minerals in their filaments and hold them together with their slimy extracellular secretions.

Archean stromatolites from Australia are commonly encrusting (Figure 7.9c) or are dome- or cone-shaped (Figures 7.9d and e and 7.10). Despite their striking resemblance to present-day stromatolites, it cannot be assumed that the microbes that built these ancient stromatolites were cyanobacteria or even close relatives. Indeed, although their shallow-water settings implies photosynthesis, it may not have been an oxygenic photosynthesis like that of today. More primitive photosynthetic metabolisms do not produce oxygen by splitting water molecules like cyanobacteria, algae, and plants. Today for example, **sulphur bacteria** perform anoxygenic photosynthesis by oxidizing sulphide, while others are capable of oxidizing ferrous iron. In any case, the strongly negative carbon isotope signatures associated with the filaments described in the Pilbara cherts strongly suggest a biological origin (Ueno et al., 2001).

The sediments of the Archean Barberton Greenstone Belt of South Africa are roughly coeval with the Warrawoona sediments of Australia and contain abundant carbonaceous remains of likely biogenic origin, as reviewed in detail by Martin

Figure 7.9 Modern and Archean stromatolites from Australia. (a) Present-day stromatolites on the foreshore of Shark Bay. The stromatolite domes average about 50 cm in height. (b) Microphotograph of cyanobacterial filaments extracted from these stromatolites. Scale bar: 10 microns. (c) 3.4 Ga encrusting stromatolites in the Strelley Pool chert, near North Pole, Pilbara, Western Australia. Scale bar: 20 cm. (d) Conical stromatolites of the same age and origin as (c). (e) Vertical section of the stromatolites from (d) showing the layered structure.

Source: (a): Paul Harrison (licence GFDL); (c, d, and e): Allwood et al. (2007) with permission of the Geological Survey of Western Australia.

Homann in 2019. Macroscopic stromatolites are rare, but microscopic fossil microbial mats are ubiquitous, occurring as 1–20 μm-thick laminations in the black cherts of the Onverwacht Group. The oldest ones (3.47 Ga) are found in the Middle Marker chert capping the thick komatiitic succession of the lower part of the Group (Figure 4.6). This thin sedimentary unit formed during a break in eruptive activity. Deposition occurred in a hydrothermal setting at depths underwater of several tens to hundreds of metres. Like their counterparts in Western Australia, sedimentary structures and fossil remnants have been preserved through early silicification. Other chert deposits above are exposed along nearly 50 kilometres of strike. They contain abundant fossil microbial with negative $\delta^{13}C$ organic matter values averaging −29.8‰, providing additional support for biogenicity. The detrital Moodies Group (3.2 Ga) also contains fossils of

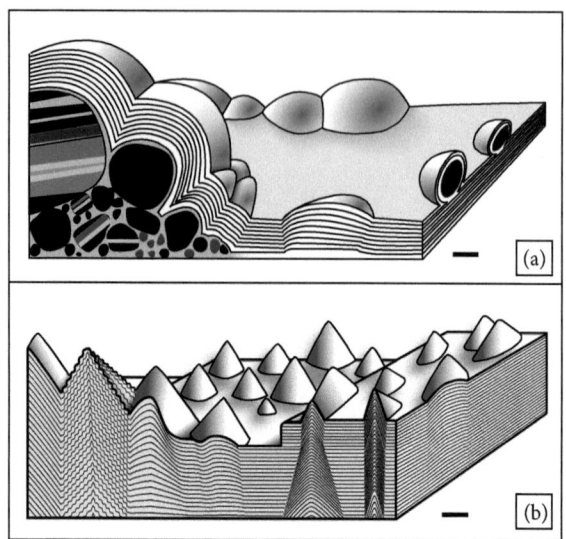

Figure 7.10 3D-reconstructions of stromatolites from Figure 7.9 c (a) and d (b). Scale bar: 5 cm.

Source: Adapted from Allwood et al. (2007).

microbial mats, these deposited in a deltaic to intertidal setting. These mats were subjected to periodic dessication and are the world's oldest traces of life on land, or more precisely along a shore. These microbial mats show striking morphological similarities to modern cyanobacterial mats. The mean $\delta^{13}C$ of organic material preserved in the mats is a negative value of –21.2‰, consistent with a photosynthetic metabolism. However, this need not imply oxygenic photosynthesis as there is no sign of free oxygen in the Barberton Greenstone Belt. The more negative isotopic signature of the older hydrothermal mats presumably reflects a different microbial metabolism.

Finally, carbonaceous material has been described in even older rocks, the Nuvvuagittuq cherts in northern Quebec, which are dated at 3,770 Ma. This discovery was made by a team of researchers led by Matthew Dodd of the University of London. Figure 7.11 shows one of these ferruginous filamentous structures. Dodd et al. (2017) propose that these structures correspond to ancient bacterial filaments preserved after impregnation with iron oxide. If this interpretation is correct, these structures would be the oldest known fossils on Earth.

Putative evidence of even earlier microbial life has been reported from 3.8 Ga rocks in Isua (south-western Greenland). The evidence is in the form of grains of graphite with highly negative carbon isotope signatures (Mojzsis et al., 1996). Unfortunately, metamorphism has transformed all organic matter in these rocks into graphite, meaning no fossils, if ever present, have been preserved. The possible evidence for life is limited to an isotopic signature. The hunt for the oldest microfossils shows its limits here. However, in spite of diminishing evidence for life in rocks older than 3.5 Ga, other arguments strongly indicate that life was present by the beginning of the Archean.

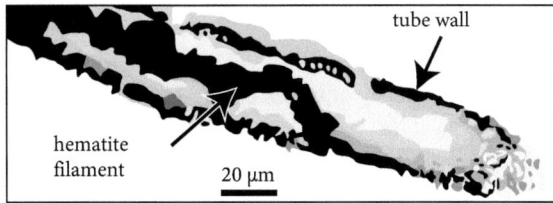

Figure 7.11 Hematite filament associated with carbon (in the form of graphite) in a northern Quebec chert dating from 3,770 Ma. This microscopic structure would be the product of the activity of iron-oxidizing bacteria (capable of oxidizing dissolved ferrous iron).
Source: Adapted from Dodd et al. (2017).

It has been argued that Early Archean life was limited by restricted access to phosphorus, an element of universal importance in all living cells. Indeed today, the main source of P is the weathering of modern continental surfaces, a source that was not available in the ocean-covered early Earth. However, this objection has been overcome by recent studies, which evidence that high amounts of dissolved P were released in the active hydrothermal alteration of the volcanic seafloor (Rasmussen et al., 2021).

The phylogenetic approach: the tree of life

Developing the tree of life was the work of Carl Woese, an American molecular biologist. The tree in question is not a simple family tree as in a family where parents and children are well known. It is called a **phylogenetic** tree, because it represents the evolutionary relationships between different organisms. When Carl Woese began his work in the early 1970s, only two domains of living beings were recognized: bacteria or **prokaryotes**, which are single-celled organisms without a nucleus, and **eukaryotes**, which are uni- or multicellular organisms, whose cells have a nucleus. Since eukaryotes are obviously more complex than bacteria, it was within the bacteria that an attempt had to be made to sort out the most primitive life forms on Earth. Carl Woese developed a quantitative approach to assess the degrees of relatedness between organisms under consideration. Specifically, he targeted ribosomes, the essential organelles present in all cells without exception, where they perform biological protein synthesis (Figure 7.3). Ribosomes are made up of two subunits, one small and one large, that fit together, and consist of RNA molecules (ribosomal RNA or rRNA) and proteins. Carl Woese compared the 16S rRNA molecule[4] which is universally present in the small ribosomal subunit of all prokaryotes. This molecule is extracted from cells by centrifugation, and then sequenced—that is, the whole suite of its 1,300 to 1,600 nucleobases is determined. The basic assumption is that the more different their 16S

---

[4] The letter S stands for a unit of measurement of centrifugation intensity, the Svedberg; centrifugation is a common method of separating cell constituents.

rRNAs are, the more distal they are from one another phylogenetically. These differences result from the gradual accumulation of random mutations, such as the chance replacement of one nucleobase by another or the repetition of a nucleobase during reproduction.

When Carl Woese began his work, microbiologists were fascinated by microorganisms living in extreme conditions, such as the recently discovered oceanic smokers or in the solfataras near volcanoes. The hunt for new thermophilic microorganisms was vividly narrated in a book by Patrick Forterre (2016). After sequencing their rRNA, thermophilic microbes were found to be very different from other bacteria. Some of them live in strict anaerobic conditions (they cannot tolerate the presence of oxygen) and have a particular metabolism that produces methane: they are therefore called **methanogens**. In fact, these methanogens are so far removed from bacteria that it was necessary to create an original group, a new domain of life, the **Archea** (capitalized here when referring to the domain; when referring to various organisms within the domain, the lower-case 'archea' is used). Thus, Carl Woese discovered that the kingdom of life was not divided into two domains, but rather three: Bacteria, Archea, and Eukaryota—a revolution in life sciences (Woese and Fox, 1977). Archea are prokaryotes like Bacteria but differ from them not only in their rRNA and their special metabolism, but also in the nature of the lipids that make up their cell membranes.

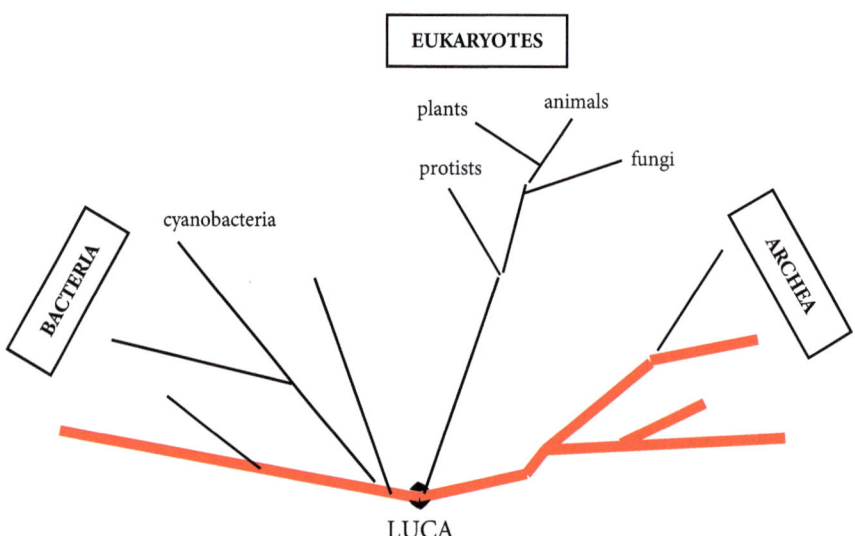

Figure 7.12 The phylogenetic tree of life and its three domains. The tree is highly simplified, but still retains its temporal dimension: the closest branches to LUCA occurred in the most distant past. The branches leading to thermophilic microorganisms, mostly archea and a few bacteria, are indicated by thick red lines and are found at the base of the tree, hence the idea that the common ancestor, LUCA, was also thermophilic.
Source: Adapted from Woese et al. (1990).

In 1990, Carl Woese proposed a modified tree of life in which eukaryotes originated from the Archea branch (Figure 7.12). Furthermore, all living beings would have derived from a hypothetical common ancestor designated by the acronym LUCA (*last universal common ancestor*, the ancestor of all living beings). Carl Woese had initially used the term 'progenitor'. The name LUCA was preferred a few years later because it is more precise and probably also because it evokes a familiar first name!

In 2003, Carl Woese was awarded the Crafoord Prize, the equivalent of the Nobel Prize (for scientific disciplines that are not eligible for it), for his remarkable contribution to the knowledge of the living world. Today, thanks to advances in computer and sequencing technology, it has been possible to sequence the ribosomal RNA of more than 270,000 species, resulting in a richer tree of life, still subdivided into three domains.

Archea in oceanic hydrothermal vents and the methane cycle

Before looking for LUCA (the ancestor of all terrestrial living beings), we need to know a little more about the archea. In December 2000, Deborah Kelley, professor at the University of Washington State in Seattle (USA), discovered a new type of oceanic hydrothermal vent, in the Atlantic Ocean at 30 N, near the Atlantis Fault. No black smokers this time, but white vents rising from the ocean floor. Some 15 km from the axis of the Atlantic ridge, this hydrothermal field evocatively named Lost City is a ghostly sight, resembling the ruins of a sunken city (Figure 7.13a). It is nearly 200 m long and about ten active vents have been discovered.[5]

Other similar hydrothermal fields have been found in the Atlantic and the Indian Oceans. The chimneys, which can reach several tens of metres in height, are made of calcium carbonate and not sulphate, as in the black smokers. They are perched upon on peridotites where mantle rocks have been exhumed and outcrop as seafloor through major faults. The fluids that flow out of the peridotites are not acidic, but rather basic, and are cooler than those of the black smokers, with temperatures of around 100 °C instead of 350 °C. Finally, they contain hydrogen and methane (CH_4) in solution.

The microbial ecosystem that thrives in these new types of smokers includes many archea that either generate methane (the methanogens) or use it (the **methanotrophs**). The life of these archea is linked to the hot reducing fluids of the white smokers, in the absence of oxygen and light. It represents a new and hitherto unsuspected living world. This deep underwater world is difficult to reproduce in the laboratory. Indeed, after twelve years of effort, Japanese microbiologists have only just succeeded in isolating and reproducing primitive anaerobic archea (Figure 7.14) from sediment samples taken near a hydrothermal vent showing the presence of methane. Many

---

[5] Washington State University has set up a website showing the findings at Lost City: www.lostcity.washington.edu.

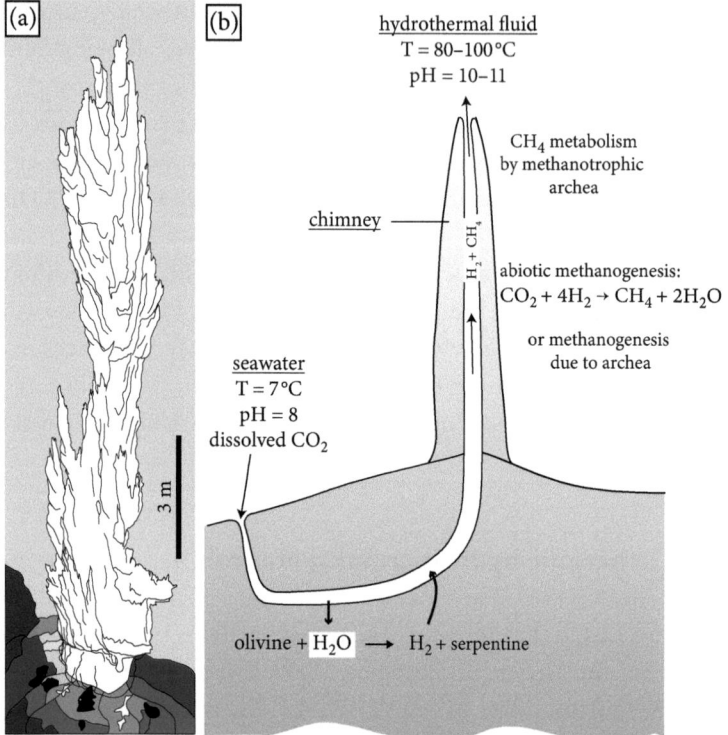

Figure 7.13 Lost City ecosystem. (a) A hydrothermal vent at Lost City. (b) Reactions between seawater and mantle peridotites, and the role of archea in the ecosystem.
Source: Adapted from Kelley et al. (2001).

archea remain uncultured. Thus, the knowledge of their biology is limited. Nevertheless, the recent progress in genome sequencing capabilities and the novel approach of sequencing genomes directly from their environment (metagenomics) has revealed the existence of a very large diversity of archea with many different metabolisms (Baker, 2020).

Let's get back to Lost City-type submarine hydrothermalism. What happens when seawater infiltrates relatively warm mantle peridotites? Mantle rocks are less rich in silica than the basalts and gabbros that make up the oceanic crust. They are mostly made of olivine. Vents on peridotites are therefore different from vents on basalts. Most importantly, the serpentinization reaction of olivine in contact with water produces H_2 (Figure 7.13b). Hydrogen in solution then allows the formation of methane by reaction with CO_2 dissolved in water:

$$4H_2 + CO_2 = CH_4 + 2H_2O.$$

This methanogenesis reaction can occur abiotically or be carried out by methanogenic archea. Methane production on its own is therefore not absolute proof of life. Although methanogenesis due to archea remains difficult to quantify *in situ*, there is no doubt that methane is used by methanotrophic archea at the outlet of the Lost City vents.

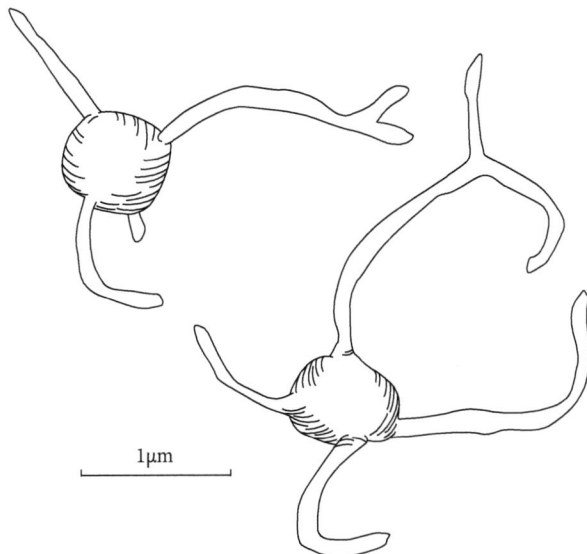

Figure 7.14 Drawing of archea obtained in culture. The cells are very small and show very long extensions that are never seen in bacteria. This morphology is not common to all archea species, many of which closely resemble bacteria morphologically.
Source: Adapted from Imachi et al. (2020).

This current ecosystem can easily be transposed to the Archean in relation with komatiites, rocks very rich in olivine, like the peridotites of the mantle. In addition, seawater was richer in CO_2 then, from which we can deduce that methanogenesis must therefore have been more active than today, where the reaction is limited by the low levels of dissolved CO_2. The early Archean environment could have supported a prolific biosphere in the vicinity of basic, reducing, oceanic hot springs. These ecosystems required no light and derived their energy from chemical reactions.

Methane is sometimes regarded as an important gas in the Archean atmosphere (Catling and Zahnle, 2020). This view is supported by the requirement to build an atmosphere enriched in gases with noticeable greenhouse effects. It is also based on the discovery of methanogens, microorganisms likely abundant in the Archean deep hydrothermal ecosystems. Personally, I do not share this view. Indeed, methane seeping at hydrothermal vents can be consumed by methanotrophic microbes before reaching the atmosphere. Today, a few ppm of methane are present in the atmosphere and this is sourced mainly from swamps and flooded soils rich in organic matter, a situation that is unlikely to have occurred in the Archean, when there were few emerged continental surfaces. In today's oxygen-containing atmosphere, methane has a very short life span: it is rapidly oxidized within a few years. Even in an oxygen-free atmosphere, methane can be oxidized to CO, then CO_2, by reaction with OH radicals in the upper troposphere. An Archean atmosphere rich in methane therefore seems unlikely. It is also unnecessary, as recent modelling shows that the greenhouse effect of methane

Who was LUCA, the ancestor of all living beings?

Before discussing the last common ancestor of living beings on Earth, it must be emphasized that LUCA is neither a fossil nor a cell whose genetic material has been sequenced. Rather, it results from the deduction that the branches of the phylogenetic tree of life are singularly rooted. The hypothesis of a common ancestor is reasonable since all living beings without exception use the same genetic code, and the same codon always corresponds to the same amino acid in the three kingdoms of life. LUCA was already using this genetic code, but this does not mean that it was the first living being on Earth. Other organisms must have preceded LUCA, and others likely coexisted with LUCA but left no descendants.

It can be assumed that LUCA had the same metabolism and habitat as the most deeply branching organisms on the tree of life. The oldest branches of the tree correspond to the methanogenic and thermophilic archea (Figure 7.13), suggesting that LUCA was also a thermophilic organism. Although not all researchers agree on this question, it is nevertheless a useful starting point for considering the environment that LUCA inhabited. One obvious candidate is in the vicinity of oceanic hydrothermal vents, with their steep thermal and chemical gradients providing a diversity of habitats and metabolic opportunities, including not just the hydrothermal chimneys but also the subsurface of nearby altered rocks. Thus, life may have been born in the darkness of the abyss. One important line of reasoning supports this scenario. What molecules make up living matter? They always consist of reduced carbon, meaning they are bonded with hydrogen atoms. The presence of hydrogen and the reducing nature of the fluids observed in hydrothermal sites associated with the serpentinization of olivine-rich rocks makes them a highly favourable incubator for the earliest life. In these deep environments, early life was also protected from harmful UV radiation. The scattered and discontinuous distribution of these first habitable oases likely did not impede these early life forms from spreading, as dilute hydrothermal plumes may rise hundreds of metres off the sea floor and disperse hundreds of kilometres off the ridge axis (Dick, 2019).

Indeed, the oldest fossils found in Archean cherts probably represent organisms that lived in the vicinity of hot hydrothermal springs, although nothing conclusive is known about their metabolism. Indeed, we cannot even be certain that they were archea. Nevertheless, we can reasonably infer that these organisms were both anaerobic and **chemoautotrophic**, meaning that they used dissolved CO_2 as a carbon source and derived their energy from chemical reactions along the steep redox gradients established between the hydrothermal fluids and seawater. However, at 3.43 Ga, microbial life was able to thrive in shallow-water environments, because adaptive evolution had provided the solutions to use light as an energy source and to protect the cells against UV. The key molecules that play these roles are pigments, such as the different types of chlorophyll (the main photosynthetic pigments), and other pigments constituting UV shields (Wynn-Williams et al., 2002). All biological pigments absorb specific wavelengths while reflecting others.

It should also be noted that, although the first traces of life date back at best to around 3.8 billion years ago, the Earth is 700 million years older. Life on Earth was possible as early as the Hadean, since water existed then in a liquid state. But nothing is known about life in the Hadean, due to the lack of a geological record. Was life already present? Did it emerge multiple times, then disappear completely during the events that led to the engulfment of the Hadean crust under torrents of lava at the beginning of the Archean? Is LUCA the only survivor of primitive Hadean life? So many questions remain open.

Before LUCA: from prebiotic chemistry to the first cell

Although we can make many inferences, we do not know precisely what LUCA was, when it appeared, or how. We do not know if other forms of life preceded it. Nor can we connect the gaps between elementary organic molecules, such as amino acids and nucleobases, and the first forms of life, although we do know that these molecules can be produced by photochemical reactions in the farthest reaches of the solar system on the surface of comets. How did these simple molecules come together to form the macromolecules (proteins, lipids, nucleic acids) that characterize living cells? Simple carbonaceous molecules seeded in the primitive terrestrial ocean by falling carbonaceous chondrites and comets would have helped, but they would have produced too dilute a solution such that the probability of random collisions resulting in larger, more complex molecules would have been very low. Thus, the primordial soup theory of the origin of life seems highly unlikely. Some researchers have proposed that mineral surfaces provided favourable niches for concentrating organic molecules and possibly catalysing their polymerization and reproduction.

Even more enigmatic is how macromolecules organized themselves into cells, and how these first cells acquired the ability to replicate. A simple experiment provides a partial answer here. Cell membranes are always made up of lipid bilayers, with their characteristic hydrophobic interiors. The hydrophobic parts naturally and spontaneously self-assemble in such a way that they are not in contact with the external or intracellular environments, which are rich in water. Experiments with lipids result in the formation of globules called 'liposomes', bounded by a double layer of lipid molecules that resembles the actual membrane of living cells (Figure 7.15). In the lab, these vesicles can capture macromolecules, including nucleic acids (RNA-like polymers). In addition, their lipid bilayer is sufficiently permeable to allow exchanges with the outside. They can fuse together and even split to produce 'offspring' in the form of small vesicles. They are not yet living cells, but could they represent a step towards the origin of life (Luisi, 2016)?

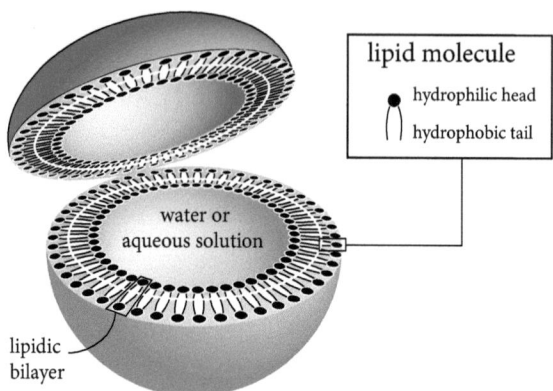

Figure 7.15 Structure of a liposome or protocell. The liposome is bounded by a lipidic bilayer resembling the membrane of a living cell.

The RNA world hypothesis

The genetic information of a living cell is enclosed in DNA molecules and translated into proteins through the mediation of diverse RNA molecules (Figure 7.3). Proteins provide the enzymes needed for metabolism and the self-replication of the living cell. Thus, nucleic acids and proteins are both essential to life. However, the discovery of short RNA molecules called 'ribozymes' (from *ribo*nucleic acid en*zymes*) that are able to catalyse specific biochemical reactions provides a new twist to this story. These molecules may represent an early step towards the origin of primordial life. This is the hypothesis of an RNA world, where RNA molecules performed the essential catalytic activities to assemble themselves in a first stage, and, once assembled, begin to synthesize proteins (Gilbert, 1986). Indeed, a few RNA enzymatic activities still exist, for instance in ribosomal RNA. Thus, a genetic world exclusively based on RNA could have been the precursor to LUCA and the following world, which relies on a combination of DNA, RNA, and proteins.

We have not yet discussed the special case of viruses. In fact, viruses are generally not considered to be proper organisms since they are incapable of self-reproduction and do not metabolize. Although they have genetic material in the form of a nucleic acid molecule that is protected by a protein coating, they have no metabolism and must inject their nucleic acid into a cell to reproduce themselves. They are therefore obligatory parasites. Isolated, they cannot carry out any biochemical reaction, grow, or reproduce. Viruses specific to methanogenic archaea have been identified, suggesting that viruses date to early in the origin of life, but their role in that origin remains an open debate.

The discovery of viroids that only consist of a circular RNA molecule containing less than 500 nucleotides without any protein envelope is also intriguing. Viroids are plant pathogens. They do not code for any protein. Their replication requires the enzymes of their host cell, which use the viroid RNA as a template to synthesize new RNAs. It has been proposed that these could be relics of a primitive RNA world (Maurel and Leclerc, 2016). However, we will say no more on the subject, returning to the classical definition of life, which conveniently excludes viruses and viroids from the living world.

> The simple carbonaceous molecules that make up living beings may have been delivered to Earth by meteorites and comets. However, it is not known how the complex organic molecules and the first cells capable of reproduction formed. The fact that all present-day living things share the same genetic code supports a single origin of life on Earth. The first simple traces of life, in the form of microbial filaments and mats, date back more than three and a half billion years; their metabolism and taxonomic affinity are unknown. Because many thermophilic microorganisms cluster near the root of the tree of life, we can speculate that life originated in the vicinity of oceanic hydrothermal vents. However, the transition from organic molecules to protocells and then to proper living organisms remains speculative.

8
Everything changes on Earth

Perhaps the most significant changes on Earth occurred about two and a half billion years ago. These changes were global in scale—the whole planet was affected. The term 'global change' is probably familiar to the reader, as these words are used to refer to ongoing climate change and its consequences. Past changes occurred on different time scales. At the end of the Archean and at the beginning of the Proterozoic, landscapes, climate, and atmosphere changed, notably under the influence of life. A different Earth, closer to our modern world, was taking shape...

Continental growth and long-term climate change

Early Archean continental domains were not very extensive, and were mostly submerged due to the abundance of dense rocks such as basalts and komatiites (Figure 6.3). This situation changed at the end of the Archean. The production of crust of a typical continental nature—let's say granitic in composition for simplicity—increased. Accordingly, the aerial surface of the continents grew while the now lower-density continents emerged above the surface of the oceans for isostatic reasons. The freeboard of the continents progressively rose, like that of a ship whose cargo is being unloaded (Eriksson, 1999).

The continental growth rate was not constant. Towards the end of the Archean, it seems to accelerate. Beginning about 3 Ga, granite production increased, as shown for example by geological studies in the Barberton region. A well-identified continental domain was formed: the Kaapvaal **craton**. A craton is a stable continental domain that formed during the Precambrian and whose extent is a few hundred to a few thousand kilometres. A thick lithosphere provides the craton rigidity and resistance to deformation, while its relatively low density prevents cratonic **lithosphere recycling** into the mantle by subduction. In addition, again thanks to its rigidity, a craton can support thick sedimentary deposits without bending under the load.

Due to the production of new granitic crust and the expansion of the oldest cratonic domains, the late Archean Earth would have contained the equivalent of 60% of today's continents, with about 20% of their present surface area, compared to only a few per cent in the early Archean. Taylor and McLennan (1985) proposed a continental growth curve (Figure 8.1) by studying the surface areas of Precambrian outcrops and their ages. Other authors propose slightly different curves, but most concur that significant crustal growth occurred at the end of the Archean.

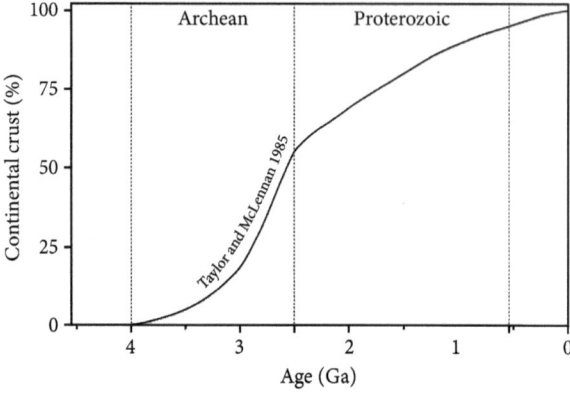

Figure 8.1 Continental growth curve as a percentage of current continental surface.
Source: Adapted from Taylor and McLennan (1985).

The landscapes of the late Archean continents were very different from those of today: no vegetation or other life inhabited their surface. Bare rock was everywhere. They were also different from early in the Archean, when the whole Earth was almost completely covered by the sea. Towards the end of the Archean, as the continents emerged, the weathering of these new land surfaces became an important process in regulating climate and ocean chemistry. Carbon dioxide dissolves in rainwater, water producing a weak acid, carbonic acid, which favours the alteration of minerals. Today rainwater pH is about 5.9, but as the Archean atmosphere was richer in carbon dioxide, ancient rain was more acidic than today—a value of about 5.2 is suggested (Hao et al., 2017). The weathering of silicate minerals outcropping on continental surfaces is a CO_2 sink; in other words, the weathering reactions reduce the amount of CO_2 in the atmosphere. It is even the main CO_2 sink in the long-term carbon cycle, such that silicate weathering has an important role in regulating global climate (Figure 8.2). Another similar sink for CO_2 is the hydrothermal alteration of oceanic crust, which would have begun early in Earth's history and contributed to some decrease of atmospheric CO_2.

The increasing continental land surface and its alteration under the effect of acidic rain progressively reduced the CO_2 content of the atmosphere (Figure 8.3). Carbon dioxide is a greenhouse gas, as we saw in Chapter 3. Therefore, consumption of CO_2 by weathering of continental surfaces should result in the decrease of global temperature. At the same time, the solar constant increased slowly over geological time, which has an opposite effect on temperature. These two factors act independently. Whether the climate is hot or cold depends on which factor prevails at a given time.

There is diverse evidence for glaciation occurring in the early Proterozoic, particularly in Canada. These are the oldest indicators of glacial activity in the geological record. Glacial sediments are made of a fine-grained matrix containing unsorted and often angular blocks of various sizes. The lack of sorting according to size is typical of ice transport. It can be observed today in glacial **moraines** or when icebergs melt at

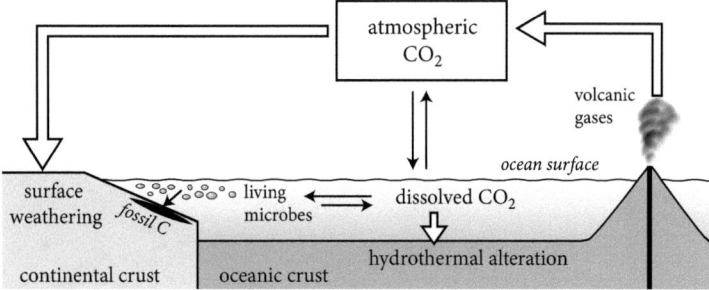

Figure 8.2 Simplified carbon cycle. CO_2 sinks: weathering of silicate minerals leads to precipitation of carbonate rocks and production of organic matter that makes up all living beings. The sources of CO_2 include volcanism and oxidation of organic matter after the death of microbes. Fossil C representing organic matter that escaped complete oxidation and was buried in the sediments represents another long-term sink for CO_2.

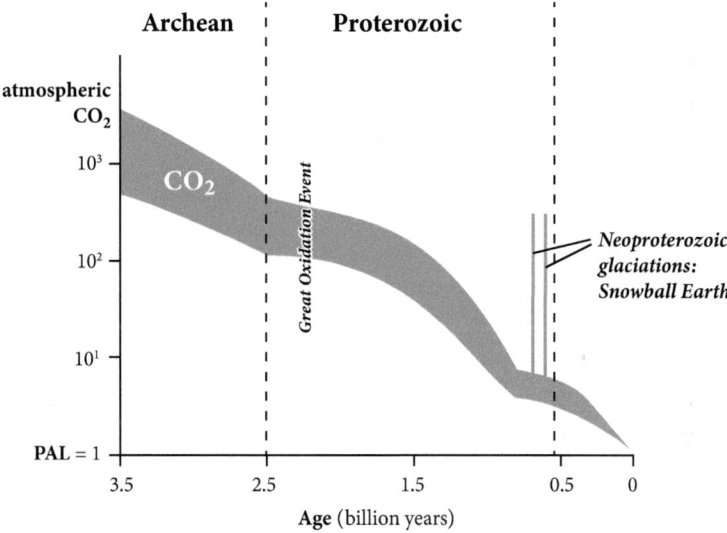

Figure 8.3 Changes in atmospheric CO_2 levels over time. Values are presented in logarithmic coordinates with respect to multiples of ten of the present atmospheric level (PAL). There has been a steady decrease since the beginning, only interrupted at the end of the Proterozoic by two peaks of short duration (see Chapter 13). The thickness of the curve reflects the uncertainties in the CO_2 estimates.

sea releasing blocks of all sizes, known as **dropstones**, that fall into laminated marine sediments (Figure 8.4). Such marine sediments made of a fine matrix and dropstones are called **diamictites**. The **glaciogenic** nature of a diamictite is confirmed if the dropstones have striae or typical abrasion traces due to their transport at the base of an ice sheet or glacier.

Four early Proterozoic ice ages were identified in Canada (Figure 8.5). The three older ones are about 2.4 Ga old (Rasmussen et al., 2013). The fourth is slightly more than 2.2 Ga old. In South Africa and Australia, evidence of glaciation of similar

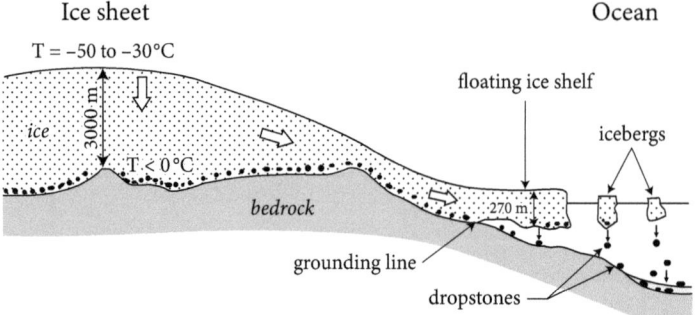

Figure 8.4 Ice sheet section from continent to ocean. From the grounding line, the ice sheet turns into a floating ice shelf. The shelf breaks and calves tabular icebergs. The floating ice melts from below, releasing rock fragments (dropstones) that fall in the marine sediment.

Figure 8.5 Paleoproterozoic glacial sediments in Canada. A dropstone (in the centre) deformed horizontal finely bedded soft marine sediments before their induration. Scale is provided by the hammer.
Source: Grant Young.

age was also found, although it is difficult to establish precise temporal correlations. Nevertheless, the early Proterozoic was a relatively cold period that contains the Huronian glaciation, so called because its traces were discovered near Lake Huron in Canada.

Gradual oxygenation of the ocean and formation of BIFs

The Archean atmosphere was rich in carbon dioxide but lacked oxygen. The ocean was also anoxic, which allowed seawater to be rich in dissolved iron of hydrothermal origin. Dissolved iron is ferrous iron (Fe^{2+}) or Fe(II), a less oxidized cation than ferric iron (Fe^{3+}), or Fe(III). Ferric iron is insoluble. As soon as dissolved ferrous iron is oxidized, insoluble iron oxides precipitate on the sea floor. Iron-rich rocks, the already mentioned banded iron formations (BIFs, Figure 6.11), bear witness to these changes. At the end of the Archean, huge volumes of BIFs were formed, as if all the iron dissolved in the ocean had precipitated. These late Archean BIFs mostly contain hematite (Fe_2O_3), a mineral where all iron is in ferric form. They indicate that the ocean was becoming oxidized.

How and when did oxidizing conditions begin in the ocean? The source of oxygen was likely photosynthesizing microorganisms, namely cyanobacteria. Oxygenic photosynthesis must have occurred before the onset of massive BIF deposits, that is, before 2.7 Ga or perhaps earlier. Owing to this reaction, the living cell can make its own carbonaceous organic matter from dissolved CO_2 in seawater using light energy. Oxygen is the waste product of this reaction. Photosynthetic microbes thrive in the sunlit photic zone (from the sea surface to a depth of 200 metres). Thus, the upper oceanic layer slowly became oxidized, while the deeper layer remained anoxic and still contained dissolved ferrous iron. At the transition zone between these water masses dissolved iron was oxidized and precipitated. Ferric oxides accumulated on the sea floor to form BIFs.

Thanks to the isotopes of iron, it is possible to determine the proportion of total dissolved iron that has been oxidized in the ocean over time. The main iron isotope, iron-56 (^{56}Fe), accounts for 92% of natural iron. The lighter iron-54 (^{54}Fe) is less abundant. The oxidation reaction of dissolved ferrous iron to insoluble ferric iron produces an **isotopic fractionation** (i.e. it changes the proportion of these isotopes), typically reported with respect to a reference standard.[1] The first iron oxide to precipitate is proportionally richer in the heavy isotope (^{56}Fe) than the dissolved iron: it has a positive anomaly compared to the reference standard. If the oxidation of the remaining dissolved iron proceeds, the positive anomaly of precipitated iron oxides will progressively decrease. If all dissolved iron is oxidized and precipitated, the isotopic signature of iron oxides is that of the initial dissolved ferrous iron.

Iron isotope measurements in BIFs can thus quantify the proportion of dissolved oceanic iron that has been oxidized as a function of the iron isotope anomaly. Data are quite scarce, as iron isotope analysis is difficult and requires very powerful instruments. However, analyses of BIFs of different ages have provided consistent results. For example, the isotopic signature of late Archean BIFs (2.7 Ga) from Carajás in Brazil corresponds to the oxidation of 40% of the iron dissolved in the surface ocean at that

---

[1] The reference standard for iron isotopy IRMM-014, that is distributed by the Institute for Reference Materials and Methods, and is close to the average composition of the Earth's crust. The analytical results are provided with the usual δ notation.

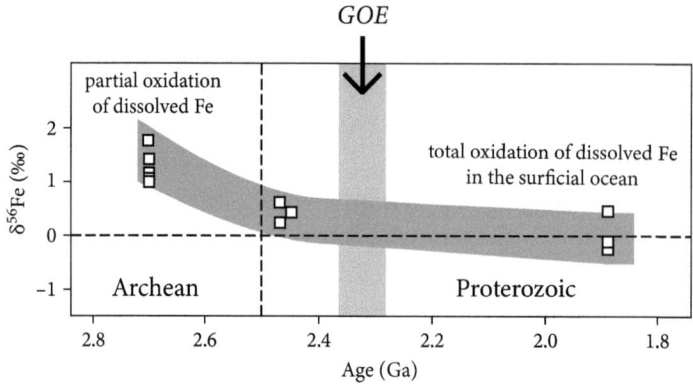

Figure 8.6 Evolution of the hematite isotopic signature of BIFs at the Archean–Proterozoic transition. Despite the small amount of data, the grey shaded curve displays a consistent evolution. It suggests a gradual increase in the oxygen content of the surface ocean, until the GOE (Great Oxidation Event in the atmosphere), which occurred at about 2.3 Ga.
Source: Adapted from Fabre et al. (2011).

time (Figure 8.6). In contrast, the early Paleoproterozoic BIFs (c.2.4 Ga) result from almost complete oxidation of dissolved iron. When there is no more dissolved ferrous iron to oxidize, oxygen from the surface ocean can then pass into the atmosphere: this is what happened around 2.3 Ga.

Archean BIFs were deposited in still waters and were mostly devoid of any continental influence. Therefore, it was assumed that they formed far from the continents. In addition, the oxidation of dissolved iron that occurred in surface seawater did not necessarily imply that deep water was oxygenated. In fact, the opposite is certain for as long as BIFs were precipitated. However, abundant iron formations also formed around 1.85 Ga. These relatively younger iron formations often occur as granular iron formations or **GIFs**, where the iron oxides form granules and not continuous beds as in BIFs. This texture indicates that the iron oxides precipitated in shallow water immediately above the stormy-weather wave base, thus implying precipitation closer to the shoreline (Figure 8.7). Despite this observation, detrital input from the neighbouring continent is limited, suggesting a dry sub-desertic climate.

Today, the world's largest deserts are located in the subtropics, because these latitudes (15–30°) correspond to the downward flow of the main global-scale atmospheric circulation called the Hadley cell. The Hadley circulation corresponds to upward motion of warm and humid air above the equator, until this air is cooled enough to lose its moisture, resulting in abundant rains. This pattern of air circulation is distributed on both sides of the equator, and is balanced by cooler dry air that descends in northern and southern subtropics. Deviation of this air circulation on Earth's surface due to the Coriolis force is responsible for the trade winds. It is also at these latitudes that **upwellings** of deep water occur along the western edges of the continents, as trade

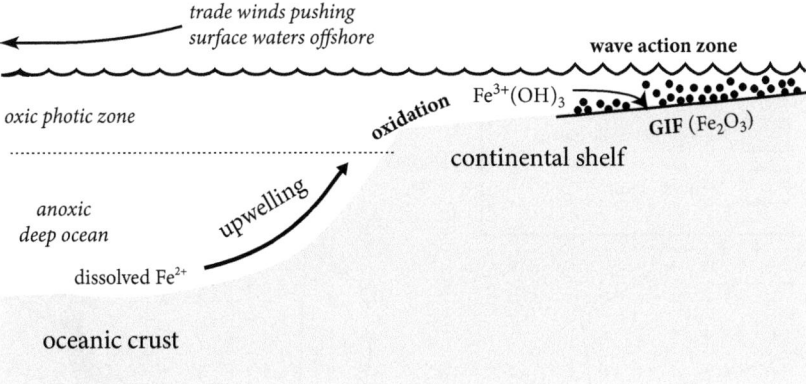

Figure 8.7 Formation of GIFs at 1.85 Ga.

winds push surface water offshore. The Hadley atmospheric circulation is a dominant process in modern climate, and likely always prevailed on the ancient Earth because it is directly related to the large amount of solar heat received at the equator, along with Earth's rotation. Thus, a similar scenario can be tested for the iron formations of the Superior craton in Canada during the Paleoproterozoic. We will try to verify this scenario by considering the iron formations of the Superior craton. The geological map of the Superior craton (so called after the Great Lake of the same name) shows that 1.85 Ga iron formations are located along the southern and eastern borders of this craton in today's coordinates (Figure 8.8).

At around 1.86 Ga, the Superior craton was not located at high latitudes nor oriented as it is today. It was part of the Columbia supercontinent (see Chapter 9) and located along its western border at subtropical latitudes (Antonio et al., 2017). As today in a similar situation, trade winds pushed the surface waters offshore and caused the upwelling of deep waters that were still anoxic and loaded with dissolved iron. Dissolved iron precipitated as oxides that collected as granules in the wave-swept coastal waters.

Deposition of iron formations ended at c.1.80 Ga. Their disappearance suggests the development of a fully oxygenated ocean during the following billion years. BIFs reappeared only at the very end of the Precambrian at a time when the Earth was completely ice-covered and the exchanges between ocean and atmosphere had been closed for a brief period, as will be described in Chapter 12.

The Great Oxygenation Event of the atmosphere

The Earth's atmosphere at the beginning of the Archean was rich in carbon dioxide but contained no oxygen. Oxygen began to be produced in the surface layer of the ocean by cyanobacteria that carried out oxygenic photosynthesis. For a very long time, most, if not all, of this oxygen had been consumed by the oxidation of dissolved ferrous iron wherever the deep waters came into contact with the surface waters. Therefore, the

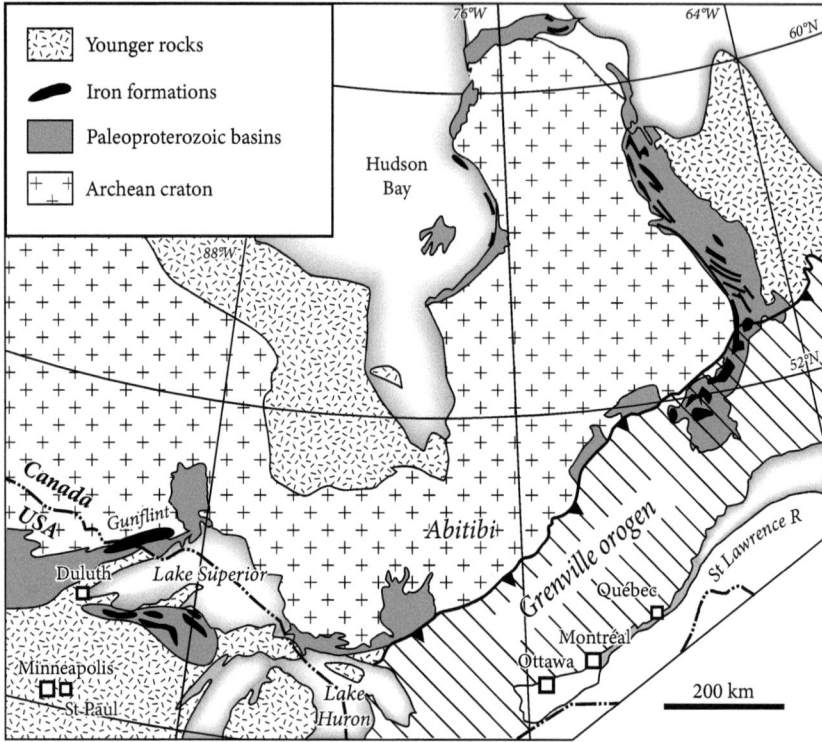

Figure 8.8 Geological map of the Archean Superior craton. Huge volumes of iron formations (mostly GIFs) were deposited in Paleoproterozoic basins along the craton borders.
Source: Adapted from Bekker et al. (2014).

atmosphere remained devoid of oxygen. It took hundreds of millions of years to purge the ocean of its dissolved iron. Huge masses of BIFs, millions of billions of tonnes (and the raw material for today's steel!), precipitated at the end of the Archean and at the beginning of the Paleoproterozoic, consuming any available oxygen in the surface waters. Oxygen began to be transferred to the atmosphere from the ocean surface at around 2.3 Ga. This major event is called the Great Oxygenation Event of the atmosphere, commonly abbreviated as the **GOE** (Figure 8.9). This name was proposed in 1984 by the geochemist Heinrich Holland, who devoted most of his career to studying the evolution of the composition of the ocean and the atmosphere throughout Earth's history.

After the GOE, conditions prevailing on the Earth's surface fundamentally changed: the ancient world turned into a modern (oxygenated) world, with immense consequences for the environment and for life itself. If the term global change was ever apt, it is for the GOE.

On the emerged continental surfaces, which represented about 20% of the total Earth surface at the end of the Archean, the landscape changed. The first red continental layers appeared, mainly sandstones with a ferruginous (hematite) cement. This

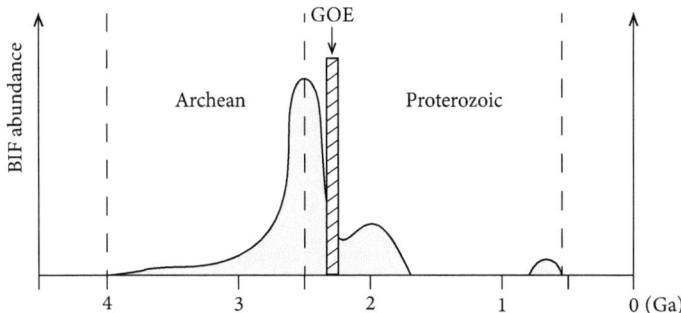

Figure 8.9 Position of the GOE with respect to deposition of BIFs. The scale of BIF deposition is a relative abundance scale.

type of sediment is well known in more recent periods and corresponds to terrestrial deposits in warm and arid (or semi-arid) climates. The red or rusty colour is that of oxidized iron, the insoluble ferric iron that remains *in situ* during continental weathering processes. It is this insoluble iron that colours all intertropical landscapes today. This characteristic colour can be seen in the photograph of the Jack Hills in Australia (Figure 2.1). Another example is the Navajo sandstone of the Colorado Plateau. But this was not the case before the GOE. Continental iron was not oxidized; in the absence of oxygen, it could be transported by water as a dissolved ion or it remained trapped in sulphide minerals such as pyrite. Before the GOE, there were no laterites, no ferruginous sandstones, no red sands as in the present-day Sahara .... Red on Earth's surface only appeared after 2.3 Ga. The world before was in black and white, or rather in shades of grey.

Even if the atmosphere became sufficiently oxidizing to trap iron on the continents, it was far from the present 20% oxygen content. Values of only 1 or 2% are suggested for the GOE. However, the oxygen level is still debated for the entire Proterozoic eon. For this reason, there remain large uncertainties in the trajectory of atmospheric O_2 evolution (Figure 8.10).

Despite uncertainties about the oxygen content of the atmosphere during the Proterozoic, the existence of the GOE is unambiguous, and its age is well constrained. The key information was obtained in a novel way by James Farquhar, then a young researcher at the University of California at San Diego. James Farquhar studied sulphur isotopes in Precambrian continental pyrite and gypsum minerals that contain sulphur atoms. Pyrite is an iron sulphide (FeS_2) and gypsum a calcium sulphate ($CaSO_4, 2\,H_2O$). Gypsum is an **evaporite**, that is, it precipitates when seawater containing dissolved sulphate is concentrated by evaporation, for example in a lagoon or an almost-closed basin. In the early Precambrian, evaporitic sulphates were rare, because seawater was low in oxygen and therefore in dissolved sulphate (Strauss et al., 2013). On the other hand, sulphides were abundant. In all cases, the sulphur atom is of volcanic origin: it comes from sulphur dioxide (SO_2), one of the main gases emitted by volcanoes.

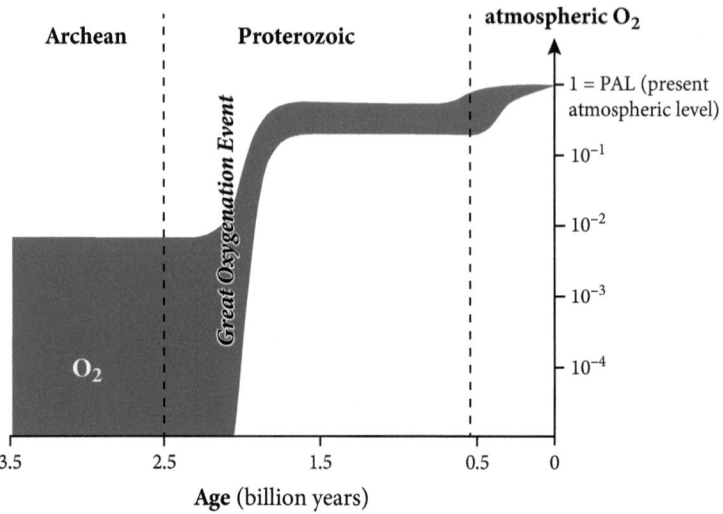

Figure 8.10 Changes in atmospheric O_2 content with respect to present atmospheric level (PAL).

In the upper atmosphere, sulphur isotopes are fractionated in an uncommon way that is indifferent to the mass differences between heavier or lighter isotopes (contrary to all the isotopic fractionations described so far, e.g. in the case of or iron). For this reason, it is called mass-independent fractionation.[2] This process affects the sulphur in sulphur dioxide gas and is the result of photolysis reactions of the SO_2 molecule, which occur only in the total absence of oxygen. Indeed, if the atmosphere contains even a small amount of oxygen, a layer of ozone appears in its upper layers, a shield against the ultraviolet radiation involved in the photolysis reactions. The isotopic distribution of sulphur, possibly modified in the upper atmosphere, is recorded in the sulphuric acid produced by the dissolution of SO_2 in rainwater. The acidified rain carries the sulphur atoms to the soil surface or to the oceans. Then, the journey of the sulphur atoms ends as they are incorporated into minerals such as pyrite or gypsum.

James Farquhar analysed many samples of sulphur-bearing Precambrian minerals. He found that mass-independent sulphur fractionation only occurred in minerals older than 2.3 Ga (Figure 8.11), thus providing the age of the GOE (Farquhar et al., 2000).

Ongoing rise of oxygen in the ocean and formation of giant manganese deposits

BIFs are evidence of the onset of ocean oxygenation, because dissolved iron requires only a very small content of oxygen to precipitate. The deposition of BIFs occurred

---

[2] The measurement of this mass-independent isotopic fractionation uses the notation Δ (capital delta), not to be confused with the usual fractionations noted δ(delta).

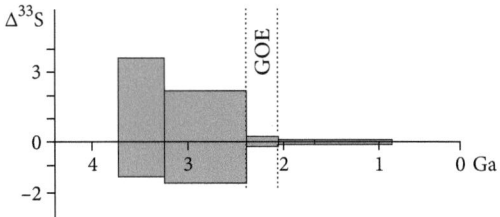

Figure 8.11 Evolution of the mass-independent fractionation of sulphur isotopes (Δ) with time. This fractionation, which is only possible in the absence of atmospheric oxygen, becomes very small at 2.3 Ga and disappears completely after 2.1 Ga. It is thus possible to date the Great Oxygenation Event in the atmosphere.
Source: Adapted from Farquhar et al. (2000).

throughout the Archean, with a maximum at the end of this eon (Figure 8.9). A little later, during the early Proterozoic, giant manganese deposits formed, for instance in South Africa, Gabon, and Brazil. As of 2022, South Africa and Gabon are the primary and secondary producers of manganese ore globally.[3] Manganese deposits also formed due to chemical precipitation in seawater, but required a higher level of oceanic oxygenation than BIFs. The manganese ion in its Mn^{2+} (reduced) form is soluble. Like dissolved ferrous iron, it was leached from the oceanic crust by hydrothermal circulation. When seawater became sufficiently oxygenated, Mn^{2+} changed to a more oxidized state (Mn^{3+} or Mn^{4+}), which then precipitated as part of insoluble manganese oxides. Mn deposition occurred between 2.4 and 2.1 Ga. The Hotazel Formation in South Africa is overlying well-dated volcanics at 2.4 Ga. It contains BIFs in its lower part and Mn deposits in its upper part (Lantink et al., 2018). Iron isotopes suggest precipitation from surface seawater that had already lost most of its dissolved iron content. This situation points to the fact that the oxygenation level of the surface ocean after 2.4 Ga up to the onset of the GOE was much higher than at 2.7 Ga. The existence of giant manganese deposits on different cratons suggests that these deposits resulted from an identical process on a global scale. The difference with the deposition of BIFs is the higher level of surface water oxygenation required. Neither BIFs nor giant manganese deposits are able to form in the modern environment. They are the witnesses of a vanished world.

Major disturbances in the carbon cycle

There is a consensus that oxygen in the atmosphere originated from an excess of oxygen in the ocean generated by photosynthetic bacteria. However, the explanation is not as simple as that. Photosynthetic microbes build their own organic matter from dissolved CO_2 in seawater. After their death, degradation of organic matter uses oxygen and regenerates CO_2. The balance is then zero from the point of view of oxygen

---

[3] Manganese is used for the manufacture of special steels.

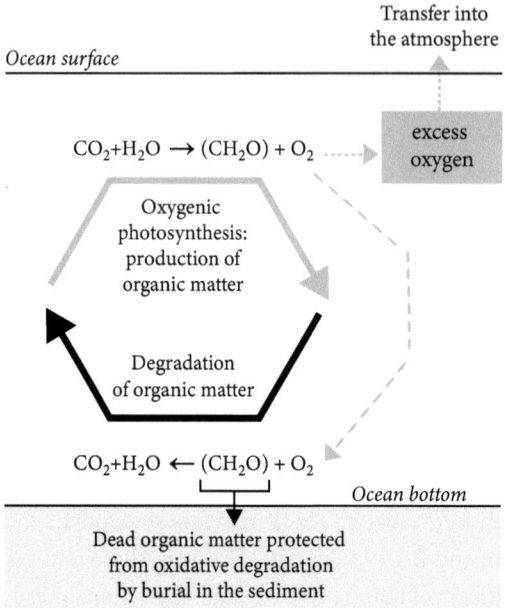

Figure 8.12 **Organic matter and oxygen cycle in the ocean.**

as well as CO_2. In order to accumulate an excess of oxygen in the ocean and then to transfer oxygen into the atmosphere, some proportion of the dead organic matter must escape complete oxidation (Figures 8.2 and 8.12).

The preservation of dead organic matter requires peculiar conditions. One favourable condition is the rapid burial of the organic matter due to voluminous detrital input from the continent. It may happen in a context of active tectonics followed by rapid erosion of newly generated topography. Such a scenario is playing out today in the Bay of Bengal as a result of the uplift of the Himalayas some 20 million years ago: abundant organic material is buried under a huge amount of continental detritus transported by the Ganges and the Brahmaputra rivers. Another condition favourable for burial of organic material occurs when the supply of dead organisms to the bottom of the ocean is so high it outstrips the remineralization capacity (i.e. the amount of oxygen available), thus preserving the remaining organic matter. This is how large reservoirs of fossil carbon accumulated at times in Earth's history, resulting in the formation of oil (in the marine environment) or, much more recently, coal (in the continental environment).

At the Archean–Proterozoic transition, around 2.5 Ga, the surface area of the emerged continents increased, and the input of detrital sediments from the continents likely also increased as a consequence. But the corresponding tectonic context is poorly known. From what we do know, tectonics seems to have been sluggish between 2.4 and 2.2 Ga. Nevertheless, continental inputs provided a global increase in nutrient supply that likely boosted biological productivity from the beginning of the Proterozoic. Remember also that the deep ocean was still poorly oxygenated at that time!

Moreover, in the absence of burrowing animals, there was no bioturbation in Precambrian marine sediments to aerate the sediments as occurs in the modern ocean. For all of these reasons, dead organic matter partly escaped remineralization, leading to the gradual accumulation of oxygen in the ocean, and then in the atmosphere.

For some organisms, the accumulation of oxygen in water can be deleterious. Oxygen is a waste product of photosynthesis, and it would have been a poison for ancient ecosystems. Many Archean microbes did not need oxygen to live: they were anaerobic, like today's archea for example. Several were even strict anaerobes, meaning they were intolerant of oxygen. Increasing dissolved oxygen content in the ocean was a real threat to them. It could have resulted in a hecatomb of anaerobic microbes, thus contributing to the accumulation of preserved organic matter in marine sediments, with the ultimate consequence of promoting oxygenation of the atmosphere.

Whatever their causes, major disturbances of the carbon cycle occurred shortly after the GOE. They are registered as isotopic anomalies. Living beings preferentially use light carbon (carbon-12) to build their own organic matter. Organic matter is therefore characterized by a very negative carbon isotope signature (i.e. it is depleted in carbon-13). Today, dead organic matter is completely remineralized in most marine settings; thus the isotopic signature of dissolved carbon dioxide (and its derived ionic species: bicarbonates and carbonates) in seawater remains close to zero. Precambrian isotopic signatures can be deciphered by analysing ancient marine carbonate rocks (limestones and dolomites), which were formed from dissolved marine carbonate ions. Around 2.2 Ga, carbonate rocks registered strongly positive values of their carbon isotopic signature worldwide, indicating a relative excess of carbon-13 in seawater. This episode was exceptional in both amplitude and duration in the history of marine sedimentation (Prave et al., 2022). It is known as the Lomagundi excursion, named after a locality in Zimbabwe. For the dissolved carbonate signature of seawater, at least near-shore seawater, to have been shifted to such very positive values, complementary light carbon with a negative isotopic signature must have been trapped somewhere. It points to an episode of extensive sedimentary burial of dead organic matter (rich in carbon-12).

First oils

Sediments containing exceptionally high levels of organic matter and aged at about 2.1 Ga have been identified in several places around the world. This is the case, for example, of the black shales of the Franceville basin in Gabon, which contain up to 15% or more organic matter. In some places near the Russian–Finnish border, up to 55% carbonaceous matter is found in the sediments! In both cases, this organic matter is bitumen, a solid product derived from ancient petroleum, which in turn was derived from dead organic matter buried in marine sediments. In Russia, the sediments were heated more than in Gabon and the bitumen is now found concentrated in pockets of insoluble pyrobitumen. In Gabon, the Franceville sedimentary basin remained nearly undeformed and unmetamorphosed after its formation and it has been possible to calculate the total mass of organic matter contained in the 2.1 Ga black shales of the whole basin. The mass of organic matter would have generated the equivalent of 84 billion barrels[4] of oil after burial! Today, an oil field containing more than 500 million barrels is called a giant oil field. The Franceville Paleoproterozoic oil field was therefore a supergiant, comparable to the current Ghawar oil field in Saudi Arabia, long considered the largest in the world with a hundred billion barrels of reserves. This suggests that marine life was proliferating in some marine places a little before two billion years ago.

> At the end of the Archean and beginning of the Proterozoic (i.e. between 2.7 and 2.1 Ga), the Earth underwent the greatest global changes in its history. Some of these changes, such as the increase of continental surfaces and the oxygenation of the atmosphere, are irreversible. The case of oxygen, a biological production, illustrates the first major interaction in the history of our planet between the living world and the mineral world. Life will have to adapt to survive this environmental turmoil for which it was partly responsible.

---

[4] A barrel, the usual unit of measurement in the oil business, corresponds 42 US gallons, that is, about 159 litres of oil.

9
Columbia

The first supercontinent

The surface area of the continents increased considerably just before two-and-a-half billion years ago, when it reached about one quarter of the surface of our planet. Thus, it is now time to tell the story of the first continents. Remember that continental crust is granitic in composition, which means that it is less dense than basaltic oceanic crust. For this reason, the continents cannot subduct. Therefore, they can preserve traces of the ancient history of the Earth. Because of their relatively low density, most continents are exposed to aerial weathering. The submerged continental margins with depths less than 200 metres belong to the continents and not to the oceans, at least from a geological point of view. This area of shallowly submerged continental crust is called the continental shelf.

Cratons at the heart of the continents

All present-day continents contain ancient cores that have been stable since the Archean eon. These cores, formerly referred to as 'shields', are now referred to as '**cratons**'. The cratons formed from Archean protocontinents, whose crust contained a variable proportion of TTG rocks of broadly granitic composition. However, to understand these cratons, we must consider their entire lithosphere (i.e. the crust and the underlying rigid mantle). Continental lithosphere is always thicker than oceanic lithosphere. Just as the continental crust is thicker than the oceanic crust (Figure 5.6), so the subcontinental lithospheric mantle is thicker than the suboceanic lithospheric mantle. This difference is particularly pronounced for Archean cratons, whose lithosphere is up to 250 km thick, as compared with 100–150 km for younger continental lithosphere. Thus, these ancient continental cores have a sort of lithospheric root or keel that is sunken into the convective mantle (Figure 9.1). Since their formation, their temperature and composition has made these Archean lithospheric keels immune to mantle convection.

What are the cratonic keels made of, and why did they remain relatively cold and therefore unaffected by convection? First we must recall that the base of the lithosphere is a thermal boundary: it corresponds to the 1,300 °C isotherm. Therefore, this isotherm defines the base of the keel. We can obtain additional information about the keels from both geophysical and mineralogical data. The geophysical data of interest here are the Earth's heat flow at the surface and the propagation of seismic waves

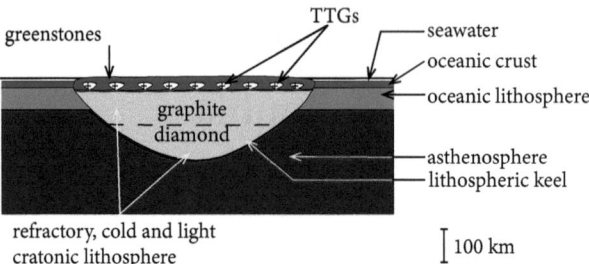

Figure 9.1 Schematic section of an Archean craton. The dashed line indicates the limit of stability of pure carbon as graphite or diamond depending on pressure and therefore depth.

deep in the mantle. The terrestrial heat flux is always low in cratons, especially in the cratonic keels, where seismic waves propagate somewhat faster than in the rest of the mantle, indicating a colder and/or denser environment.

Mineralogical data can tell us more. Volcanic rocks derived from magmas sourced very deep in the mantle are rare rocks known as **kimberlites**, but they are actively sought after, as they may contain diamonds. Kimberlites can be quite young, although no historical eruptions are known. A well-known example is from Kimberley in South Africa, whence the word is derived. The Kimberley kimberlite is only 85 Ma old, and like other kimberlites of all ages, it is found on an Archean craton. Kimberlite magmas come from 150 to 250 km below the surface. It is at these depths that pressures are high enough to produce diamonds from pure carbon; at shallower depths the carbon occurs in rocks as the mineral graphite (Figure 9.1). Kimberlite magmas form via the melting of a very low percentage of cratonic peridotites, made locally fusible by percolation of siliceous and carbonic fluids, possibly derived from subduction zones at the craton periphery.

In addition to diamonds, kimberlite magmas transport fragments of the cratonic mantle, up to several centimetres in diameter. These stowaway enclaves from the mantle always display the composition of a peridotite depleted of its most easily melted minerals. Early partial melting episodes extracted minerals rich in Ca, Al, and Fe and depleted the initial peridotite constituents by 40% or more. Such extensive partial melting is only known to have occurred in the Archean, when they formed the high-temperature komatiite magmas described in Chapter 4. Therefore, the Archean subcontinental mantle is made of refractory peridotite, mainly containing magnesium-rich olivine. These refractory peridotites are called **harzburgites**. Due to their relative depletion in iron, refractory peridotites have a somewhat lower density than ordinary peridotites, which, as stated earlier, are known as lherzolites.

During partial melting of the peridotite, radioactive elements that normally contribute to the heat flow—mainly uranium, potassium, and thorium—were also extracted. The result is an Archean cratonic lithosphere that is both colder and slightly more buoyant than younger lithosphere. Finland, at the heart of the European Baltica craton, is an example of an old, cold craton (Figure 9.2). We can think of these old cratons as having floated on the Earth's surface since the Archean.

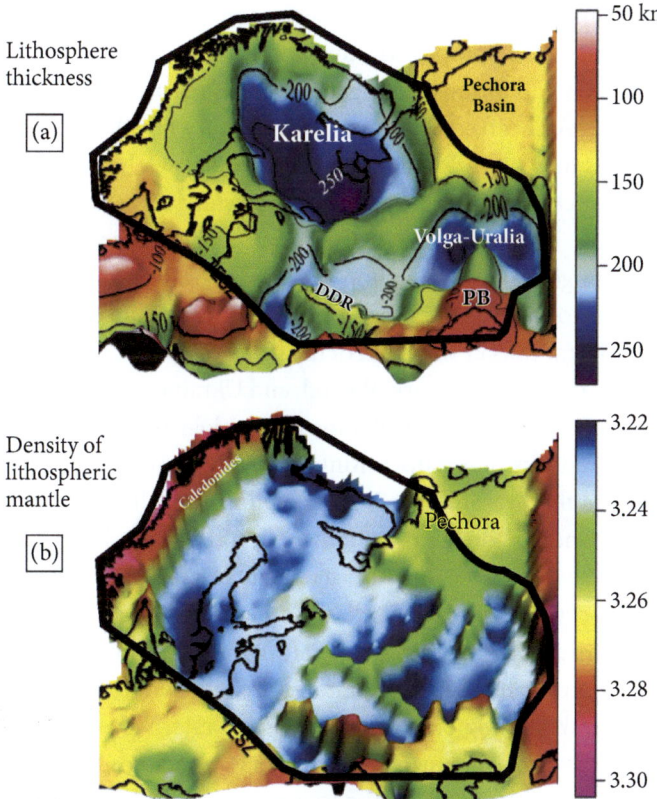

Figure 9.2 The northern European Baltica craton (thick black outline).
(a) Lithosphere thickness map: a maximum of about 250 km is reached in Archean domain in Karelia near the border between Finland and Russia. To the south, the Archean lithosphere was thinned to form the Dniepr–Donetsk Rift (DDR) and the Pericaspian Basin (PB) during the Paleozoic. (b) Density map of the lithospheric mantle, based on geophysical data: the very old lithosphere is less dense due to the refractory (magnesian) nature of the mantle. TESZ is the trans-European suture zone, also known as the Teisseyre–Tornquist line.
Source: Adapted from Artemieva (2007).

The growth of cratons

The oldest cratons, which are only a few thousand kilometres in width, may have collided as a result of plate movements early in Earth's history. The region where two cratons collided and welded together to form a larger continent is like a geological scar, bearing signs of heating and intense deformation. Thus, the Kaapvaal craton in South Africa now appears to be joined to the Zimbabwe craton to its north. Both cratons are approximately the same age (over 3 Ga) and their geological weld is known as the Limpopo Belt, a deformed metamorphic zone. Since this ancient collision over two billion years ago, these cratons have travelled together as a single, larger craton known as the Kalahari craton.

Making up much of north-eastern Europe, Baltica is a composite of three Archean cores dating back to more than 3 billion years, which together form a craton that is about 5 million km². The larger of the cratonic cores, known as Karelia, underlies Finland and extends to the border with Russia. It is in this Finnish craton that Hervé Martin studied TTGs and proposed his geodynamic interpretation for the genesis of these rocks. Another Archean core outcrops in Ukraine, while the third one extends to the Ural Mountains in western Russia but is not well known because it is covered by thick sedimentary rocks. These ancient cores were welded together towards the end of the Paleoproterozoic, between 2 and 1.8 Ga, meaning that Baltica, while still ancient, is considerably younger as a single craton than Kalahari. The south-western edge of Baltica defines a major step at the base of the lithosphere. Indeed, the Baltica lithosphere is nearly 100 km thicker than that of the nearby regions. This major boundary, which runs straight through Denmark, Poland, and Ukraine, is known by geologists as the Teisseyre–Tornquist line, named after the two geologists, one Polish and one German, who recognized this major discontinuity, in 1893 and 1908, respectively. This boundary separates the Precambrian Baltica craton from the younger terrains that make up southern and western Europe. However, today Baltica and its surrounding terrains belong to the same plate, the Eurasian plate (Figure 5.5).

About two billion years ago, another large craton formed as part of what would become North America. The Laurentia craton (Figure 9.3) encompasses (at least) four smaller Archean cratons: the Superior craton (already mentioned, Figures 5.12 and 8.8), along with the Wyoming craton in the north-western United States, the Slave craton near Great Slave Lake in north-western Canada, and the Greenland craton that makes up southern Greenland, where the Isua series is found. These last two cratons contain the oldest rocks known to date, with respective ages as old as 4.0 and 3.9 Ga. Throughout most of Laurentia's history, Greenland was welded to Canada; it was only separated along the Baffin Sea during the early Cenozoic because of the opening of the North Atlantic Ocean.

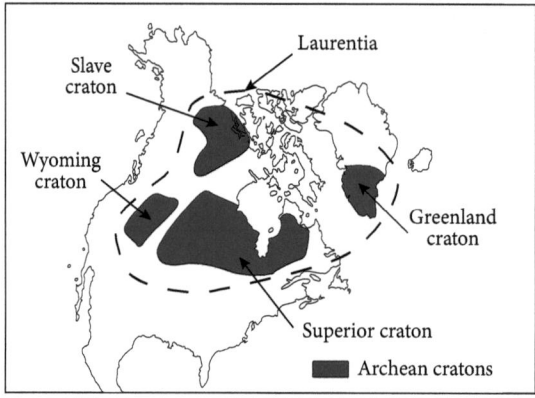

Figure 9.3 The North American craton, or Laurentia, formed by the assembly of four smaller Archean cratons. The dashed line shows the extent of Laurentia after its initial assembly between 2.0 and 1.8 Ga. The remainder of North America was added by subsequent episodes of crustal growth.

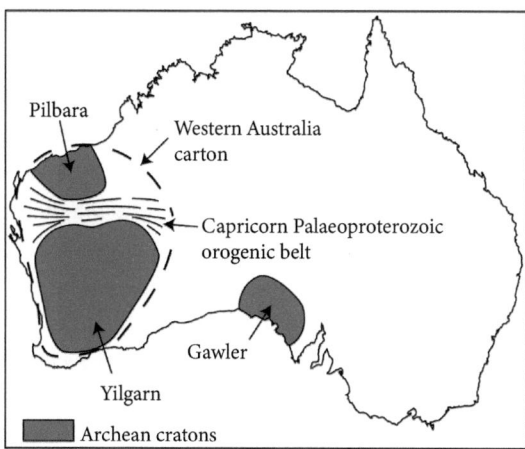

Figure 9.4 The Australian cratons. The Western Australia craton results from the collision of the older Pilbara and Yilgarn cratons along the Paleoproterozoic Capricorn belt. The eastern half of present-day Australia was mostly added during the Phanerozoic.

In Australia, the Archean Pilbara and Yilgarn cratons are welded together by the c.1.9 Ga Capricorn orogen. Together these two cratons form the Western Australian craton (Figure 9.4).

Other cratons also came together between c.2.2 and 1.8 Ga, giving the impression that that was a tectonically active interval of the Paleoproterozoic. Indeed, it has been proposed that these cratons assembled to form the first **supercontinent** in the history of the Earth. A supercontinent is a composite of cratons or continents that accounts for at least 75% of the total continental surface area at the time. In the more recent past, Pangea formed 300 million years ago, at the end of the Paleozoic era, when virtually all of the continental pieces had joined into a single continental land mass. The assembly and subsequent fragmentation of Pangea were the consequences of plate tectonics—the drift of the continents. Plate tectonics had probably initiated by 3.2 Ga, and as the area of continental land masses increased considerably at the end of the Archean, the probability of continental collisions also increased, making the formation of a Paleoproterozoic supercontinent plausible at least. However, reconstructing these ancient paleogeographies based on geological similarities alone is difficult, as the cratons are mobile and have drifted across the globe for the past two billion years. Fortunately, paleomagnetic data provide an additional tool for putting together the pieces of these supercontinental puzzles.

Paleomagnetism: a paleogeographic tool

Paleomagnetism is the study of the Earth's past magnetic field. The Earth's magnetic field exists as the result of the circulation of liquid iron in the outer core, which generates electric currents that in turn induce a magnetic field. This magnetic field has existed at least since the Archean, but its intensity has varied over geological time. The

Earth's magnetic field is thought to have always been **dipolar**, with a magnetic north and south pole, as if a large bar magnet lies at the centre of the Earth. However, this dipole is somewhat restless, because the magnetic poles migrate quite rapidly, by a few kilometres or tens of kilometres per year, around the geographical poles, such that the magnetic poles rarely coincide exactly with the geographic poles. Fortunately for paleomagnetists, this secular variation in the location of the magnetic poles averages out over a few thousand years, so that, on the scale of geological time, the location of the magnetic poles approximates to that of the geographic poles. After being located for a decade in the far north of Canada, the north magnetic pole, where the magnetized needle of a compass is pointing to, is currently migrating towards Siberia through the Arctic Ocean. If left free, the needle of a compass would also show a variable tilt from horizontal, because it aligns parallel to the magnetic field lines. The **inclination** of the magnetic field is vertical at the poles but horizontal at the equator and varies with latitude between these extreme positions (Figure 9.5). A simple trigonometric equation can be used to calculate **paleolatitude** from the inclination of measured magnetic minerals.[1]

The magnetic field is recorded by certain magnetic minerals in rocks when they crystallize from a magma or settle out in sediments. Magnetite, which is a very common iron oxide,[2] is unsurprisingly a particularly important magnetic mineral. The record of past magnetic alignment of minerals can persist over geological time: it is therefore called **remanent magnetization**. In magmas, remanent magnetization is acquired during cooling of the already solidified rock, when the temperature cools below a threshold known as the Curie temperature, which is 580 °C for magnetite.

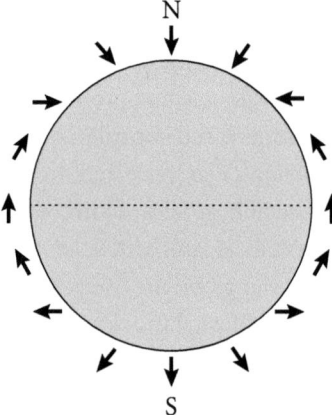

Figure 9.5 Variation of the inclination of a magnetized needle with latitude. Magnetic and geographic poles are superimposed at the geologic scale.

---

[1] Tangent (I) = 2 tangent (λ), where I and λ are the angular values of the inclination (I) and the paleolatitude (λ) in degrees.
[2] The formula of magnetite is Fe_3O_4.

Rock crystallization and magnetization are assumed to be of the same age, as cooling is quick, at least on geologic timescales. This magnetization persists if the rock is not reheated above the Curie temperature, which may happen as a result of a younger magmatic or metamorphic event (or a lightning strike!). The remanent magnetization retains both the **declination** (i.e. the direction of the magnetic north pole at the time) and the inclination of the magnetic field at the time and place where the rock was formed.

Earth's magnetic field has another feature that has proven especially useful in geology. In 1906, the French geophysicist Bernard Brunhes discovered that some ancient lava of the French Massif Central was magnetized in a direction opposite to that of the current magnetic field. He concluded that the north magnetic pole of the time of the lava's formation was at the south geographic pole, meaning the magnetic field of the Earth must have reversed in the past. Indeed, **geomagnetic reversals** are characteristic of the magnetic field and have occurred randomly throughout Earth's history (Channel et al., 2004).

When a paleomagnetist collects a rock in the field for study (either a volcanic rock, because it can be well dated, or a sedimentary rock, because its layers provide an indication of the original horizontal position), he must note precisely both its location and the orientation of the rock (i.e. if it has been tilted). In the lab, the paleomagnetist measures the orientation of the remanent magnetization in the rock sample, and then uses the location and orientation data to reconstruct the declination and inclination of the magnetic field with respect to the geographical coordinates at the time of the rock's formation. The declination indicates the direction of the North Pole (or South Pole in times of a reversed magnetic field). This paleopole generally does not point to the current pole, suggesting an apparent drift of the pole from the Earth's spin axis. In fact, this magnetic signature does not reflect that the magnetic pole moved with respect to the spin axis, but rather that the tectonic plate carrying that sample had drifted across the surface of the globe.

For the purpose of paleogeographic reconstruction, the craton where the studied sample comes from is first oriented to superimpose the identified former magnetic pole with a current geographical pole (Figure 9.6). Both the declination and inclination are used, which together provide the paleolatitude of the location from which the sample was collected and an indication of any rotation the continent has experienced during its drift.

Paleogeographic reconstruction using paleopoles is hindered by two sources of uncertainty. The first is the result of the regular reversals of the Earth's magnetic field. Since it is generally not known whether the field was normal (identical to the present-day field) or reversed when the magnetization was recorded, the continent in question might be in either hemisphere, with opposing orientations. The second uncertainty is that paleomagnetism yields no direct information on paleolongitude: the studied craton could therefore be located anywhere along the parallel of its paleolatitude. Thus, it is necessary to return to geology to test different solutions and narrow down the most viable configuration (Figure 9.6). Obviously, this is not an easy task, especially as data

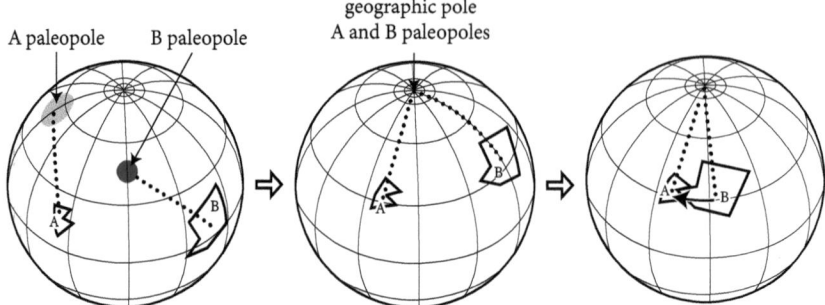

Figure 9.6 Paleogeographic reconstruction from paleomagnetism. Two samples coming from two different cratons, but of the same age, provide the magnetic paleopoles A and B in present-day geographic coordinates. First, the cratons are moved so that their paleopoles are in a polar position. Second, craton paleolongitudes are not known, but geographic and/or geological arguments allow a plausible solution to be proposed by moving one of the cratons along a parallel for a better fit of both cratons.

from Precambrian times are relatively few and difficult to obtain. As a result, different researchers commonly generate contrasting paleogeographic reconstructions for the same times in Earth's past.

Reconstructing the first supercontinent

As noted, geologists have proposed that a supercontinent formed at the beginning of the Proterozoic, based in part on an abundance of rocks dating to between 2.1 Ga and 1.8 Ga. These ages include magmatic rocks that are often associated with deformation and metamorphic events related to collision belts between Archean cratons. In 1988, Paul Hoffman, a Canadian geologist then at the Geological Survey of Canada, proposed that the formation of the Laurentian continent in the middle Paleoproterozoic was part of a global network of collisions that assembled a supercontinent. The existence of this supercontinent, now called Columbia (or sometimes Nuna), is supported by paleomagnetic evidence (Meert and Santosh, 2022). Paleomagnetic data, coupled with geological data, make it possible to propose a paleaogeographic reconstruction of Columbia (Figure 9.7a). The details are debated among specialists, and further paleomagnetic data are required for a better consensus—therefore, this paleogeographic reconstruction should be regarded as a work in progress.

Sedimentation on the surface of Columbia

The existence of the Columbia supercontinent has several consequences. First, a supercontinent is characterized by large land areas far from coastlines, and therefore far from sources of evaporation and precipitation. Increased aridity of the Earth's global climate is a direct consequence of the existence of a supercontinent, a finding

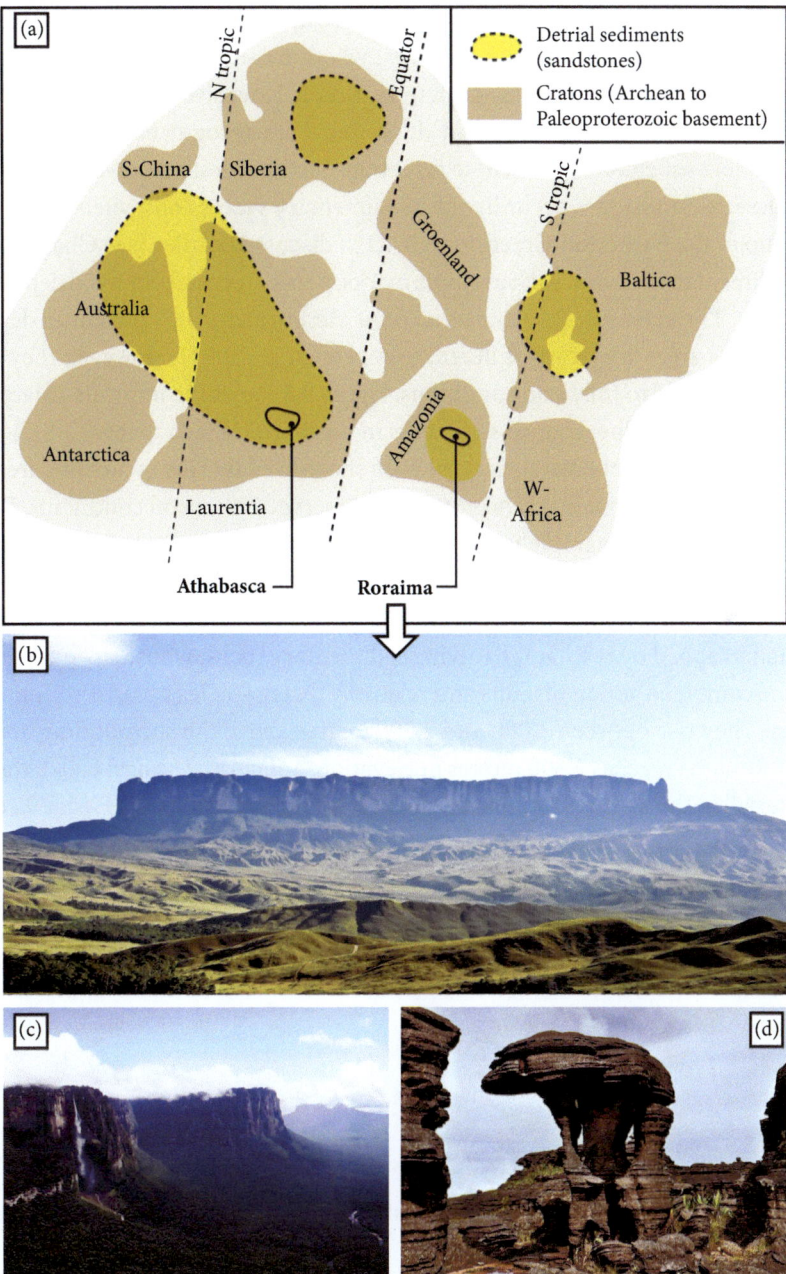

Figure 9.7 (a) Location of Proterozoic continental sandstones and sands (yellow) on a reconstruction of the Columbia supercontinent. Note that north faces to the left of the map. (b) View of Mount Roraima from Venezuela. (c) The Auyan tepui with the Salto Ángel (Angel Falls) waterfall on the left. (d) Close-up view of the Proterozoic sandstones at the top of the Roraima plateau.

Source: (a) Adapted from Zhang et al. (2012). (b) and (d) Paolo Costa Baldi (licence GFDL/CC-BY-SA 3.0). (c) Heribert Dezeo.

also documented for the more recent Pangea supercontinent. Furthermore, aridity is most intense in subtropical latitudes, where most large modern deserts, such the Sahara, Kalahari, and Atacama deserts, occur. Extensive continental sandstone deposits representing large ergs and dune fields are recognized from the time of the Columbia supercontinent. These represent Earth's oldest preserved former large, sandy deserts. Sandstones occur in older sedimentary successions (such as the Moodies Group of the Barberton Greenstone Belt) as discussed earlier (see Chapter 6), but their sedimentary features, such as ripple marks, point to coastal or fluvial depositional conditions. The extensive Paleoproterozoic to Mesoproterozoic sandstones deposited on Columbia areas are clearly different (Simpson et al., 2004). However, they do not always correspond to fully arid conditions. Figure 9.8 shows the imprints of large raindrops in interdunal fine-grained sediments in northern Canada that are 1.8 Ga in age. Elsewhere, traces of ephemeral rivers or even short-lived marine incursions are found.

The Columbia continental sandstones are preserved on several continents. Perhaps the most spectacular example is in South America, at the border between Brazil and Venezuela, where they form spectacular tabletop **mesas**, called 'tepuis' ('houses of the gods') by the people native to these areas (Figure 9.7b). These tepuis appear as isolated mountains capped by resistant, flat-lying sedimentary rocks, whose steep edges result from the combined action of faults and relatively recent—at least post-Precambrian—erosion. They rise between 1,000 and 2,000 metres above the surrounding area, and Mount Roraima reaches 2,800 metres in elevation. *Salto Ángel* (Angel Falls), the highest waterfall in the world (979 metres), drops from the top of a tepui (Figure 9.7c). The

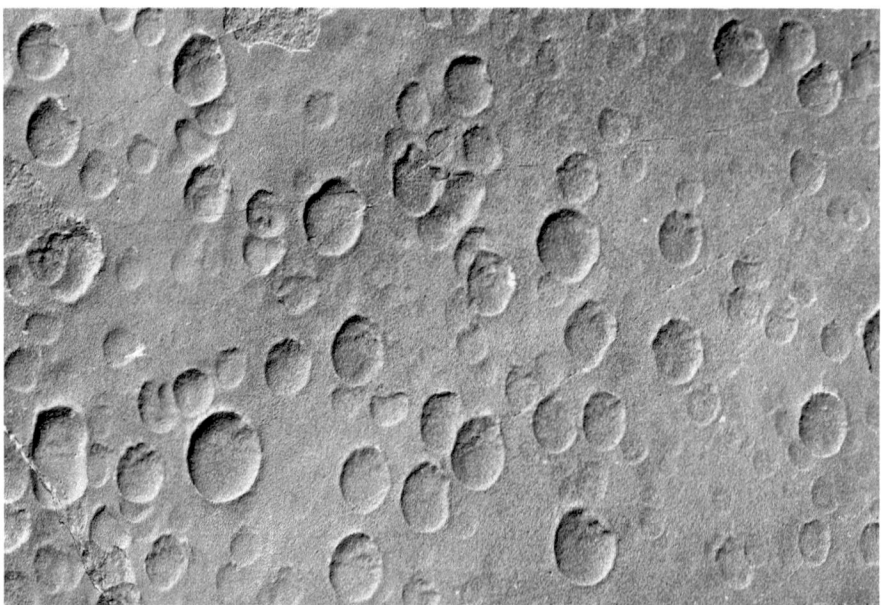

Figure 9.8 **Large raindrop imprints** (largest one: 1.8 cm) from 1.8 Ga fine interdunal sediments of northern Canada (Nunavut province): traces of a temporary storm over a sandy desert.
Source: Robert Rainbird.

summits of these high plateaus are often hidden by thick clouds. Erosion has carved out fantastic figures in the sandstone (Figure 9.7d). The steep topography and intemperate meteorology provoke the imagination and make the tepuis difficult to explore. Indeed, Sir Arthur Conan Doyle's *The Lost World* is set on a tepui, and the imaginative animated film *Up*, directed by Pete Docter for Pixar Studios in 2009, visits such a plateau.

Thermal effect of the supercontinent

A continental lithosphere is always thicker than the oceanic lithosphere. Transfer of the Earth's internal heat by conduction is therefore less efficient in continental lithosphere. And rocks are generally poor heat conductors. These factors have consequences for the temperature of the underlying mantle over long timescales.

Numerical models show that a supercontinent causes the temperature of the underlying mantle to increase. The two-dimensional model of Lenardic et al. (2011) considers a vertical section of crust and mantle (Figure 9.9). Cold plumes are the

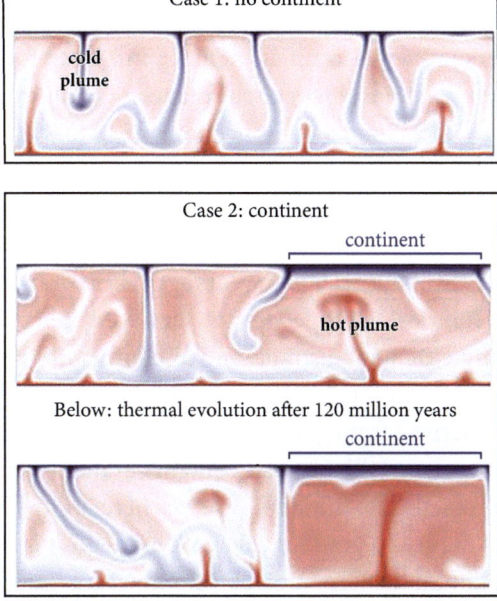

Figure 9.9 **Two-dimensional numerical models of the temperature distribution in the mantle with or without a supercontinent.** The mantle is simulated as a rectangular box representing a vertical section. In case 1, the cold (blue) and hot (red) plumes are manifestations of mantle convection; these plumes detach from the upper (lithosphere) or lower (D″ layer at the core–mantle boundary) boundary layer, respectively. In case 2, a thermally insulating supercontinental lid induces a progressive rise in the temperature of the underlying mantle, which is particularly pronounced if the supercontinent is bordered by subduction zones.
Source: Adapted from Lenardic et al. (2011).

equivalent of the subducting oceanic lithosphere. According to this model, the thermal effect is most significant if a significant part of the supercontinent is bordered by subduction zones, which act as a thermal barrier between the subcontinental mantle and the neighbouring suboceanic mantle. The temperature under the continental lithosphere thus rises some 100 °C after 100 million years after the formation of the supercontinent, and by 200 °C after 300 million years.

Under these conditions, the mantle is expected to begin to melt under the continental lithosphere. Indeed, the Columbia supercontinent is characterized by significant intracontinental magmatism throughout its existence. These hot mantle-derived magmas caused partial melting of the continental crust at their contact, resulting in abundant granites and their volcanic equivalents at the surface, **rhyolites**. The intense volcanic activity would have strongly impacted the atmosphere and had metallogenic consequences, which are discussed further below. However, the magmatic consequences of a supercontinental thermal blanket were less intense for younger supercontinents since the mantle was then cooler.

Formation of giant uranium deposits

Uranium is a characteristic element of the continental crust, though the average uranium content of the continental crust is only 3 ppm. Therefore, the formation of large U deposits requires extraordinary circumstances. Uranium ore, or **pitchblende**, a uranium oxide (UO_2), was mined in France from 1948 to 2001 in modest deposits associated with the 300 Ma old Limousin granites in the French Massif Central. In the last 30 years, the discovery of giant deposits in Australia and Canada has radically changed the geology of uranium. These deposits were formed on the surface of the Columbia supercontinent about 1.5 billion years ago. These uranium ores occur in two different modes with contrasting properties: low-grade but high-tonnage polymetallic deposits such as Olympic Dam in Australia, and **unconformity-related deposits** such as Cigar Lake in Canada, with very high grade and a lower tonnage. In both cases, the uranium was initially derived from granitic rocks.

The Olympic Dam polymetallic deposit contains uranium, copper, gold, and silver, all of which are mined. The uranium reserves here are the largest in the world (over 1 billion tonnes of ore at a grade of 0.04%). This deposit is linked to the granitic magmatism which followed the formation of Columbia as a result of the thermal conditions generated by the supercontinent. Large volumes of granites and associated rhyolitic lavas were emplaced around 1,590 Ma in the Gawler craton in southern Australia (Figure 9.4). The very hot granitic magmas rose high in the continental crust and produced hydrothermal fluids rich in dissolved metals that fractured the rocks above the granite. Hydrothermal circulation continued for some time in these breccias, and when pressure and temperature decreased, metals, including uranium, precipitated.

Unlike polymetallic deposits that occur close to their granitic source, unconformity-related deposits form in sedimentary basins. The Cigar Lake and McArthur River

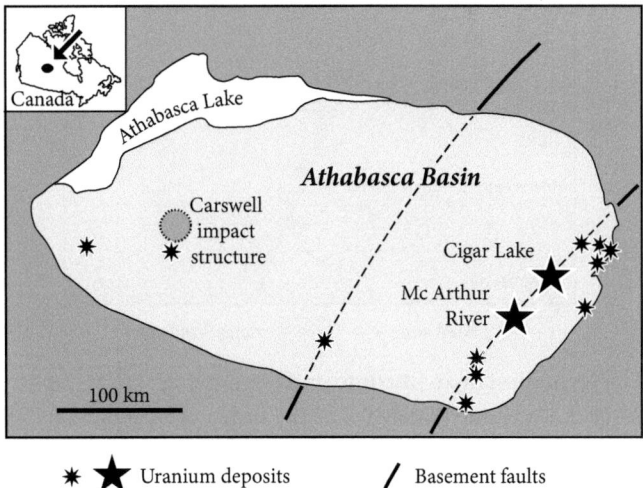

Figure 9.10 Unconformity uranium deposits in the Athabasca Basin, Canada. The U deposits are located along a basement fault, which was partly hidden under the sediments.
Source: Adapted from Jefferson et al. (2007).

deposits are located in the Athabasca Basin in Saskatchewan, Canada (Figure 9.7a). This basin contains continental sandstones derived from the erosion of the surface of the Columbia supercontinent. At the base of these sandstones, along faults, are unconformity-related uranium deposits. Cigar Lake and McArthur River reach fantastic grades of 19 and 24% uranium (Figure 9.10). These deposits are so rich, and therefore so radioactive, that mining operations have to be automated. These deposits have made Canada the second-largest producer of uranium in the world, after Kazakhstan.

Environmental conditions on the continental surface of Columbia allowed uranium to be transported and concentrated at the boundary between the ancient rocks and their sedimentary cover. It is the oxidation state of the element uranium that is important, for uranium is highly soluble in its most oxidized form (U^{6+}), but precipitates if it is reduced to U^{4+}. This is the opposite of iron! On the surface of Columbia, rocks were exposed to weathering over huge areas since there was no vegetation cover. The Proterozoic atmosphere already contained oxygen, and rainfall was more acidic than today because of the still high CO_2 content of the atmosphere, which was augmented by relatively large episodic volcanic emissions. Therefore, continental rocks at the surface of Columbia were likely subjected to aggressive, oxidizing chemical alteration that efficiently dissolved their minerals and released their elemental constituents, including uranium. Indeed, Fabre et al. (2021) tested a thermochemical and climatic weathering model based on the chemical composition of the Roraima sandstones and found that weathering intensity was up to one hundred times greater than the tropical current weathering intensity during Proterozoic times. Sandstones deposited across Columbia show similar characteristics.

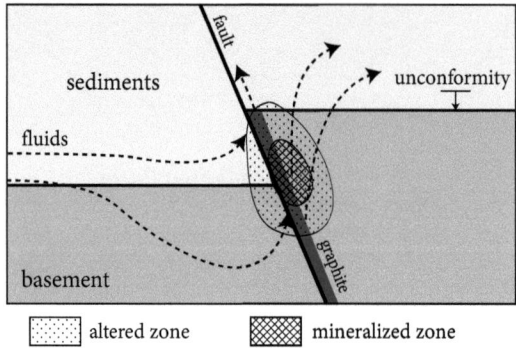

Figure 9.11 The formation of an unconformity-related uranium deposit. The U mineral deposit is located immediately below the Proterozoic sediments, where graphite generated reducing conditions.
Source: Adapted from Jefferson et al. (2007).

The uranium liberated from granitic rocks was transported in solution by runoff and then groundwater flowing through porous surficial sediments, such as sands. Locally, a fault might have raised a rock compartment, bringing the basin fluids into contact with older rocks of a reducing nature, such as organic- or graphite-rich black shales. Dissolved uranium (U^{6+}) in the fluids was then reduced and precipitated as pitchblende (where uranium is in the U^{4+} valence state). Precipitation occurred along the ancient erosion surface, or unconformity, which separates the sedimentary strata from the underlying, older rocks (Figure 9.11).

These conditions on Columbia represented an unprecedented situation in the history of the Earth. During the Archean eon, the atmosphere was not oxidizing enough for uranium to be mobilized in this way, and there were not yet large exposures of granitic continental crust. Formation of giant uranium deposits on the surface of Columbia during the Proterozoic was a result of the intersection between the internal evolution of the Earth and changes in its atmosphere composition. The changes in atmosphere itself were partially the consequence of life on Earth, specifically oxygenic photosynthesis. Once again, as in the case of iron and manganese, metallogeny had been influenced by the evolution of the biosphere.

> Archean cratonic nuclei gradually coalesced to form larger continental fragments, culminating in the formation of the Paleoproterozoic supercontinent Columbia, some two billion years ago. The isolated interior of the supercontinent was arid, and the first sandy deserts, comparable to the present-day Sahara, formed at this time. The gradually oxidizing Proterozoic atmospheric composition, combined with high carbon dioxide levels, resulted in intense weathering of the continental surface. Uranium released from granitic rocks was mobilized, then concentrated to form giant uranium deposits.

10
Diversification of life in the Proterozoic

After the many changes in the environment that took place about two billion years ago, what happened to the relatively simple microbial life of the Archean? Proterozoic life appears diverse and widespread in comparison, if only because there are nearly 3,000 fossil sites compared to only 30 or so in the Archean! Even if there is a bias due to better preservation of fossils and more abundant rocks from this time, microbial life likely proliferated in the Proterozoic. Moreover, Proterozoic fossils indicate the appearance of new life forms. Living cells became larger and more complex with the development of eukaryotes. Finally, the first fossils of multicellular organisms are observed: algae dating back a little over one billion years, and animals at the end of the Precambrian.

Multiplicity and diversity of stromatolites

From the beginning of the Archean, 3.5 billion years ago, evidence of microbial life was preserved in the form of special structures, the stromatolites, which are not fossils *sensu stricto*, but sedimentary structures induced by the presence of microbes, including the important oxygen-producing cyanobacteria. At the beginning of the Proterozoic, growth of the continents offered more and more surfaces that could be colonized by microbial mats all along the coastline; in the foreshore; and, immediately below, in shallow submerged conditions. Lagoons and lakes were also possible habitats. Any area sheltered from heavy storms was suitable for the development of microbial mats. Moreover, in the absence of foraging animals, the sedimentary support of microbial life was never disturbed. There was nothing to fear from grazers, as these would not appear for almost another two billion years! Thus, microbial mats proliferated in the Precambrian. However, life remained water-bound: there was no purely terrestrial life on emerged continental lands in the Precambrian, neither in the Archean nor in the Proterozoic.

The greatest variety of stromatolites date from the Mesoproterozoic era, between 1,600 and 1,000 Ma. Several hundred different forms have been described for this time interval and have given rise to a true classification with Latin names based on the Linnaean classification principles,[1] though it is important to remember that these stromatolite forms do not represent a single organism. In addition to simple domes and columns, large, complex, sometimes branched forms are found. Complex stromatolites are particularly abundant between 1,200 and 1,000 Ma, as exemplified by stromatolites in the Adrar region of Mauritania (Kah et al., 2009), near the town of Atar, as shown in Figure 10.1.

Beautiful Mesoproterozoic stromatolites have also been described on other continents, for example in Brazil, Siberia, Australia, and China. The Mesoproterozoic is regarded as a golden age of stromatolites, but the profusion of complex stromatolites in the Proterozoic is not simple to explain. Microbial mats that produced stromatolites would have consisted of a consortium of species, with cyanobacteria on the surface and other bacteria below with varying requirements, for example for light and other energy sources. Stromatolites were also strongly shaped by their environment. Flat, low-relief stromatolites could have withstood periodic emersion in the tidal zone, whereas large, branched stromatolites protruding above the seafloor would have formed in deeper waters where they were never emergent, possibly forming large three-dimensional constructions on the model of those of the Great Barrier Reef in Australia. The comparison has its limits because coral reefs are the constructions of animals: they do not result of microbial activity alone. Nevertheless, coral reefs today occupy similar environments to those represented by the stromatolitic buildups characteristic of the Mesoproterozoic. Long-lived stromatolite columns would require an episode of relative sea level rise to provide accommodation space while the microbial construction was rising in search of light.[2] In the case of encrusting stromatolites, only the top and younger part of the stromatolite building was possibly exposed at the sediment–water interface. Actually, little is known about the growth rate of stromatolites. Whatever their external shape, stromatolites are always made up of a succession of multi-millimetre laminae, each lamina corresponding to the presence of a microbial mat and the sedimentary particles trapped or precipitated by the microbial filaments. When one mat dies, another one settles on top. But how long did the formation of a lamina last: days, months, years? We don't know. Finally, growth and preservation of stromatolites depends on sedimentary inputs: if detrital input increases too much, the microbial mats are buried and disappear. Conversely, in the absence of detrital input or precipitation, the indurated layers do not form.

---

[1] The Linnaean classification is the systematic way of classifying living beings. It is due to the Swedish botanist Carl Linnaeus (1707–1778). The basic unit is the species, a group of identical and potentially interbreeding individuals. Similar species are grouped into genera. Each species is given a Latin name consisting of two terms: the genus name with a capital letter, followed by the species name with a small letter. The next hierarchy is the grouping of genera into families, then orders, classes, phyla, and finally kingdoms. Thus, for example, the panther, *Panthera pardus*, and the tiger, *Panthera tigris*, are two different species of the genus *Panthera* and both belong to the Felidae (feline or cat) family, in the order Carnivora, the class Mammalia, and the phylum Chordata.

[2] Sea level changes can have several causes. After the last Quaternary ice age, the melting of continental ice sheets led to a sea level rise of about 120 m. Other causes are tectonic: the bottom of the ocean rises or sinks, resulting in a relative change in sea level, without any change of the volume of the ocean.

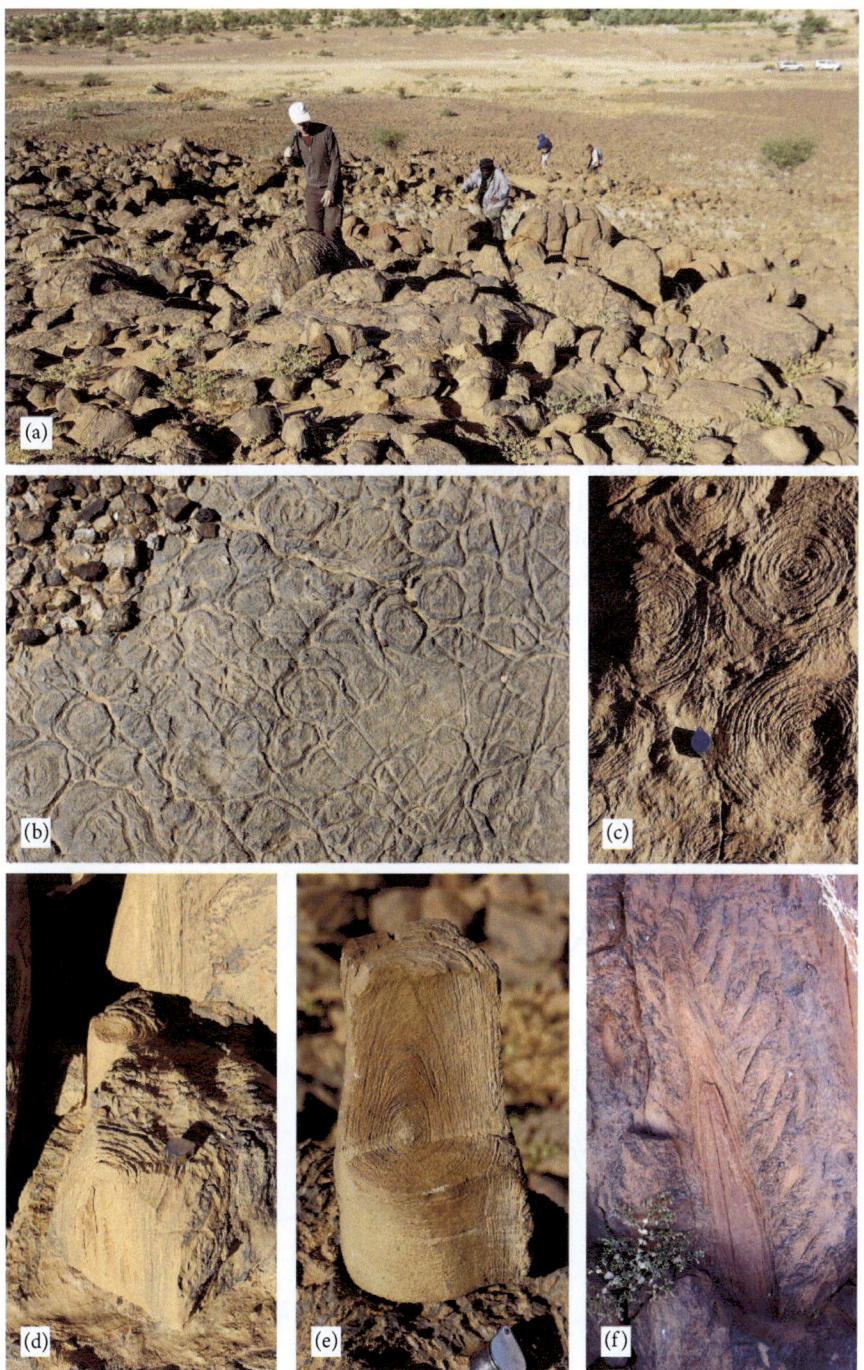

Figure 10.1 **Stromatolites near Atar town (Mauritania).** (a) Overview of stromatolitic forms sticking up from an outcrop. (b) Plan view showing horizontal cross-sections across numerous stromatolite columns. (c) Close-up plan view of stromatolitic columns showing the layers that correspond to growth stages of the stromatolites; hand lens for scale. (d) View showing the 3D structure of a stromatolite column. (e) Conical stromatolite column of the form-genus *Conophyton*, which is about 20 cm in diameter. (f) Branched stromatolith.

Source: Hervé Diot.

The Franceville Basin in Gabon: oil, biomarkers, and enigmatic fossils

In 2010, a small bombshell exploded in the world of paleontologists and geologists: the discovery of large fossils in the Paleoproterozoic sediments of Gabon by Abderrazak El Albani, professor at the University of Poitiers (France). These sediments are a little over two billion years old and contain a very high percentage of organic matter, preserved as oil or bitumen. They belong to the Francevillian Formation, named after Franceville, the second largest city in Gabon, which is nearby. They were deposited directly on the Archean basement, part of which underwent significant deformation around 2.1 Ga. At the time of deposition of the Francevillian sediments, mountains must have stood in the west of Gabon, where the Atlantic shoreline lies today. At that time, the Atlantic Ocean did not exist, and Gabon was adjacent to north-eastern Brazil such that Port-Gentil and Salvador de Bahia were twin cities (Figure 10.2). Indeed, the Archean continental cores of Gabon and Brazil collided between 2.2 and 2.1 Ga and formed the Congo–Sao Francisco craton in an early phase of the progressive assembly of the Columbia supercontinent.

Over the millions of years subsequent to the collision, the mountain range eroded, yielding sediments that were transported by rivers to the sea. The sedimentary sequence begins with conglomerates and sandstones, referred to as Francevillian A, deposited in a fluvio-deltaic environment. The overlying sediments are finer, more heterogeneous, and marine; they are referred to as Francevillian B to E. **Shales**

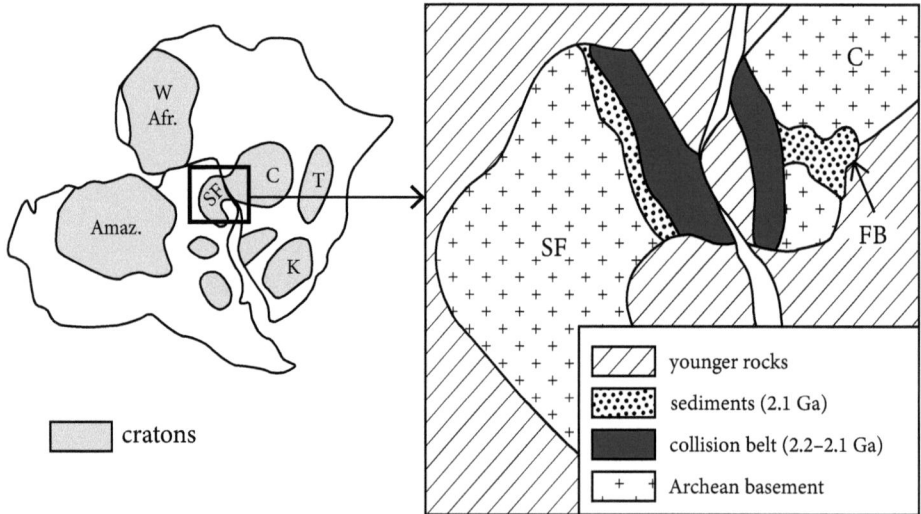

Figure 10.2 Location of the Francevillian Basin (FB). The Congo (C) and Sao Francisco (SF) cratons were welded by a collision that occurred about 2.1 Ga as the Columbia supercontinent began to form. Africa and South America are shown with their present-day contours in a reconstruction that has closed the Atlantic Ocean, which is a much younger geological feature. Other cratons shown are Amazonia (Amaz.), Kalahari (K), Tanzania (T), and West Africa (W Afr.).

and mudstones dominate. Shales are fine-grained rocks containing abundant clay minerals, along with fine grains of quartz, feldspars, and micas. To the naked eye, they appear homogeneous, but finely laminated. Their colour is variable, but they are grey to black when they contain organic matter. They are low-permeability rocks, meaning that once formed, they do not easily allow fluids to pass through them. While the lower sandstones are red in colour, indicating an oxidizing depositional environment (they were deposited after the GOE), the overlying sediments are black shales, due to their high organic content. It is in this series that the giant manganese deposits mentioned in Chapter 8 are found. The Francevillian deposits are exceptionally well preserved and represent one of the oldest basins in the world that has remained unaffected by major deformation and metamorphism.

The Francevillian series is up to 2,000 metres thick. However, numerous sedimentary figures, such as ripple marks, indicate that the water depth in the Franceville basin was never very great: 40 m at the most. The accumulation of sediments was due to the progressive sinking of the basin floor: a process named **subsidence**. Subsidence generally affects thinned areas of the lithosphere which may correspond to an oceanic margin or an extending continental rift, but also occurs adjacent to mountain ranges, whose immense mass causes the crust to flex downwards.

The weathering of the nearby mountain range delivered nutrients to the sea, helping to support a high level of biological productivity. Francevillian mudstones contain fragments of microbial mats and other organic remains. The remains of these prolific microbes consumed the oxygen in the water column below the surface, thus supporting the accumulation of organic-rich, fine-grained sediments. The content of organic matter preserved in these sediments reaches very high levels (locally up to 15%). As sediments are buried, they are gradually heated due to the geothermal gradient, and this heating slowly converts some of the organic matter into oil. At the base of the black shales and through faults that offset the sedimentary layers, some of the oil migrated down into the Francevillian A sandstones, which are porous. At the same time, the sandstones were infiltrated by oxidizing fluids that had dissolved the uranium-bearing minerals in the conglomerates below. These fluids lost their oxidizing potential when they encountered oil—a reducing liquid—causing the precipitation of uranium as insoluble oxides (Figure 10.3). This is how the uranium deposits at the top of the Francevillian A sandstones formed.

In Gabon, uranium was locally concentrated enough for nuclear chain reactions to have been triggered. This discovery was made by engineers and technicians from the French CEA (*Commissariat à l'Energie Atomique*, Atomic Energy Commission) who found small but significant differences in the proportions of the isotopes of uranium 235 and 238 in the Francevillian ores. Specifically, the isotope ^{235}U, which decays the fastest, is found in lower concentrations than would be expected (0.6 instead of 0.72% of the total uranium, Neuilly et al. 1972). This intriguing situation could only be explained by the occurrence of nuclear fission in the Francevillian sandstone, which has been verified by the detection of other isotopic signatures of nuclear fission. This fossil nuclear reactor, named Oklo, is the only known natural nuclear reactor. The

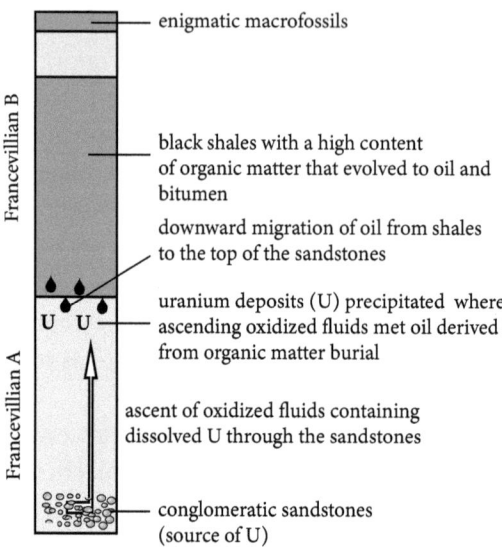

Figure 10.3 Stratigraphic column of the Francevillian Basin showing the succession of Francevillian sediments (sandstone in light grey, mudstone in darker grey), formation of uranium deposits, and location of the fossils discovered by El Albani and his collaborators.

Source: Adapted from Gauthier-Lafaye and Weber (1989).

Oklo reactor was small in volume (about 10 m³) and presumably could only have formed due to relatively higher concentrations of ^{235}U (~3%) two billion years ago, accompanied by exceptional uranium enrichment in Francevillian sandstones. Such a process would be impossible today under natural conditions, as too little ^{235}U remains to be concentrated by natural processes. The Oklo reactor has been operating for 800,000 years.

Let us return to the macroscopic structures discovered in the black clay at the top of the Francevillian B sediments. These are centimetre-scale nodules of variable appearance—round, folded, or lobed—concentrated in certain beds (Figure 10.4) and revealed during the excavation of a quarry. Sediment is draped over the nodules, implying they formed on the seafloor.

This discovery in such ancient rocks inevitably raised a question: were the nodules true fossils or biogenic structures, or rather abiotic mineral concretions? The diversity and complexity of the forms virtually reconstructed in three dimensions suggest an organic origin (Figure 10.5).

Moussavou et al. (2015) extensively studied the mineralogy, microstructure, and geochemistry of the nodules. The nodules contain microporous and microcrystalline silica that replaced early calcite. Several microscopic remnants were identified, including a few multicellular clusters 50 to 250 micrometres in size. The δ^{13}C of the organic matter is −24‰ on average, in the usual range of cyanobacterial mat. δ^{34}S values of pyrite grains match those expected from bacterial sulphate reduction at that time.

Figure 10.4 **Top view of the surface of a Francevillian clay bed** showing moulds and imprints from one to a few centimetres long.
Source: Abderrazak El Albani.

Figure 10.5 **The Francevillian fossils from Gabon.** 3D reconstruction (left) and partially transparent virtual image (right) of a fossil; the curved structure in the centre is not a particular organ: it is simply due to folding of the whole structure.
Source: Abderrazak El Albani and Arnaud Mazurier.

These data are indicative of former microbial consortia composed of microorganisms with different metabolisms, living on the surface of the Franceville Basin muds. They must have been rapidly smothered by a thin layer of sediment to have been preserved. This mud, saturated with sea water, quickly became depleted with oxygen, at which point sulphate-reducing bacteria that use the dead organic matter as a source of carbon began to generate sulphide. These sulphide ions combined with iron in the sediment to form small pyrite crystals, which gradually replaced the buried organic matter. These pyrite crystals form a fossil template of the organic material that is perfectly preserved if the sediment remains undisturbed and protected from the oxygen-containing air.

The Francevillian nodules are unlikely to represent multi-cellular animals as initially proposed by El Albani. Microscopic observations, along with molecular phylogenetic estimates on when animals originated, suggest that these provocative fossils represent microbial colonies. The debate about what these unique fossils represent continues (El Albani et al., 2019). Another interesting clue from these rocks comprises **biomarkers** found in small oil inclusions preserved at the top of the Francevillian A sandstone as well as in solid bitumen. Biomarkers are organic molecules derived from living organisms that have been preserved in sediments. Organic geochemists know how to extract and identify biomarkers in sediments that have not been heated too much. This is a common technique for recognizing oils from different times and origins. The samples from Gabon contain many biomarkers typical of bacteria, especially cyanobacteria, but also, more unexpectedly, many biomarkers typical of eukaryotes, the **steranes** (Dutkiewicz et al., 2007). These molecules were derived from sterols or steroids that are produced by eukaryotes and not by bacteria. We must conclude that eukaryotes were already abundant at 2.1 Ga in the Franceville Basin and that their appearance dates back at least to this period. So, what were the first fossil eukaryotes?

Acritarchs: the first fossil eukaryotes

Discreet and enigmatic but evolutionarily important microfossils appeared in the Proterozoic: the **acritarchs**. They are made of organic matter and are therefore certainly the remains of microscopic organisms, but lack modern equivalents. They were protected by an envelope, a sort of flexible but resistant shell, formed of polysaccharides (i.e. complex carbohydrates) such as chitin or cellulose.

The earliest known acritarch fossils date back to just over 1,600 Ma. They have been found in India and China. Figure 10.6 shows a specimen called *Shuiyousphaeridium*. The delicate protuberances attached to the surface of the shell protecting the cell suggest planktonic living conditions: the protuberances probably helped the acritarch to float. This small carbonaceous sphere measures just over 100 microns in diameter, which is considerably larger (about fifty times) than a bacterial cell. The combination of their large size and the protuberances suggests that at least some of these acritarchs represent early eukaryotes. While the taxonomic affinity of acritarchs is uncertain, it has been proposed that some may be related to the peridinians or dinoflagellates, which are protists (unicellular eukaryotes) that feed on microscopic algae and that are also often protected by a tough cellulose shell.

Diverse acritarchs have been found in Mesoproterozoic and Neoproterozoic sediments, ranging in age from 1,600 to 540 million years. Their sizes vary between 50

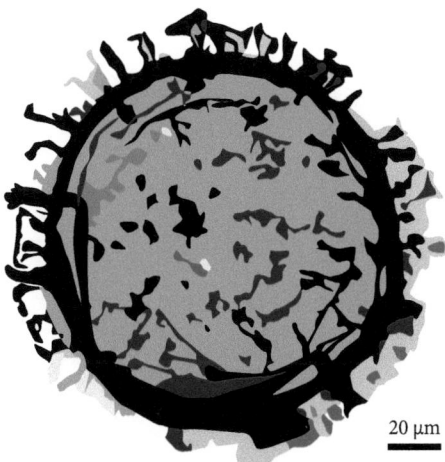

Figure 10.6 *Shuiyousphaeridium*, **one of the oldest known acritarchs**, which has been found in *c*.1,600 Ma rocks in China.
Source: Adapted from Yin (1997).

and 200 microns. Acritarchs are often spherical, with various ornaments or extensions. Others have a more elongate shape. Acritarchs are particularly abundant in rocks around 800 Ma.

The nature of the organic matter associated with acritarchs provides very important information. Acritarchs are sometimes associated with a type of biomarker characteristic of eukaryotes, the steranes. This argument, together with their size, leads to the conclusion that acritarchs were likely eukaryotes.

The origin of eukaryotes

The Eukaryota, the third domain in the living world after the Archea and Bacteria, appeared a little over two billion years ago. Note that human beings belong to the eukaryotic group, like all animals and plants. Eukaryotes are very different from archea and bacteria, as their cells are much more complex. The eukaryotic cell has a nucleus, bounded by a membrane, which contains the DNA (the genetic material of the cell) in the form of chromosomes. In contrast, the cell of an archaeon or a bacterium does not contain any nucleus. The DNA of the bacterial chromosome is a single circular molecule, located in the cytoplasm of the cell. In contrast, the eukaryotic cell is about twenty to one hundred times larger than the bacterial cell and is compartmentalized with many organelles that play specific roles (Figure 10.7).

Among these organelles, two deserve close consideration: **mitochondria** and **chloroplasts**. They contain a circular DNA molecule, similar to the DNA of bacteria. Thus, mitochondria and chloroplasts have their own genetic material, independent of the cell genetic material contained in the nucleus. They also possess their own ribosomes, which enable them to make specific proteins. These ribosomes are smaller

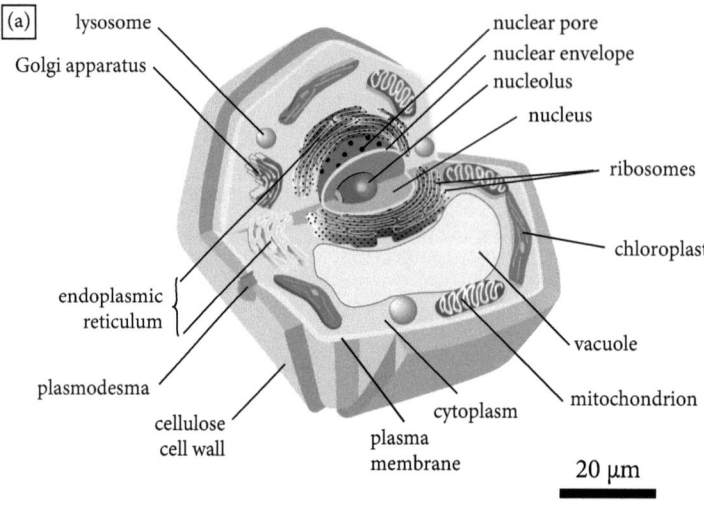

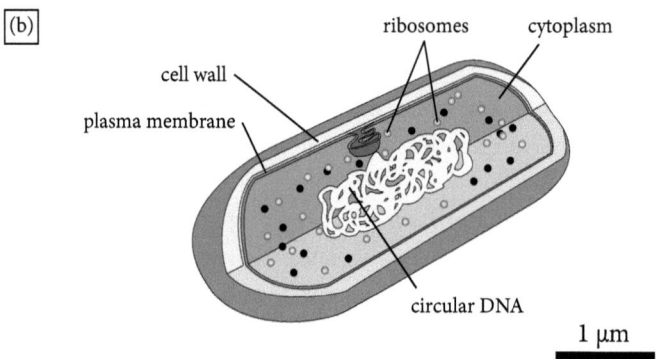

Figure 10.7 Comparison of eukaryotic and prokaryotic cells. (a) Plant cell. (b) Bacterial cell.

than the ribosomes in the cytoplasm of the eukaryotic cell; they resemble bacterial ribosomes.

Chloroplasts are found in algae, green land plants, and many protists. They contain the chlorophyll that enables them to carry out photosynthesis. Mitochondria are present in all eukaryotes. They produce the energy needed by the cell by transforming organic molecules from food, such as glucose. The breakdown of glucose in the presence of oxygen produces CO_2 and energy. Mitochondria are tiny, but very numerous in a single cell: their number ranges from several hundred to several thousand. In size, morphology, and function, as well as owing to the presence of DNA, a mitochondrion (Figure 10.8) closely resembles species of the Alphaproteobacteria, a class of bacteria.

When a cell divides into two daughter cells, the stock of mitochondria is divided and distributed between both daughter cells. Then the number of mitochondria increases in each cell, because mitochondria can replicate by simple fission, independently of cell division. In the case of sexual reproduction, which involves the fusion of an egg

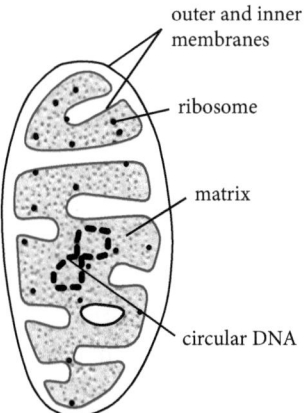

Figure 10.8 Schematic cross-section of a mitochondrion, as seen under an electron microscope (its length is a few microns).

and a sperm, the mitochondria are brought by the cytoplasm of the egg. We therefore inherit our mitochondria and their DNA from our mothers and maternal grandmothers! In contrast to this mitochondrial inheritance, the DNA of the chromosomes in the nucleus is inherited from both parents in equal parts.

In 1967, a 28-year-old American microbiologist, Lynn Margulis (born Lynn Alexander and first married to Carl Sagan), defended the idea that mitochondria represent ancient bacteria that lived in symbiosis in the cells of primitive eukaryotes (i.e. the hypothesis of an endosymbiotic origin of mitochondria).[3] Symbiosis was not a new idea at that time, because this phenomenon is not rare in the living world. For example, a lichen is a symbiotic association between a fungus and an alga. Unlike parasitism, which involves the exploitation of one living being by another, symbiosis implies that each partner finds an advantage in the association.

At the time of the Margulis' initial publication on the endosymbiotic origin of eukaryotics, the idea was deeply unpopular among biologists, many of whom scorned the hypothesis. But after more than a decade of struggle, the hypothesis of the endosymbiotic origin of mitochondria finally gained full acceptance in the scientific community, in particular with the discovery and study of the mitochondrial DNA (Gray and Doolittle, 1982). Lynn Margulis became a professor at Boston University in 1977 and later moved to Amherst University (Massachusetts) in 1988. She passed away in 2011. Her contribution to our understanding of the origin of complex life was extraordinary.

Analogous to the origin of mitochondria, the chloroplasts of algae and land plants are also considered to have an endosymbiotic origin, in which the host cell was a protist, and a cyanobacterium became the chloroplast. Upstream of these endosymbiotic processes, an essential point could be the capacity for phagocytosis. Bacteria and

---

[3] The endosymbiotic hypothesis was first put forward in 1905 by Constantin Mereschkowski, a professor of microbiology at the University of Kazan in Russia, but his hypothesis was forgotten shortly afterwards.

archea can only take up dissolved substances that pass through the cell wall and cell membrane. They are not capable of phagocytosis, unlike many eukaryotes, which have a flexible and deformable cell membrane without any cell wall allowing them to ingest food particles of different sizes by phagocytosis (Figure 10.9). Normally, food absorbed in this way is digested in the cytoplasm. One can imagine that the micro-organisms that form the endocellular organelles started their symbiotic life in eukaryotic cells after a phagocytosis event without digestion.

When did these processes occur? The biomarker records suggest that eukaryotes may have appeared by 2.1 Ga ago. Moreover, with possibly exceedingly rare exceptions, no eukaryotes are known to lack mitochondria. Did primitive eukaryotes already have mitochondria? If so, the origin of mitochondria would therefore also date back to at least 2.1 Ga, though the timing of this event remains controversial. The acquisition of mitochondria results in cells that have a considerable energetic advantage over other cells. Later, likely some time between about 1.6 and 1.2 Ga, a second endosymbiotic event gave rise to chloroplasts and enabled algae and ultimately plants to carry out photosynthesis (Figure 10.10).

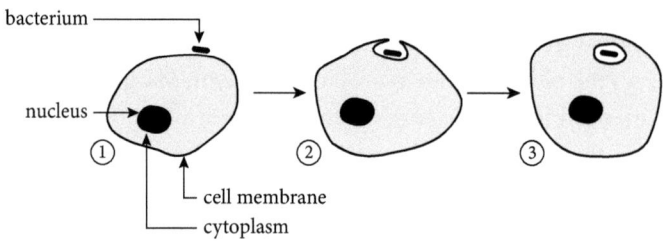

Figure 10.9 **Steps in the phagocytosis** of a bacterium by a eukaryotic cell.

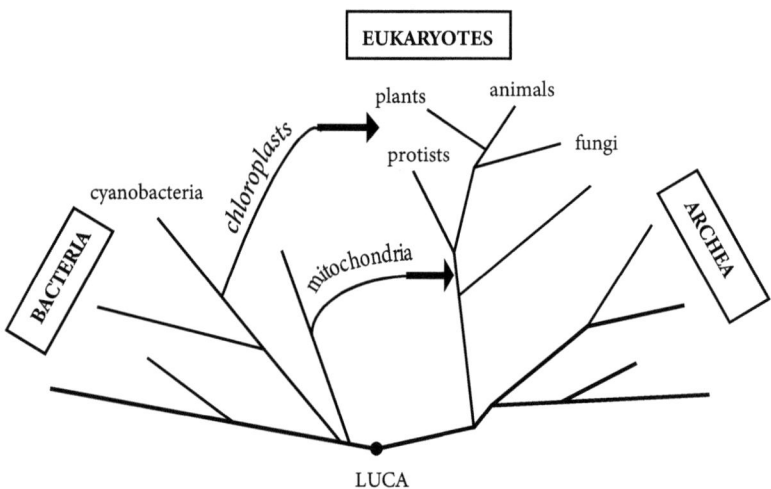

Figure 10.10 **The place of mitochondria and chloroplasts in the tree of life.** Compared to Figure 7.13, this version of the tree of life suggest that Eukaryotes derived from the Archea and not directly from LUCA, the Last Universal Common Ancestor.

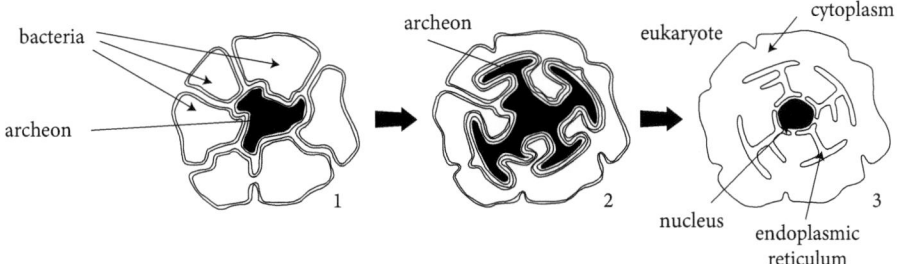

Figure 10.11 Origin of the eukaryotic cell. (1) An archaeon (in black) is surrounded by five or six bacteria (in white) with which it forms a symbiotic association. (2) The cytoplasms of the bacteria merge, while the archaeon emits long extensions into the common cytoplasm. (3) This leads to the eukaryotic cell with its nucleus and endoplasmic reticulum (the interconnected network that serves as the transportation system in the cell).
Source: Adapted from Moreira and López-García (1998).

The question now arises as to the origin of the eukaryotic nucleus. For the interested reader, the origin of eukaryotes is discussed at length in the very accessible book written by John Maynard Smith and Eörs Szathmáry (2000). Most biologists assume that eukaryotes derive from archea—that is, an archaeon was the host cell in the endosymbiotic event giving rise to the mitochondria. This is the assumption that is made in Figure 10.10 where eukaryotes branch off from archea and not directly from LUCA.

Several symbiotic associations involving methanogenic archea and sulphate-reducing bacteria have been identified in nature. Both partners benefit from the association. When isolated, each partner grows only poorly. In 1998, David Moreira and Purificación López-García considered such an association to explain the genesis of eukaryotes (Figure 10.11). In this hypothesis, the archaeon would provide the nucleus and the associated bacterium would constitute the cytoplasm of the future eukaryotic cell. Thus, eukaryotes are actually chimeras.[4]

A central controversial question regarding the origin of eukaryotes concerns the role that oxygen may have played. If eukaryotes did indeed appear sometime just prior to 2.1 Ga, then it would be tempting to connect the origin of the mitochondria, and hence, eukaryotes, to the GOE.

The emergence of algae

The first multicellular fossil, *Bangiomorpha pubescens*, found so far only in northeastern Canada and shown in Figure 10.12, is interpreted to belong to the red algae (Butterfield, 2015). It was dated at 1,050 Ma (i.e. the close of the Mesoproterozoic

---

[4] A chimera is a mythological half-goat, half-lion creature with the tail of a snake. In the same way, a unicellular eukaryote would be a composite being resulting from the obligatory association of an archaeon and several bacteria.

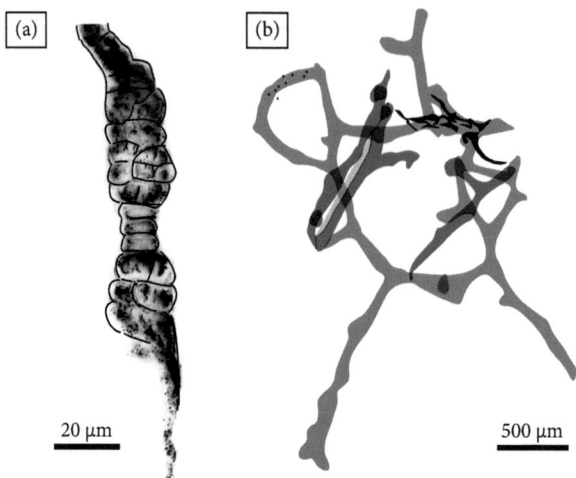

Figure 10.12 Fossil algae. (a) *Bangiomorpha pubescens*, the oldest fossil, from the late Mesoproterozoic (*c*.1,050 Ma) in Nunavut, north-eastern Canada. Note the arrangement of cells in groups, some of which may be specialized for a particular role. (b) A slightly younger, larger, and more branched algal fossil, found in Siberia.
Source: Adapted from Butterfield (2015).

(Gibson et al., 2018)). Using a molecular clock evolution model, the latter authors also calculated that photosynthesis emerged within the eukaryotes at around 1.25 Ga, representing the minimum age for chloroplast acquisition by endosymbiosis. The new multicellular organisms were quite different from bacterial colonies. Bacteria are prokaryotes, and in some groups the cells remain closely associated in colonies after replication. This tendency is favoured by the production of sticky, mucilaginous substances which help bind them together. Nevertheless, a bacterial colony is not multicellular. The bacterial colony grows according to the resources of the environment and not according to a precise genetically programmed plan. It is made of identical cells, with no differentiation. All cells retain the property of developing and reproducing independently of the other cells. By contrast, multicellular eukaryotes have several cell types organized as different tissues or organs and reproduced from specialized germ cells.

Of course, the appearance of the first macroscopic algae likely preceded the preservation of the first algal fossil. And other, older putative red algal fossils have been described. However, it is broadly accepted that algae appeared in the Mesoproterozoic. Sterane molecules with 27 carbon atoms, characteristic of the red algae, first appear in the geological record around 800 Ma. The oldest fossils of green algae date to about 1.0 Ga, but their characteristic steranes with 29 carbon atoms do not appear until about 650 Ma.

It seems that once macroscopic algae appeared and started to play an important role in the marine environment, stromatolites started to decline. One possible explanation is competition for the substrate: the stromatolites no longer had as much

EARTH AND LIFE 165

space available in which to live and attach themselves. Microbial mats necessarily need a support to grow. Their growth rate may also have been lower than that of the algae.

First marine animals: the Ediacara fauna

After the algae, a new stage in evolution is marked by the appearance of other multicellular organisms, the animals. Unlike plants, which produce their own organic molecules from carbon dioxide through photosynthesis, animals are heterotrophic, meaning they must feed on existing organic matter.[5]

The oldest animal fossils date to around 575 Ma, in the Ediacaran period—the final period of the Precambrian. These first animal fossils are locally abundant and diverse (Figure 10.13) and belong to an association called the Ediacaran fauna (named after a locality in Australia). The Ediacaran fossils include soft-bodied animals, whose traces

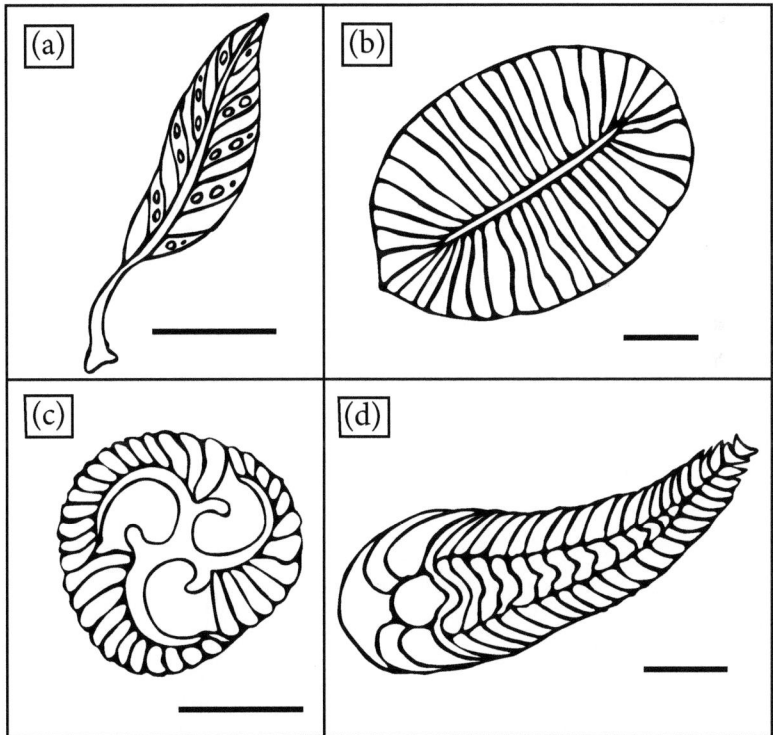

Figure 10.13 **Some representatives of the Ediacara fauna.** Each scale bar measures 1 cm.

(a) An animal attached in the form of a frond. (b) A flattened disc-shaped animal with bilateral symmetry: *Dickinsonia*. (c) *Tribrachidium*, with its enigmatic triradial symmetry. (d) The fossil *Spriggina*, named after Reginald Sprigg, who first discovered Ediacaran fossils in South Australia. *Source*: Adapted from Laflamme et al. (2013).

---

[5] The same applies to fungi, which are neither plants nor animals.

have been preserved in fine sediments. Some of the most spectacular examples occur in Australia, Newfoundland, Namibia, and Siberia. In any case, the Ediacaran fauna was diverse, and some living beings must have been quite sophisticated. The ancestors of these animals may have appeared before the Ediacaran period, but they were too few or too small, or were not preserved in the sediments. Or perhaps they just have not yet been discovered?

The relationships of the Ediacaran fossils with currently known animals are not straightforward. Figure 10.13a shows an animal that resembles a pennatula, a kind of feather-shaped sea anemone. But what about the enigmatic discoid forms (Figure 10.13b and c)? In truth, we know little or nothing about their lifestyle. Were they filter feeders? Or grazers of algae or microbial mats? Even morphological similarities can be misleading in assigning a systematic affinity. Indeed, neither head nor tail can be recognized in the disc-shape *Dickinsonia* (Figure 10.13b). First regarded as a sort of jellyfish and variously interpreted as many different classes of animal, it could even be related to an unknown metazoan group.

Finally, a discrete fossil from the very end of the Ediacaran, around 550 Ma, heralds further changes in the biological environment. This is *Cloudina*, named after the American geologist Preston Cloud, a small fossil protected by a shell made of small, nested cones and measuring no more than a few millimetres in length (Figure 10.14). It is the first animal capable of making its own shell as an external mineralized structure, which is distinct from the soft-bodied Ediacara fauna. What is the biological advantage of a shell? It seems that *Cloudina* was a small, bottom-dwelling animal living in shallow waters. Its shell may have protected it from waves, but perhaps also from predators. This protection was not always sufficient, as *Cloudina* shells are sometimes found perforated! The regular outline of these perforations suggests that they were made by other unknown animals that feasted on the little *Cloudina*. The ecosystem of 550 Ma ago therefore included a complex food web with carnivores at the top.

At the end of the Precambrian, the biosphere began to resemble life in today's oceans: plankton, algae, animals, some attached to the seafloor and others mobile,

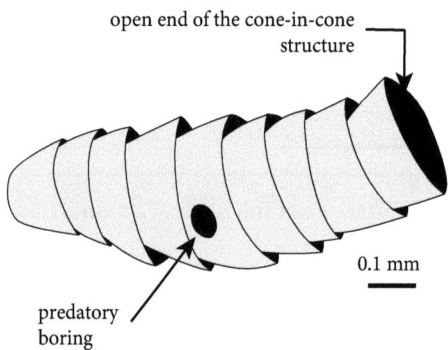

Figure 10.14 ***Cloudina* shell** with a predatory boring.
Adapted from Becker-Kerber et al. (2017).

and even carnivores among them. This was very different from the purely microbial ecosystems of the Archean and even most of the Proterozoic. Moreover, the new fauna began to play an important role modifying their ecosystem, for some of these animals began to scour and burrow in the sediment as they moved and fed, which in turn aerated the sediments. This burrowing, which is characteristic of Phanerozoic sediments, ended the proliferation of continuous, poorly oxygenated laminae that were common in many Precambrian sediments. In addition, the ability to produce a calcareous shell, which appeared at the end of the Precambrian and was later found in other groups of animals that appeared shortly thereafter (e.g. mollusks) modified the global carbon cycle. Indeed, recent carbonate rocks formed dominantly through the accumulation of skeletons and shells.

> The Proterozoic was a time of major diversification in life. Eukaryotes, or cells with a true nucleus, appeared, possibly in response to the oxygenation of the marine environment. Their cells also contain organelles, most importantly the mitochondria, which resemble bacteria. This suggests that eukaryotes derived from a symbiotic association between archea and bacteria. Eukaryotes started out as single-celled organisms and later became more complex, sexually reproducing, multicellular organisms. The fossil record allows us to recognize the first algae in the Mesoproterozoic era, followed by the first animals in the Ediacaran period, at the close of the Precambrian. Nevertheless, no animals or plants had yet colonized the continental surface.

11
From Columbia to Gondwana

The ballet of the continents

Columbia, the first supercontinent, formed about two billion years ago. A second supercontinent, Rodinia, formed later during the Proterozoic. The sequence of tectonic events began with the fragmentation of the first supercontinent, leading to the formation of new oceans, followed by the reassembly of continental fragments in collisions that gave rise to mountain ranges. This perpetual shuffling of the continents across Earth's surface is like a ballet at a huge scale, driven by plate tectonics. For convenience, we only refer to continents. Continents are part of lithospheric plates, and these plates move because new oceanic crust is created and then disappears back into the mantle by subduction. While the oceanic part of the plates has a geologically transient existence, the continental part is preserved on Earth's surface for eons, participating in different configurations, like pieces of a puzzle that are put together in different ways. The drift and collision of the continents result in successive paleogeographies—ancient maps of the world that geologists strive to draft to understand ancient environments.

The Wilson cycle

The Canadian geologist J. Tuzo Wilson (1908–1993) was one of the most fertile minds in launching new ideas in the early days of plate tectonics. In 1963, he hypothesized that the Hawaiian volcanoes originated from a fixed hotspot under the moving Pacific plate. Two years later, he proposed the existence of a third type of plate boundary, complementing divergent (oceanic ridge) and convergent (subduction zone and collisional chain) boundaries. These vertically sliding boundaries came to be known as 'transform' faults and include the San Andreas Fault[1] in California, which separates the Pacific and American plates.

Finally, in 1966, Wilson proposed that the displacement of plates gives rise to a cyclical repetition of tectonic processes with a succession of oceanic openings and closures. More precisely, he considered that the Atlantic Ocean was preceded by another ocean at approximately the same location, which he called the proto-Atlantic. He placed the

---

[1] Sliding along this fault causes recurrent earthquakes from Los Angeles to San Francisco. The great 1906 earthquake that heavily damaged the city of San Francisco occurred on the San Andreas Fault, and surely other large earthquakes will occur in the future.

proto-Atlantic in the location of the present-day North Atlantic, with one small difference: western Scotland was separated from the rest of the UK and belonged to Laurentia (the large Proterozoic North American craton). Today, this ancient ocean basin is reasonably well understood and accepted among geologists and is known as the Iapetus Ocean. It formed at the very end of Precambrian and then closed in the first half of the Paleozoic era during the Caledonian **Orogeny**,[2] which formed the mountain range known as the Caledonian Range. This collision welded Laurentia and Baltica together along the eastern edge of Greenland on one side to Norway and to Scotland on the other side, forming a larger continental mass, formerly referred to as the Old Red Sandstone continent. Later, towards the end of the Paleozoic era, some 300 million years ago, the Old Red Sandstone continent collided with Gondwana to form supercontinent Pangea during the Hercynian/Alleghenian orogeny. The Atlantic Ocean only began to open at the beginning of the Mesozoic era, around 200 Ma, with the beginning of the break-up of Pangea. The Atlantic Ocean is still a relatively young ocean. The Pacific Ocean, on the other hand, is an older ocean, surrounded by subduction zones that have consumed more-than-200-million-year-old oceanic crust. An active ridge in its eastern part, the East Pacific Rise, is constantly renewing the oceanic crust.

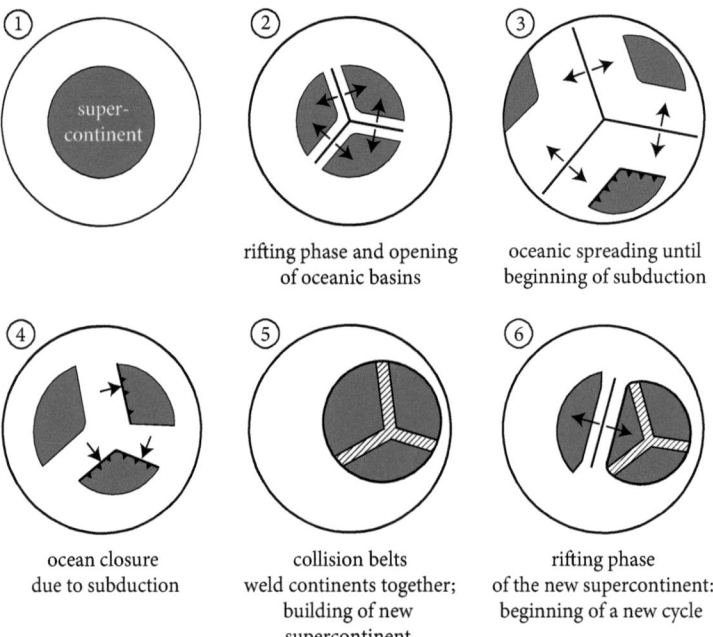

Figure 11.1 **Schematic representation of the Wilson cycle.** The individual events in a Wilson cycle are not always so well synchronized. For example, some collisions may still be in progress while a new ocean is already opening elsewhere.

---

[2] The adjective Caledonian comes from *Caledonia*, the Latin name for Scotland.

The idea of cyclic tectonism has gained favour, and the phenomenon is aptly referred to as the Wilson cycle: ocean opening, then subduction and continental collision, followed by new ocean opening, and so on (Figure 11.1). The Wilson cycle operates at the speed of plate movements (i.e. a few centimetres per year). The total duration of a Wilson cycle varies, but it is generally in the order of several hundred million years. Thanks to the progress of absolute dating and the contributions of paleomagnetism, we can extend the record of Wilson cycles to the Precambrian. In this chapter, we'll attempt to reconstruct the Wilson cycles that have taken place since the formation of Columbia, and describe the supercontinents that succeeded.

The role of hotspots in the breaking of supercontinents

The fragmentation of a supercontinent is usually triggered, or at least facilitated, by the effect of several hotspots. A hotspot is the surface expression of a deep plume of hot material that rises through the mantle. Recall that the material in plumes is solid but has sufficiently low density and viscosity to rise buoyantly through the mantle. The hottest plumes originate at the core–mantle boundary. The quasi-adiabatic ascent of a plume (i.e. without heat loss), delivers very hot mantle to the base of the lithosphere. The effect is like a blowtorch: the base of the lithosphere rises above the buoyant plume, since it is only a thermal boundary (Figure 11.2). This hot mantle at the base of the lithosphere also weakens the lithosphere, making it more prone to breaking apart.

At shallow depths (200 to 250 km below the surface), the hot mantle plume begins to melt, producing basaltic magmas. In the case of a hot plume of deep origin, the thermal anomaly is sufficient that large volumes of magmas are produced under the thinned lithosphere. Some of these magmas reach the Earth's surface, where effusive volcanic eruptions form piles of extensive volcanic floods that may be several kilometres thick: these flood basalts (also known as traps) form the carapace of large igneous provinces (Figure 5.4). The huge volumes of magma ascend through vertical fractures, leaving behind vertical basaltic veins called **dykes**. The dykes that fed the traps are generally

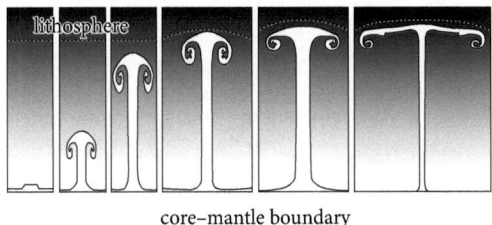

core–mantle boundary

Figure 11.2 Theoretical evolution of a hot plume. The plume forms at the base of the mantle, moves up, and finally spreads under the lithosphere, which is thermally thinned. The duration of the whole process is a few hundred million years. White dashed line: isotherm corresponding to the base of the lithosphere. Real plumes may not be perfectly vertical and can have variable widths (see Figure 5.7).

a few tens of metres wide and several hundred to several thousand kilometres long. In addition, they commonly form a radiating pattern that points to the central position of the mantle plume under the lithosphere.

Thus, flood basalts and/or dyke swarms indicate the position of the main hotspots that caused the break-up of Pangea (Figure 11.3). The oldest flood basalts have been almost completely eroded, as in the case of the hotspot between West Africa and Laurentia that opened the central Atlantic 200 Ma ago. However, the dykes that fed this giant magmatic province are still recognizable.

Volcanism associated with a hotspot lasts for several tens of millions of years, but the largest volumes of lava are released early, when the plume reaches the base of the lithosphere. The Icelandic hotspot, which triggered the opening of the North Atlantic Ocean about 55 Ma ago, is still active today. When it appeared, it extruded huge quantities of basalts in Greenland, Scotland, and Ireland, including the Giant's Causeway basalts in Northern Ireland.

The formation of the Atlantic Ocean and the initial break-up of Pangea was triggered by the hotspot between West Africa and Laurentia 200 Ma (Figure 11.3). Subsequently, the Karoo–Ferrar hotspot initiated the formation of the Indian Ocean between Africa

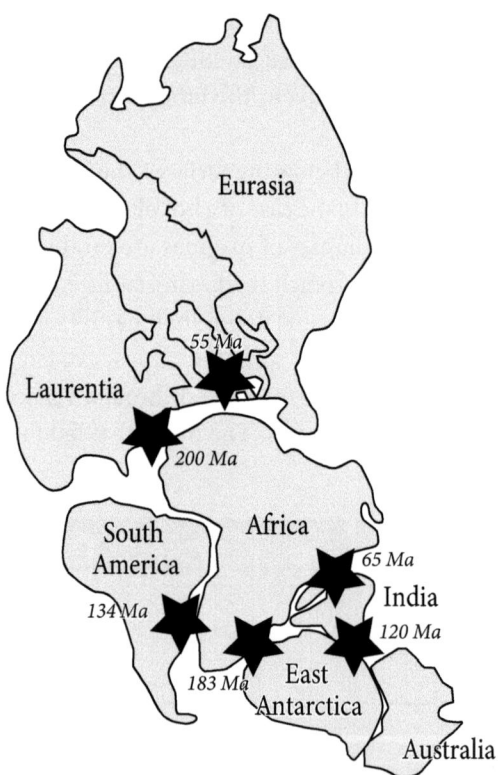

Figure 11.3 Reconstruction of the Paleozoic-era Pangea. The black stars correspond to the hotspots that triggered the fragmentation of the supercontinent and the opening of the Atlantic and Indian oceans during the Mesozoic and early Cenozoic eras.

and Antarctica. Gradually, the Atlantic and Indian oceans widened, as oceanic crust was continuously produced at the mid-oceanic ridges. As the continents carried by the two plates diverge, the ocean basin accretes. As noted by Wegener, this continental drift elegantly explains why the western coast of Africa and the eastern coast of South America appear to fit together like puzzle pieces.

When a mantle plume first impinges upon a continent, the thinned lithosphere eventually gives way: a fracture appears and widens to form a divergent rift system. This process is playing out today in Africa, where the formation of the Ethiopian traps at around 30 Ma initiated rifting. The result is the Great African Rift, which stretches from Malawi to Djibouti via Kenya and Ethiopia, illustrating the onset of the break-up of the African plate. The nearby Red Sea, which separates the African plate from the Arabian plate, illustrates the next phase of break-up, when new oceanic crust begins to form. Indeed, seismic tomography has identified a superplume, like that shown in Figure 5.7, in the mantle beneath north-eastern Africa. Thus, the Red Sea and the Great African Rift together herald the formation of a future ocean in a few tens of millions of years.

In search of lost oceans

We have seen that the oceanic crust has a geologically short lifespan on the Earth's surface. In fact, as oceanic lithosphere pulls from the ridge, it cools and becomes increasingly dense. It is for this reason that after about 200 million years, oceanic lithosphere is so dense that it will begin to subduct (although it may subduct much earlier given the right tectonic conditions). This oceanic lithospheric lifespan results in the inevitable disappearance of the oceanic lithosphere and eventually the closure of the whole ocean if the production of new oceanic crust at the ridge is not efficient enough to compensate for subduction. As the opposing continental margins begin to collide with one another, a new mountain belt forms.

A recent example of this process is the Himalayas, a mountain range that resulted from the disappearance of the Tethys[3] Ocean, which existed between Eurasia and Gondwana in the Mesozoic Era. The closure of the Tethys Ocean led to the collision of Eurasia in the north and India in the south. The vanished ocean left traces in the form of scattered relics of oceanic crust and upper mantle from the oceanic lithosphere, the **ophiolites**, which were recognized in Tibet, all along the Himalayan chain. It also shows up in seismic tomographic images beneath the current Indian Ocean.

Finding extinct Proterozoic oceans is a more difficult challenge. Ancient ophiolites and other traces of ancient subductions are rarely preserved. Nevertheless, ancient mountain ranges, although long since erased, can be inferred by linear belts of highly deformed metamorphic rocks that can be traced over hundreds to thousands of

---

[3] Tethys was a sea goddess in the Greek mythology. The hypothesis of an inland sea called the 'Tethys Sea' between Laurasia and Gondwana was suggested by Eduard Suess.

kilometres. Numerous field studies, backed up by increasingly precise radiometric dating and increasingly abundant paleomagnetic data, now make it possible to outline the tectonic history of the last two billion years.

The evolution and fragmentation of Columbia

One possible method of reconstructing this rifting stage of the break-up of Columbia is to look for traces of hotspot magmatism that would have preceded the opening of an ocean. As discussed previously, even if old flood basalts are fragmentary in their preservation, the same is not true for the swarms of dykes that fed them, for they are difficult to erode away. This is the case, for example, with the Mackenzie dykes, a giant swarm of basaltic dykes that cut across much of Canada (Figure 11.4). Many of these dykes can be mapped over thousands of kilometres. They converge towards a place that is naturally interpreted as the arrival point of their parent mantle plume under the lithosphere and where some of the original flood basalts are still preserved.

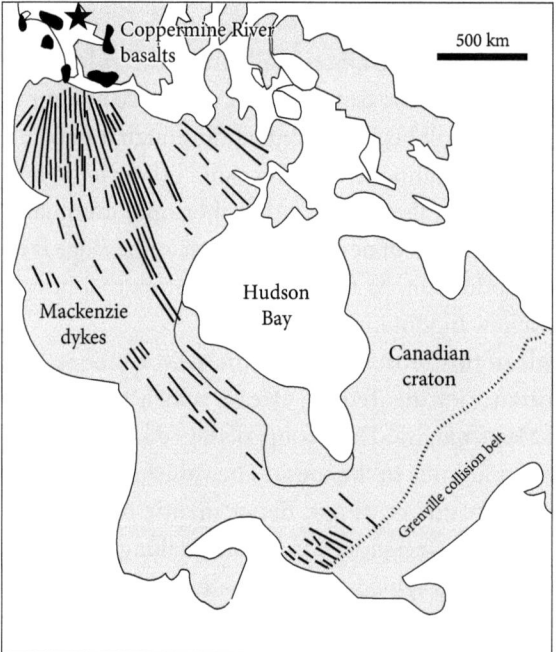

Figure 11.4 Mackenzie dykes in northern and eastern Canada. These dykes converge at a point indicated by a star and around which the coeval Coppermine River flood basalts still exist.
Source: Adapted from Baragar et al. (1996).

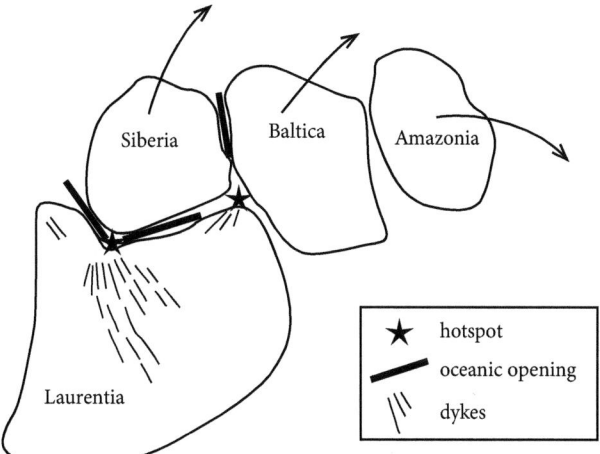

Figure 11.5 Location of the two hotspots and corresponding dykes that separated Laurentia, Siberia, and Baltica from the Columbia supercontinent.
Source: Adapted from Evans and Mitchell (2011).

The Mackenzie dyke basalts are 1,270 Ma old. The corresponding hotspot is thought to have contributed to the break-up of the Columbia supercontinent by rifting apart the Laurentia and Siberia cratons, which were previously welded within Columbia. Nearby, another slightly older dyke swarm (about 1,340 Ma) contributed to the separation of the Baltica, Laurentia, and Siberia fragments (Figure 11.5).

Other dyke swarms of ages around 1,300 Ma have been reported in China, Australia, and Baltica, suggesting that the Columbia supercontinent began to break up at this time. Since the assembly of Columbia began at around 1,900 Ma, this supercontinent remained intact for at least 600 million years, which seems a long time. Indeed, the stability of the last supercontinent, Pangea, appears to have been no more than a hundred million years. This contrast is striking and remains unexplained. The period beginning towards the end of the Paleoproterozoic, after the assembly of Columbia, is often referred to as the 'boring billion'. In fact, this period covered less than one billion years: from the late Paleoproterozoic, *c.*1,900 Ma, to the beginning of the Rodinia assembly 1,100 Ma.

Columbia experienced unique intracontinental magmatism due to its thermal blanket effect (Figure 9.9). In addition, the intense alteration of its surface under oxidizing conditions favoured the genesis of giant uranium deposits. Finally, the break-up of Columbia would have resulted in the development of continental shelves along all newly created margins. The continental shelf is the submerged part of the continent along the coast. It is a shallow area (less than 200 m deep) where the sea is penetrated by sunlight almost to the bottom and which receives nutrient inputs from the weathering of the continent. It is therefore an environment where marine life usually proliferates. It is perhaps no coincidence that the golden age of stromatolites took place around 1,200 Ma. Stromatolites would have benefited from this favourable paleogeography before competition from algae became too intense. However, ancient passive sedimentary margins corresponding to the continent–ocean transition are rather difficult to identify, because their fate is generally to be involved in younger collisions. Nevertheless, relevant Mesoproterozoic sedimentary successions have been identified, along Siberia for instance (Bradley, 2008).

The Grenville orogeny and the formation of Rodinia

The rather quiet conditions described above did not last very long, geologically. Continental fragments on the Earth's surface were involved in numerous collisions between 1,100 and 1,000 Ma. For example, Laurentia collided with Amazonia, resulting in the formation of the Grenville orogen (or mountain range) in eastern Canada. The European Baltica craton was also involved in this collision (Figure 11.6). The landforms of the original Grenville mountain range have long since disappeared, though remnants of the collision remain in the form of rocks deformed by the high pressure and temperature conditions that prevailed in the crust during the collision. Erosion of that ancient mountain chain has brought those deep, deformed rocks to the surface, where geologists can trace the trend of the original mountain belt using the fabric of metamorphic minerals that formed during the Grenville orogeny.

Several chains, roughly coeval with the Grenville chain of North America, have been recognized in the world, such as the Kibaran chain located between the Tanzanian craton and the Congo–São Francisco craton. It is inferred that a new supercontinent resulted from these collisions at the end of the Mesoproterozoic: the Rodinia supercontinent. The existence of Rodinia, as an end-Proterozoic Pangea, is now widely accepted. There are several reconstructions of Rodinia. These differ in detail, but still have many similarities, in particular on the location of Laurentia at the centre of the supercontinent. Figure 11.7 presents a reconstruction by the paleomagnetist Zheng-Xiang Li, professor at Curtin University in Perth, Western Australia.

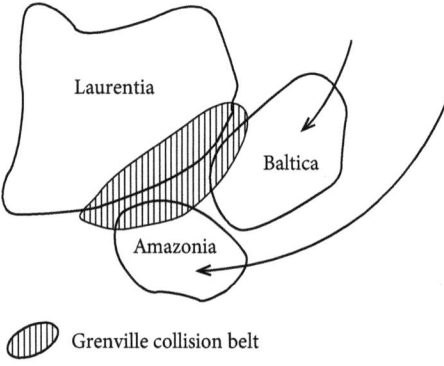

Figure 11.6 Formation of the Grenville collision belt between 1,100 and 1,000 Ma. Note that the direction of movement of the cratons (arrows) is curved on this map. This is because these arrows are map-plane representations of the movement of the cratons on the surface of a globe.

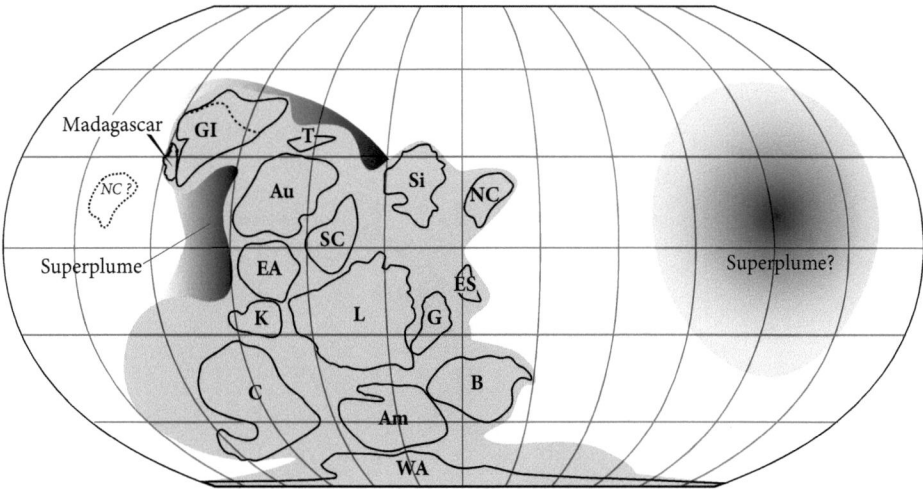

Figure 11.7 **Reconstruction of Rodinia at 780 Ma, just before its fragmentation.**
Cratons: Am: Amazonia, Au: Australia, B: Baltica, C: Congo, EA: East Antarctica, ES: East Svalbard, G: Greenland, GI: India, K: Kalahari, L: Laurentia, NC: North China, SC: South China, Si: Siberia, T: Tarim, WA: West Africa. I suggest placing North China south-west of Madagascar rather than next to Siberia or NW Laurentia, based on geological and geochronological arguments derived from 780 Ma old granites in Madagascar (Nédélec et al., 2016). Note that the suggested position is compatible with paleomagnetic data, since the paleolatitude remains unchanged. The abundance of magmatic rocks dated to 780 Ma supports the hypothesis of a (very) hot spot or mantle (super)plume between Australia and Madagascar at that time. In contrast, the existence of the superplume on the right is only inferred for reasons of symmetry of the mantle convection system.
Source: Adapted from Li et al. (2013).

Starting from the beginning of the fragmentation of Columbia around 1,300 Ma, it took 300 million years for the new supercontinent Rodinia to finish forming around 1,000 Ma. The subsequent Wilson cycle starts with the fragmentation of Rodinia and leads to the formation of Pangea.

Fragmentation of Rodinia

Emplacement of huge volumes of basaltic magmas dated between 800 and 750 Ma in several places of the globe testify to the onset of Rodinia's break-up. These large magmatic provinces, still recognizable today, indicate the position of the hotspots that facilitated the break-up of the supercontinent. The reconstruction by Zheng-Xiang Li emphasizes the presence of a hotspot around 780 Ma in south-eastern Australia. Another hotspot had been identified north-west of Laurentia (Figure 11.8). It heralded the formation of an ocean that separated North America from Australia and southern China, which we can consider as the proto-Pacific Ocean.

The reader may be surprised that the Pacific Ocean, a present-day ocean, is said to have been formed over 700 million years ago given that we have learned that oceanic

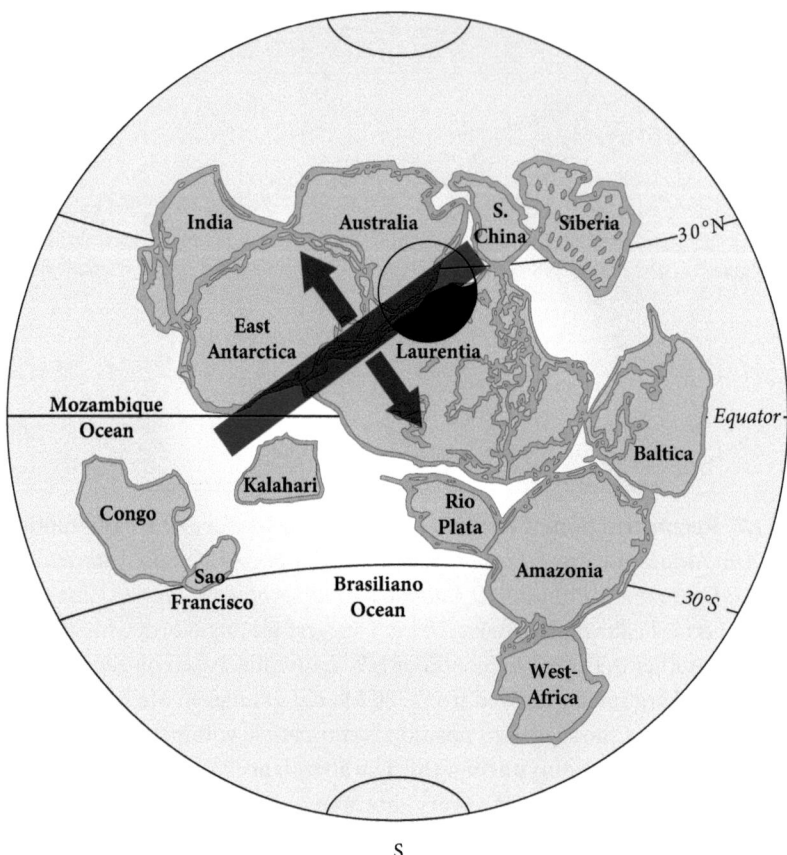

Figure 11.8 Another reconstruction of Rodinia at 750 Ma. In black: magmatic rocks of the Laurentia hotspot; dark grey: location of the future proto-Pacific Ocean. In this reconstruction, which is older than Li's, North China was not included due to lack of paleomagnetic data. The place of South China close to Laurentia and Australia is no longer accepted. Finally, India and Australia were not positioned as they are in the younger reconstruction. However, the important point remaining undisputed is the opening of the proto-Pacific along western Laurentia.
Source: Adapted from Meert and Torsvik (2003).

plates should sink into the mantle by the time they reach 200 million years old. Indeed, this is what is currently happening all along the Pacific Ring of Fire. But the oceanic crust that disappears at subduction zones is continuously replaced by new oceanic crust formed at the East Pacific Rise (Figure 5.5). As this ridge is very active, the Pacific Ocean remains expansive despite the subduction of its Asian and American margins.

The Pan-African orogeny and the formation of Gondwana

As early as 650 Ma, fragments of Rodinia began to reassemble to form a new supercontinent. Geologists working in Africa during the previous century had noted that

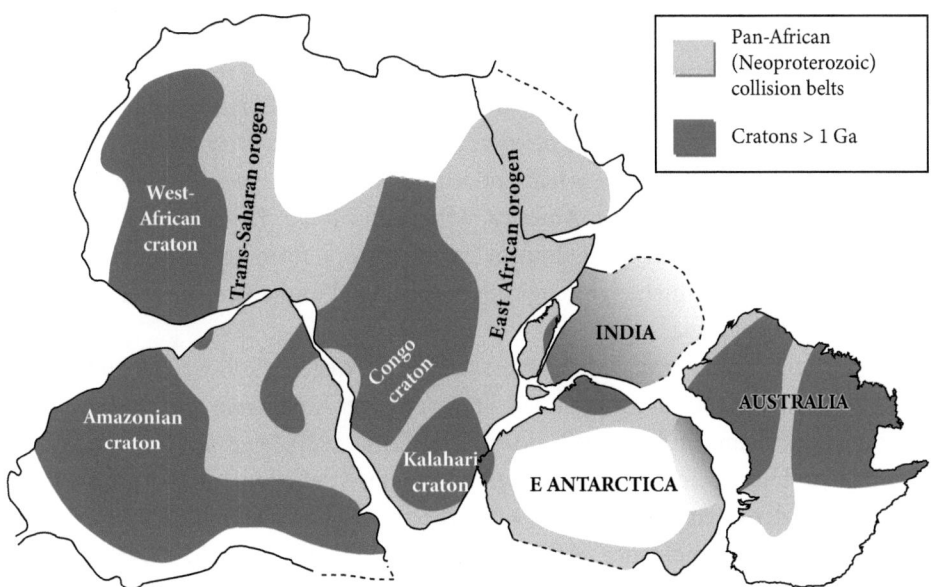

Figure 11.9 Reconstruction of Gondwana at the end of the Precambrian.

the period between 650 and 500 Ma corresponded to intense deformation and metamorphism in many parts of Africa. In 1964, in a two-page abstract presented at the first African Geology Colloquium in Leeds (UK), William Kennedy suggested the existence of a tectonic event which he called 'Pan-African'. At that time, plate tectonics was still in its infancy and geologists had not developed a global reference frame in which to cast tectonic events. However, William Kennedy was right and, within a few years, the Pan-African orogeny was broadly accepted. More precisely, two main collision chains were identified (Figure 11.9): one that stretches along the length of eastern Africa, the other that crosses the Sahara from north to south: the trans-Saharan chain. Several French geologists contributed to the study of this Pan-African chain in the Hoggar in the second half of the twentieth century (Boullier, 1991). The trans-Saharan chain continues in Brazil, where Brazilian geologists call it the 'Brasiliano' chain (Brito-Neves et al., 1999). These two collisional mountain chains, which must have been as high as the present-day Himalayas, together with minor orogenic belts welded different cratons and formed a very large continental entity: the Gondwana supercontinent.

Gondwana may be considered a supercontinent in terms of its size, since it includes about half of the Earth's continents. However, it is not a **Pangea** in the etymological sense (from the Greek 'all land') as it would have to contain at least 75% of the continents. The Wilson cycle, which began with the fragmentation of Rodinia around 750 Ma, was not yet complete at the end of the Precambrian. It was not until the collision of Gondwana with the Old Red Sandstone continent (i.e. the Laurentia–Baltica association) and the Siberia craton that the full Pangea supercontinent finished assembling around 300 Ma.

The existence of Gondwana, as an association of most southern-hemisphere continents, was suggested more than 100 years ago by the Austrian geologist Eduard Suess in his global synthesis *Das Antlitz der Erde* ('The face of the Earth'), which presented a comprehensive view of Earth's geology at that time (Durand-Delga and Seidl, 2007). Long before plate tectonics, Suess had noticed the similarity of fossil fern assemblages, the *Glossopteris* flora, in South America, Africa, and India. He proposed that these continents may have been contiguous in Earth's history, forming a landmass that he named Gondwana,[4] which etymologically means 'land of the Gonds', after the name of a native tribe of India.

These observations provided the basis for Alfred Wegener's theory of continental drift, exposed at length in his book *Die Entstehung der Kontinente und Ozeane* ('The Origin of continents and oceans'). Born in Berlin in 1880, Wegener, an astronomer and meteorologist by training, imagined continents (the continental crust or 'sial') floating on a denser mantle (the 'sima'). His theory was based on multiple arguments: geographical (the coincidence of the coastlines on either side of the Atlantic), paleontological (the similarity of fauna and flora), paleoclimatic (the presence of ancient glacial sediments in tropical regions), and structural (the similarity of older mountain chains on either side of the South Atlantic Ocean). It was the first attempt of a global understanding of the Earth. However, Alfred Wegener's ideas were strongly opposed at the time, especially in North America. He died tragically in 1930 during an expedition on the Greenland ice sheet. Although some individuals, such as the great South African geologist Alexander du Toit, also believed that Africa and South America had once been contiguous, Wegener's hypothesis of continental drift was not accepted in his lifetime. Plate tectonics is sometimes considered to be an avatar of continental drift. But there are some differences. The current theory never considers continents in isolation: they are part of plates that can be both continental and oceanic. Moreover, plates are not limited to the crust, but also include the upper part of the mantle, the whole constituting the lithosphere. The lithospheric plates move across the surface of the planet, dragging the continents in their motion. However, Alfred Wegener is now rightly considered to be a precursor of current ideas.

The Cadomian chain on the northern edge of Gondwana

For the French reader, the question arises as to where metropolitan France was located at the end of the Precambrian. The continental fragments that make up France were on the northern edge of Gondwana. France was then in Africa! So was much of western Europe, except for Scotland, which belonged to Laurentia.

Late Precambrian rocks outcrop in northern Brittany and western Normandy. Studied for a long time, these rocks are grouped under the name Brioverian, a name that is only used in France and comes from *Briovera*, the Latin name of the small town of

---

[4] Eduard Suess initially used the term 'Gondwana-Land', but this name is redundant because etymologically, 'Gondwana' means 'land of the Gonds.'

Saint-Lô in Normandy. The Brioverian corresponds to the end of the Neoproterozoic, between 670 and 540 Ma.

The Brioverian terranes tell a story that can be read from the age and nature of the rocks. A little over 650 Ma, a volcanic archipelago bordered the northern coast of the Land of Trégor, a historical province of Brittany, along the Channel. The chemical composition of the Trégor rocks indicates that they formed above a subduction zone, possibly in a tectonic context similar to that of Japan. The Trégor archipelago was separated from the continent by a narrow sea that closed, leading to the formation of the Cadomian chain, named after *Cadomus*, the Latin name of the city of Caen in Normandy. The Cadomian chain formed at the same time as the great Pan-African chains that welded together the different parts of Gondwana, but it is only a coastal chain of modest scale, at least compared to continental-scale mountain ranges.

Recent research has documented the existence of several continental fragments that originated on the northern periphery of Gondwana, that were later involved either in subduction or rifted from the Gondwana mainland. Such fragments (Avalonia, Armorica . . .) span from western Africa to the Middle East. The rest of the history of Gondwana and its peripheral fragments took place during the Paleozoic era and will be discussed in Chapter 14. For now, we shall conclude by noting that Gondwana and the related continental blocks were welded to the northern continents to form Pangea during the Hercynian/Alleghanian orogeny, which left a pronounced imprint in western and central Europe as well as in North America.

> Plate tectonics is responsible for a changing geography, with continents colliding violently, only to separate again and then collide once more. This rhythmical ballet of the continents is the result of plate tectonics, which opens and closes oceans through the production of oceanic lithosphere at ridges and destruction in subduction zones. Three supercontinents formed during the Proterozoic eon: Columbia, Rodinia, and Gondwana.

12
The Snowball Earth

Today, ongoing global warming is a matter of grave concern, and media coverage of climate change sometimes takes on a frightening tone. Indeed, the anthropogenic contribution to the warming of the planet and acidification of the oceans is irrefutable. However, Earth's climate has varied naturally over its long history, and life has had to cope with environmental change. Some 21,000 years ago (virtually yesterday in the eyes of a geologist!), prehistoric humans experienced cold conditions during the peak of the last ice age. Nearly all the British Isles and northern Europe were covered by ice. A lower sea level meant that Britain was not an island, and in place of the English Channel was a land bridge to mainland Europe. Herds of reindeer and mammoths roamed France and Spain. In North America, the situation was more extreme: the Laurentide ice sheet covered most of Canada and extended to the latitude of New York, c.40°N, the same latitude as Naples, Italy. In fact, the climate of the Quaternary era experienced many regular oscillations between relatively warm and cold. In order to understand the impressive climate changes that occurred at the end of the Precambrian, it helps first to review the more recent climatic oscillations.

The causes of the last ice age

Counterintuitive though it may seem in this time of global warming, the Earth is in a glacial state today. Indeed, climatologists define a glaciation as a time interval characterized by a permanent ice cap on a part of the globe. This is currently the case with two ice caps, one in Greenland and another thicker and more extensive one in Antarctica. The Antarctic ice cap likely first appeared around 35 million years ago. Three factors account for its initial appearance and growth. First, the middle Cenozoic era was a time of falling atmospheric carbon dioxide concentrations, which resulted in cooling global temperatures. Furthermore, Antarctica straddled the South Pole, making it naturally vulnerable to glaciation. However, it had occupied this austral position for over a hundred million years, and fossil forests and other evidence indicate temperate conditions in Antarctica. It turns out that the initiation of the Antarctic ice cap coincided with the opening of the Drake Passage between the Antarctic Peninsula and the southern tip of South America. Antarctica thus had no connections with any landmass and was surrounded by a cold oceanic current driven by the strong westerly winds of the high latitudes. The Antarctic Circumpolar Current, as it is known, isolated the continent thermally, causing average temperatures to fall and triggering the rapid development of a permanent ice cap.

The current Greenland ice cap is much more recent: it dates back only a few million years (i.e. to late in the Cenozoic era). The geography of the North Polar region is different from that of the Antarctic because it is mainly oceanic. Around the Arctic Ocean, only Greenland is covered by an ice sheet, although it is not the region with the lowest recorded temperatures. So how did the Greenland ice cap come about? Several causes led to the inevitable global cooling during the Cenozoic era, ultimately leading to the freezing of Greenland and the formation of huge ice sheets in Europe and North America.

The first cause is the consequence of the continental collision between India and Eurasia, which began around 50 Ma. Substantial topographic relief appeared at around 20 Ma, forming the Himalayan chain, which includes the current highest peaks in the world, and the Tibetan plateau. These high mountains naturally became susceptible to mechanical erosion, and the fragmentation of the rocks facilitated chemical weathering and the trapping of atmospheric CO_2. In addition, these high mountains in the very heart of the Asian continent modified the atmospheric circulation: the Indian monsoon appeared. The monsoon brings heavy rainfalls to India and the Himalayan range in summer, conditions that increase the chemical weathering of rocks. Thus, the rise of the Himalayas accelerated the precipitous decline in atmospheric CO_2 that had already begun earlier in the Cenozoic.

An additional factor played a role in the formation of the Greenland ice cap. As in the case of Antarctica, it was a change in ocean circulation. The opening of the Fram Strait between Greenland and the Spitsbergen archipelago facilitated communication between the Arctic Ocean and the North Atlantic (Figure 12.1). Circulation in the Arctic Ocean was no longer limited to a circular transit around the North Pole. Instead, cold waters were carried southwards along the elevated east coast of Greenland. They maintained very cool temperatures in summer on this continent and, as the winter snow did not melt completely in summer due to the fairly high latitudes, an ice cap appeared and gradually expanded. The situation is different in eastern Siberia, although it is the coldest region in the Circum-Arctic area in winter. Indeed, Siberia receives less precipitation than Greenland and has higher summer temperatures, limiting the development of permanent ice.

In summary, the important factors in climate variability over geological time are the CO_2 content of the atmosphere, the distribution of land masses, the oceanic circulation, and regions of high topographic relief. All these factors link back to plate tectonics, which results from convection of the mantle within the globe. Thus, much of the long-term climate change during the Cenozoic era resulted from the inner dynamic of the Earth.

Origin of Quaternary climate oscillations

In addition to these climatic variations operating on the scale of millions of years, other, shorter period fluctuations are known to modulate Quaternary climate. These

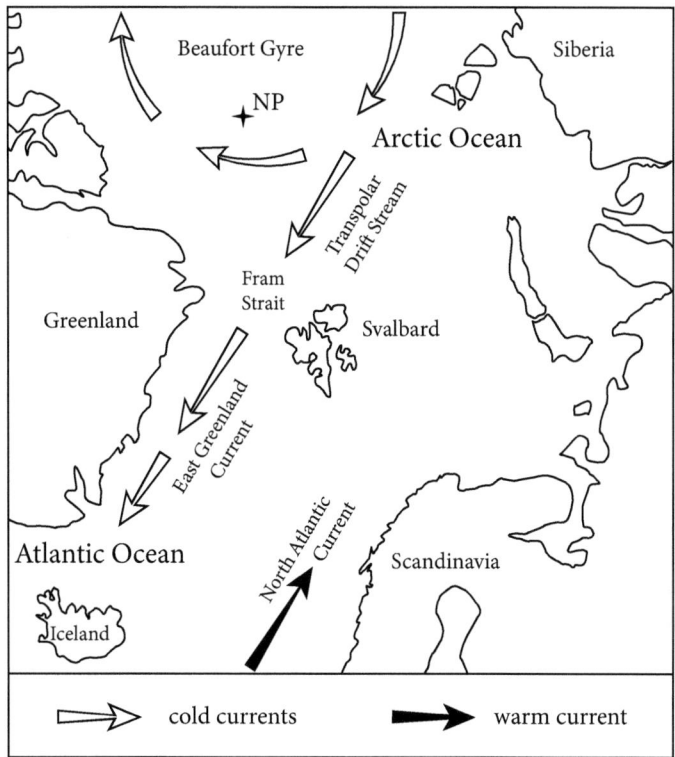

Figure 12.1 The East Greenland Current carries cold waters from the Arctic southwards through the Fram Strait. On the other side of the Atlantic, a warm current, the North Atlantic Drift, maintains somewhat warmer temperatures. *NP: North Pole.*

periodic changes are related to small fluctuations of the astronomical parameters that control Earth's rotation and orbit around the Sun. These subtle but important changes arise from the gravitational attraction of the other planets. The Earth completes one revolution around the Sun in one year, tracing an almost circular ellipse, with the Sun at one of its foci. However, this ellipse shows small changes in its elongation or eccentricity, which occur over periods of c.100,000 and 400,000 years (Figure 12.2). Variations in the eccentricity of the Earth's orbit influence the amount of solar energy received by the Earth. Finally, the Earth's axis of rotation itself rotates, like a spinning top. Earth's axis of rotation currently points to the Pole Star in the Little Bear constellation, but this was not the case 12,000 years ago. It is the axial precession (or precession of the equinoxes) occurring at a periodicity of 19,000 and 23,000 years. The result of axial precession is that the timing of the seasons relative to Earth's orbit around the Sun continually changes. Eccentricity modulates the effect of precession, which may drive relatively strong variations in latitudinally distributed solar insolation.

Finally, the Earth rotates, giving rise to our familiar 24-hour day (note the duration of the day has steadily increased over time as the moon has migrated away from Earth). The Earth's axis of rotation is currently tilted by just over 23° from the perpendicular to the plane of the Earth's orbit. This angle, called obliquity, varies between 22 and 24.4°

ECCENTRICITY of the Earth's orbit:
Departure of the orbit from nearly circular to mildly elliptical
Variation cycles of 400,000 and 100,000 years

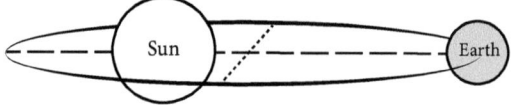

OBLIQUITY (axial tilt): from 22.1 to 24.4°
Angle of the axial tilt with respect to orbital plane
Variation cycle of 41,000 years

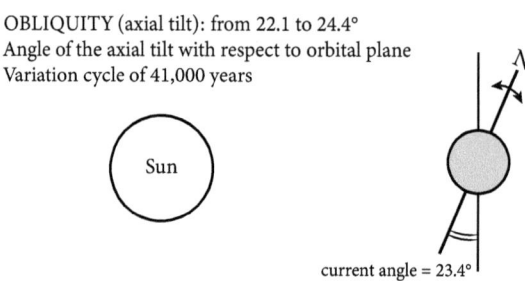

current angle = 23.4°

Axial PRECESSION (or precession of the equinoxes):
Direction of the Earth's rotation axis relative to fixed stars
Variation cycles of 19,000 and 23,000 years

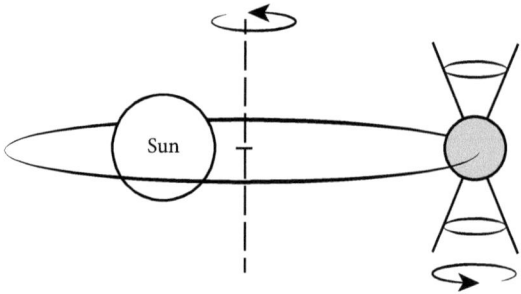

Figure 12.2 **Astronomical parameters and their periodicity.**

with a periodicity of 41,000 years. The greater the obliquity, the greater the difference between the seasons.

The Serbian geophysicist and climatologist Milutin Milanković accurately calculated the variations in solar insolation in the northern hemisphere over the past hundreds of thousands of years as a result of the variations in astronomical parameters. He compared his results with what was known at the time about the advance and retreat of ice in Europe, which he could explain by astronomical theory. However, Milanković died in 1958, before his theory was fully confirmed or accepted. It was eventually brilliantly validated by several independent observations in the following decades, most importantly the variations in oxygen isotope ratios in the shells of calcareous microorganisms, recovered from deep-sea sediment cores, as well as similar fluctuations recorded in ice cores taken from Greenland and Antarctica.

These isotopic variations provide a precise record of the Quaternary climate. Ocean water evaporates in large quantities near the equator. The water molecule (H_2O) contains oxygen, mainly in the form of ^{16}O, but also ^{18}O in very small quantities (and ^{17}O

in even scarcer quantities). With two more neutrons, ^{18}O is heavier than ^{16}O. Water molecules containing this heavy isotope evaporate more slowly than those containing the lighter isotopes and are therefore concentrated in seawater. The opposite happens when water vapour clouds condense into rain. The net result is that the water vapour that reaches the high polar latitudes and feeds the polar ice caps in the form of snow is very depleted in heavy isotopes, such that it has low $^{16}O:^{18}O$ ratios, which are typically reported in $\delta^{18}O$ notation. During cold periods, the volume of the ice caps increases significantly, stocking huge quantities of frozen water depleted in ^{18}O. As a result, ocean water (whose total volume also decreases) becomes relatively enriched in ^{18}O, and its $\delta^{18}O$ signature increases (Figure 12.3).

Cesare Emiliani, a student in Urey's laboratory in Chicago, had the idea of measuring the variations in the ratio of oxygen isotopes in the calcitic shells secreted by small ubiquitous marine creatures, the foraminifera, which are unicellular, heterotrophic eukayotes. When foraminifera die, their shells fall to the bottom of the sea and are preserved in sediment, which can be sampled and dated. The analysis of these shells makes it possible to track variations in the $\delta^{18}O$ signature of seawater through time. In fact, the $\delta^{18}O$ signature encoded in these shells also reflects a temperature contribution. Fortunately for us, cooler temperatures drive the $\delta^{18}O$ values in the shells higher, such that both changing ocean temperatures and ice volumes work in concert to control the variations in the $\delta^{18}O$ signatures of foraminifera shells. Emiliani observed that this ratio changed rhythmically over the past hundreds of thousands of years and concluded that the changes tracked the advance or retreat of continental ice caps. The development of isotope studies and precise dating of sediments has

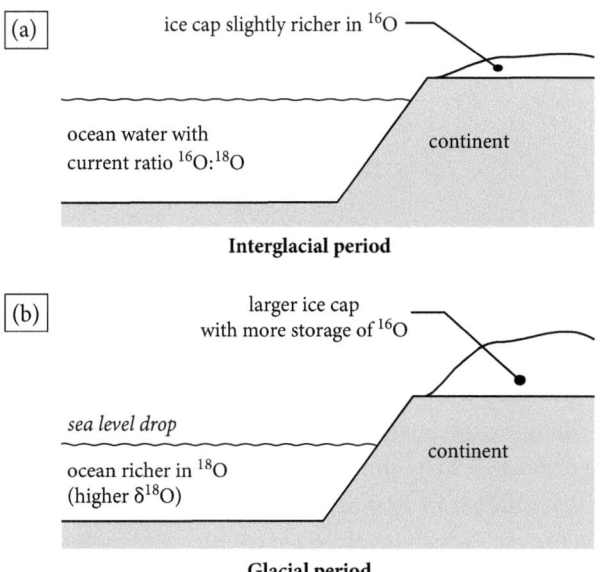

Figure 12.3 Relationship between continental ice volume, sea level, and the isotopic signature of seawater.

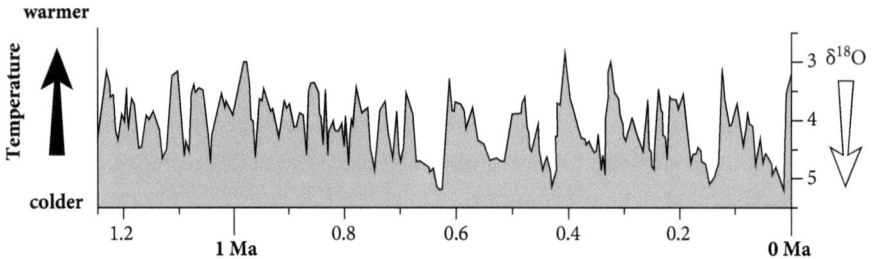

Figure 12.4 Oxygen isotope record of carbonate shells of foraminifera from deep-sea sediment cores. Note that temperature and oxygen isotopic ratios vary in opposite senses. Recent variations are particularly well-defined and show that warming occurs faster than cooling.
Source: Adapted from Railsback et al. (2015).

made it possible to clearly identify the oscillations that punctuate the overall cold climate of the Quaternary: an alternation of cooling, leading to an increase in the volume of continental ice (the glacial stages), and warming causing the ice retreat (the **interglacial** stages). These oscillations occur with a frequency that can be related to variations in astronomical parameters, notably to the c.100,000-year eccentricity cycles (Figure 12.4). Thus, 21,000 years ago there was a temperature minimum that corresponded to the maximum extent of the ice sheets—the Last Glacial Maximum (**LGM**), whereas today we are in a warm interglacial. The last interglacial (called the Eemian) was 125,000 years ago, when temperatures were slightly warmer and sea level was 6 metres higher than today. It is important to note that glacial and interglacial stages represent climatic oscillations within an overall period of glaciation. Thus, the LGM is sometimes misleadingly referred to as 'the last ice age', though it is more aptly considered the last glacial stage of the Quaternary glaciation (or Quaternary ice age).

Ice core data

At the end of the twentieth century, several deep boreholes were drilled into the heart of the continental ice of Greenland and Antarctica, where snow, transformed into ice, had been accumulating for tens of thousands of years in layers akin to those found in sediments. These boreholes provided exceptional climate archives for the Quaternary. They were obtained under very difficult climatic and technical conditions; the first Antarctic drilling by the Soviet Union in the 1970s took place at Vostok, where the temperature drops to −90 °C during the southern winter (Figure 12.5). The first Vostok ice core was studied in France (Jouzel et al., 1987). In the early 2000s, the European EPICA project,[1] which involved ten countries, obtained a 3,270 m-long core sample from the Concordia station in the heart of Antarctica. This core extends back

---

[1] European Project for Ice Coring in Antarctica.

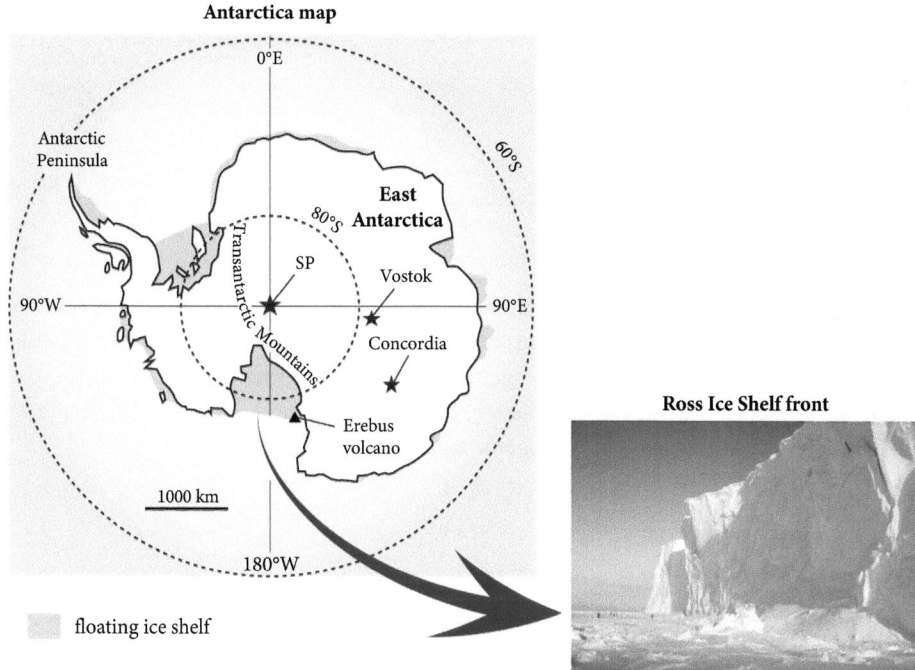

Figure 12.5 Map showing the location of Antarctic drill cores referred to in the text, and floating ice shelves that buttress the sheets. SP: south pole ice core. Right: the edge of the Ross Reef and the pack ice, or sea ice, at its foot.
Source: NOAA.

to 800,000 years ago, spanning eight glacial–interglacial cycles. The EPICA project has resulted in numerous publications.

Oxygen isotope analyses of the ice cores support the data from the ocean sediments. The ice cores also provide additional information, as they contain air bubbles that were trapped when the snow turned to ice: fossil air, so to speak! The carbon dioxide content of these air bubbles has been determined to reconstruct the composition of the atmosphere over the last few thousand years (Figure 12.6). The CO_2 content of the atmosphere has varied in parallel with temperature changes. This was partly the result of CO_2 exchange between the ocean and the atmosphere. Henry's law is used to calculate the amount of atmospheric gas that can dissolve in the ocean.[2] This law depends on temperature. For example, from 2 °C to 40 °C the solubility of carbon dioxide in the ocean is reduced by a factor of three. In other words, the lower the water temperature, the more easily CO_2 dissolves. Conversely, if the temperature rises, CO_2 is released into the atmosphere by the ocean. The CO_2 content of the atmosphere was slightly less

---

[2] Henry's law calculates the fraction x of gas that passes into solution in a liquid with which the gas is in contact. This fraction depends on the proportion p (or partial pressure) of the gas in the atmosphere. The equation, $x = p / H$, applies to all gases in the atmosphere. The constant H depends on the temperature and the gas concerned. Its value is much lower for carbon dioxide than for oxygen or nitrogen, which means that CO_2 is much more soluble in water than the other two gases. Furthermore, the constant H increases with temperature for all three gases mentioned.

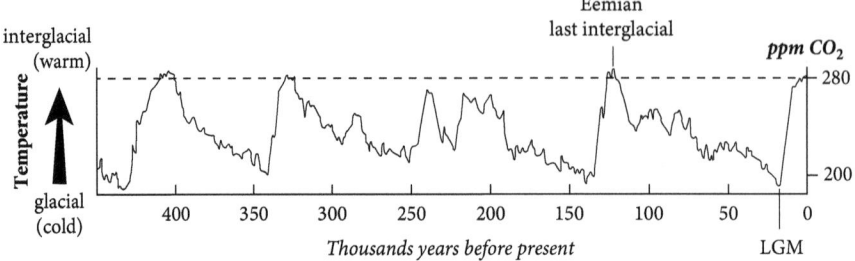

Figure 12.6 Variation in the CO_2 content of air bubbles from Antarctic ice cores over the last 450,000 years. These variations are synchronous with the temperature variations deduced from the isotopic data. Once again, we see that warming phases leading to interglacial stages are always shorter (faster) than cooling phases. Note that this figure does not capture the recent, anthropogenic rise in pCO_2.
Source: Adapted from EPICA (2004).

than 200 ppm (0.02%) during the LGM and reached about 280 ppm (pre-industrial value) during the last interglacial before the recent climate change. The industrial era has strongly perturbed atmospheric composition due mainly to the combustion of coal and petroleum, which add CO_2 to the atmosphere, enhancing the greenhouse effect. Today, atmospheric CO_2 exceeds 400 ppm (0.04%), a higher level than at any other time in the past 3 million years.

The geological record of the late Precambrian glaciations

There is no trace of glaciation during the Mesoproterozoic between 1.6 and 1 Ga, a period when life proliferated. It seems that equable climate prevailed on the planet. The amount of carbon dioxide in the atmosphere was much lower than in the Archean, but at the same time the Sun was more intense than at the beginning of our planet's history. These two main factors, or **climate forcings**—namely the composition of the Earth's atmosphere and the brightness of the Sun—contributed to the generally warm climatic conditions. However, the climate changed considerably at the end of the Precambrian. Indeed, in many parts of the world, geologists have long recognized sediments that were deposited by or influenced by glaciers (glaciogenic sediments) in the late Precambrian, during what is now referred to as the Neoproterozoic era, which lasted from 1,000–541 million years ago. The earliest identification of glaciogenic sediments was in 1871 with the description of ancient moraines, or **tillites**, at Port Askaig in western Scotland. Other late-Precambrian glaciogenic deposits were subsequently recognized in Australia, Norway, Svalbard, China, India, and Africa.

The geologists who first identified these ancient glacial deposits relied on comparison with recent glacial deposits. Indeed, glaciers are efficient agents of abrasion and transport. This is evidenced by the width and U shape of alpine glacial valleys and the accumulation of moraines in front of and alongside the glacier tongue (Figure 12.7).

Figure 12.7 The Aletsch glacier (Switzerland) and its lateral moraines. The rock is freshly scoured on the slopes on either side of this glacier, the largest in the Alps, indicating a recent decrease in ice volume.
Source: Dirk Beyer.

Glacial moraines are characteristically unsorted, displaying a wide range of sizes and abundant fine-grained sediment (mud). Many of the clasts have grooves, or striations, which result from rubbing against the hard substrate as the huge mass of ice carried them downstream, slowly but inexorably. No transport agent other than glaciers can produce such features. Torrents of flowing water can move large blocks, but these will roll and become progressively rounded. Moreover, flowing water tends to sort fluvial deposits and wash away the mud. Larger boulders remain upstream, close to their sources, while the smaller pebbles are carried downstream. As the current weakens, rivers deposits sands and finally fine particles, silts and clays—where in contrast glaciers tend to leave behind a jumble debris of all sizes.

Therefore, geologists can recognize ancient moraines even when the ice has disappeared. In rare cases, the remains of a striated floor (i.e. the rock surface on which the blocks transported by the glacier left traces) have been observed under the tillite. A particularly convincing example was described by Reusch in 1891 in the Varangerfjord in Norway (Figure 12.8).

Another type of deposit of glacial origin is formed in marine settings and is characterized by debris released by icebergs as they drift and melt. Icebergs carry blocks of all sizes at their base, which are plucked from the continent and embedded in the ice sheet. These clasts are transported towards the shoreline with the flowing glaciers, and if the glaciers reach the sea, they form floating ice shelves. These ice shelves calve icebergs, which gradually release the debris they are transporting as they

192 THE SNOWBALL EARTH

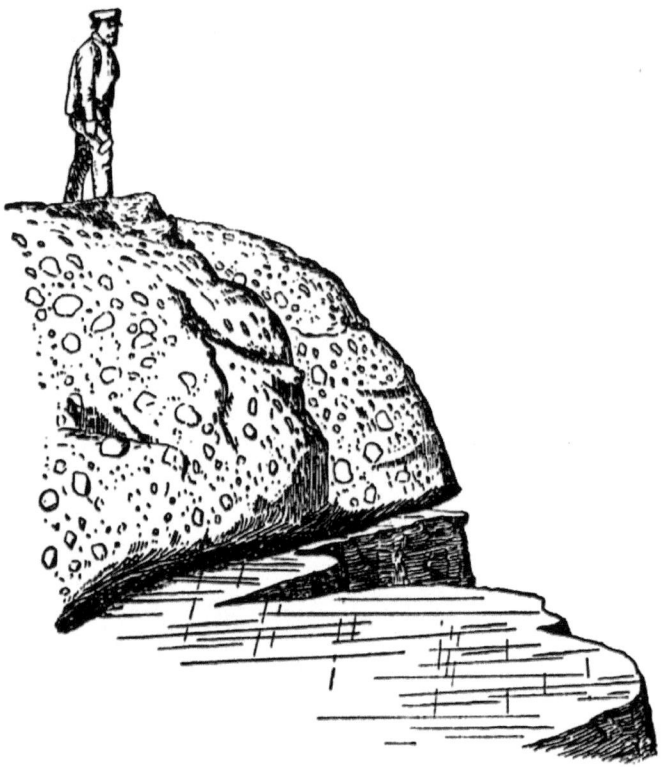

Figure 12.8 Late Precambrian moraine deposits resting on a striated surface of the underlying sandstone. Drawn by Reusch in 1891. The 'Reusch moraine' (or Bigganjarga tillite) outcrops in the inner part of Varangerfjord in north-eastern Norway.
Source: Adapted from Bjørlykke et al. (1967).

melt (Figure 8.4). These ice rafted blocks tend to fall into soft, fine-grained, laminated marine sediment. Where these blocks clearly deform the soft layers of marine sediment, they are known as *dropstones*. The fine, bedded matrix, the unsorted sizes of the blocks, and evidence of deformation of the sediment before consolidation are all clues to identify this type of glaciomarine deposit within ancient sedimentary series. These deposits are referred to as **diamictites**.

The floating ice shelves that form when glaciers reach the sea, and their calved icebergs, should not be confused with pack ice, which is the ice that forms from freezing of seawater. Pack ice is much thinner than the floating ice shelves from the continent. Huge areas of ice shelf are found along the Antarctic coast. The largest of these expanses of floating ice, the Ross Ice Shelf,[3] is almost the size of France! At its seaward margin it rises about 50 m above sea level and extends about 450 m below sea level. The Ross

---

[3] James Ross led a British expedition to Antarctica from 1839 to 1843 with two ships, HMS *Erebus* and HMS *Terror*. This expedition discovered the large ice shelf, later known as the Ross Ice Shelf, the Transantarctic Mountains, and the active volcano Erebus, named after one of the two ships of the expedition.

Figure 12.9 Map of the present location of Neoproterozoic glaciogenic deposits (white circles). The large number and wide distribution of these deposits on all continents is striking.
Source: Adapted from Harland (1965).

Ice Shelf was previously called the Ross ice barrier, because it prevents any southward navigation (Figure 12.5).

In 1965, the University of Cambridge geologist Brian Harland, a specialist in geological expeditions to the Arctic, in particular to the Svalbard archipelago, compiled all deposits of supposed glacial origin of late Precambrian age (Figure 12.9). He concluded that these deposits were indeed glaciogenic, but that they were not necessarily all coeval. This conclusion raised the question of how many glaciations occurred at the end of the Precambrian. The very wide distribution of these deposits on the surface of the globe was, according to him, proof of continental drift (*sic*). Nevertheless, Brian Harland made a premonitory remark by suggesting that paleomagnetism would provide the necessary data to reconstruct the Neoproterozoic paleogeography and thus establish the extent of the glaciations of this period.

Evidence of global glaciations

A first intriguing paleomagnetic result was reported in Harland's article. Red sandstones located just below the oldest of Norway's two late-Precambrian glaciogenic units recorded a magnetization that corresponded to a paleolatitude of about 10° (i.e. close to the equator at the time!). However, during the last glacial maximum, 21,000 years ago, ice sheets did not extend beyond 40° latitude. The first paleomagnetic result reported by Harland would have been easy to question if it had remained unique. Indeed, paleomagnetism is an arduous and time-consuming discipline that requires many conditions to obtain a reliable result. One could also question the quality and precision of the equipment available at the time, or whether the rocks are faithful recorders of late Precambrian paleolatitude.

But similar results of the same type—namely a low paleolatitude for Late Precambrian glaciogenic deposits—were subsequently obtained with improved reliability between 1986 and 2005 in Australia, Canada, China, Brazil, and Oman. The inescapable conclusion was that the late Precambrian glaciations had affected the entire surface of our planet. The hypothesis of a late Precambrian global glaciation was formulated by the paleomagnetician Joe Kirschvink in 1992, using a poetic comparison, as he evoked the Earth as white as a snowball in interplanetary space. This beautiful image of the Snowball Earth could have gone unnoticed, as it was the title of only a short, two-page abstract in a huge collective work devoted mainly to Proterozoic fossils!

In 1998, Paul Hoffman, a Canadian geologist and professor at Harvard University (USA) at the time, resurrected the term in an article documenting geological and geochemical evidence in northern Namibia in support of the hypothesis. This short scientific paper unleashed a decade of excitement (and controversy) on the international geological community. The proponents of the bold hypothesis, that the entire globe was encased in ice for millions of years at a time, searched for additional evidence in support of the theory. The detractors, commonly invoking the uniformitarian principles that have largely guided geology for the past two centuries, looked to the glacial record itself to refute the hypothesis. Other curious scientists joined the intellectual fray, eager to explore the implications of the extreme climates that accompanied a Snowball Earth. The discussions were sometimes lively, as at a 2006 meeting in Ascona, Switzerland, devoted to the topic, which highlighted that scientists are often passionate and not necessarily models of objectivity and fair play.

Since that conference, a great deal of new, significant data have accumulated. In addition to new and more robust paleomagnetic data confirming the tropical distribution of Neoproterozoic ice sheets, great progress has been made in dating these glaciations. It is now established, based on radiometric dates obtained from across the globe, that there were three Neoproterozoic glaciations: the Sturtian glaciation (the longest, which lasted from 717 to 661 Ma), the Marinoan glaciation (from c.640 to 635 Ma), and the Gaskiers glaciation (which lasted only a few hundred thousand years around 580 Ma (Pu et al., 2016)). A distinct prediction of the Snowball hypothesis is that the glaciations were long-lived and synchronous. Thus, only the Sturtian and Marinoan glaciations were snowballs. The Gaskiers, it seems, was more like the familiar glaciations of the Phanerozoic eon, including the present one.

I was happy to take part in this scientific adventure in search of the Snowball Earth by coordinating a project funded by the French CNRS (*Centre National de la Recherche Scientifique*) from 2000 to 2007. This project involved a multidisciplinary team of field geologists, paleomagneticians, geochemists, astronomers, and climate scientists from several French laboratories. Brazilian and American scientists also participated in this project, for scientific research has no boundaries. For readers who are not familiar with the way research projects materialize, there follow some characteristics of the processes involved.

Research projects begin as a proposal, commonly by a team of scientists with complementary expertise, which is subjected to a critical evaluation. Many, and commonly most, are rejected without funding. In the event of success, the funding granted is not necessarily a large amount, but these sums are nevertheless necessary for setting up expeditions in the field, transporting samples, carrying out chemical analyses, acquiring software, recruiting students to participate in the project, and allowing participation in national and international meetings, which are essential for the exchange of ideas. In addition to the financial aspect, such projects have another critical advantage. They force scientists to work together, no easy task when they belong to different disciplines—and often not even when they belong to the same discipline. Coordinating a large, interdisciplinary project is challenging, but how much enthusiasm, how many fruitful exchanges, and how much joy you experience when you succeed in advancing our understanding of the planet![4]

The comeback of BIFs

In addition to exceptional climatic conditions, the end of the Neoproterozoic was also marked by the return of BIFs (i.e. marine deposits very rich in iron oxides, which are evidence of an anoxic deep ocean). These marine iron formations had disappeared more than a billion years ago. Indeed, the last deposits, mentioned in Chapter 8, date back to 1.8 Ga. BIFs returned, associated with glacial deposits from the end of the Precambrian. They are best known from the very long-lived Sturtian glaciation (see e.g. Halverson et al., 2011). The recurrence of BIFs during the Neoproterozoic glaciations convinced paleomagnetician Joe Kirschvink that the Neoproterozoic glaciations were global. Indeed, their formation suggests a completely frozen ocean, thus isolated from the atmosphere by a thicker ice pack than today. It is assumed that the oxygen content of the ocean then gradually decreased until it reached such low values that iron from hydrothermal alteration of the oceanic crust could remain in solution as reduced ferrous iron in seawater, just as it did earlier in the Archean and Paleoproterozoic. Assuming that Sturtian BIFs were deposited at the end of glaciation, the retreat of the glaciers and melting of sea ice would have allowed oxygen to dissolve back into the ocean. The mixing of the shallow oxygenated waters with deep ferruginous waters would have rapidly oxidized the dissolved iron, which deposited iron oxide minerals on the seafloor. Continued warming would have accelerated the collapse of ice sheets at the same time, accounting for the abundance of dropstones found within the BIFs (Figure 12.10).

The recent discovery of a 'bleeding' glacier in Antarctica illustrates a similar situation on a smaller scale. These so-called blood falls correspond to an occasional discharge of subglacial water loaded with dissolved iron at the front of the Taylor Glacier, not far from the Ross Ice Shelf (Mikucki et al., 2009). As soon as this water

---

[4] I am grateful to French CNRS *Centre National de la Recherche Scientifique* (National Center for Scientific Research) for having provided the support for this and many other projects over the course of my career.

Figure 12.10 Examples of ice rafted boulders dropped into Sturtian BIFs from Iron Creek, Yukon, Canada.
Source: Galen Halverson.

comes into contact with the oxygenated atmosphere, dissolved iron precipitates as iron hydroxides (i.e. rust), which imparts a spectacular red hue that contrasts with the gleaming glacier in the background (Figure 12.11). This is a very special situation that is not seen at the front of other glaciers. Indeed, the subglacial water of the Taylor Glacier is a brine which derives from ancient sea water that was trapped under ice several hundred metres thick during a period of glacial advance. The brine, which is highly concentrated in dissolved elements, does not freeze, despite sub-zero temperatures. Isolated from the atmosphere, the brine became anoxic, and the iron, presumably sourced from weathering of the bedrock below the glacier, remains in solution in the ferrous state. These observations are interesting because they provide an example of a cold glacier (its surface is at $-17\,°C$) producing chemical deposits not unlike those seen during Snowball Earth. However, whether they provide an actual analogue to the Neoproterozoic BIFs is another question.

Causes of the Snowball Earth

How did the Earth become a completely ice-covered planet in the Neoproterozoic? Let's compare Snowball Earth with the LGM 21,000 years ago. In addition to the Antarctic sheet, huge ice sheets covered northern Europe, Greenland, and a vast part of North America. Large erratic blocks left behind after the retreat of the North American glaciers can still be seen in New York's Central Park. New York City's latitude corresponds to the maximum southward advance of the Quaternary ice sheet in North America. There is additional evidence of ice rafted debris reaching $30\,°N$ latitude in the Atlantic (Figure 12.12). The glaciation was therefore impressive, but it was far from being global.

Figure 12.11 Blood Falls, a subglacial outflow at the front of the Taylor Glacier in Antarctica.
Source: J. C. Priscu team (Montana State University).

Using numerical models of Earth's energy balance, climatologists show that if continental ice expands down to 30° latitude, a global glaciation cannot be avoided, because the extent of ice dramatically increases Earth's **albedo**. The albedo, a word derived from the Latin *albus* meaning 'white', is the reflectivity of a surface. White surfaces reflect the sun's rays more efficiently than darker surfaces, thus having a cooling effect on Earth. On a scale of zero to one, the average terrestrial albedo today is around 0.3. This value rises to 0.6 for ice and even higher for fresh snow. Hence, beyond a certain percentage of ice-covered surfaces, a positive feedback process occurs: the albedo increases and cools the climate even more, which leads to more ice and cooler temperatures. Once initiated, this albedo runaway will rapidly freeze the whole Earth's surface. While mathematically relatively simple to comprehend, this process requires that the continental ice has already reached the vicinity of the tropics. How could such a situation have occurred at the end of the Precambrian?

One of the first explanations proposed for the Snowball Earth ice ages was that the Earth's rotational axis had tilted significantly, which would have considerably reduced the amount of solar radiation reaching the lower latitudes, while increasing it at high latitudes. This hypothesis was quickly dismissed on geophysical grounds: because Earth and Moon are gravitationally coupled, the Moon stabilizes the Earth, preventing major variations in the Earth's obliquity (Laskar et al., 1993). Obliquity variations are therefore restricted to between 22 and 24.5°, as discussed earlier in the context of the Milanković cycles (Figure 12.2). In contrast, Mars, which lacks a large satellite, experiences chaotic variations in the tilt of its rotation axis between 11 and 49°.

We need to return to the other climate forcing mechanisms: the solar constant and the atmospheric greenhouse effect. Since the birth of the solar system, our star, the

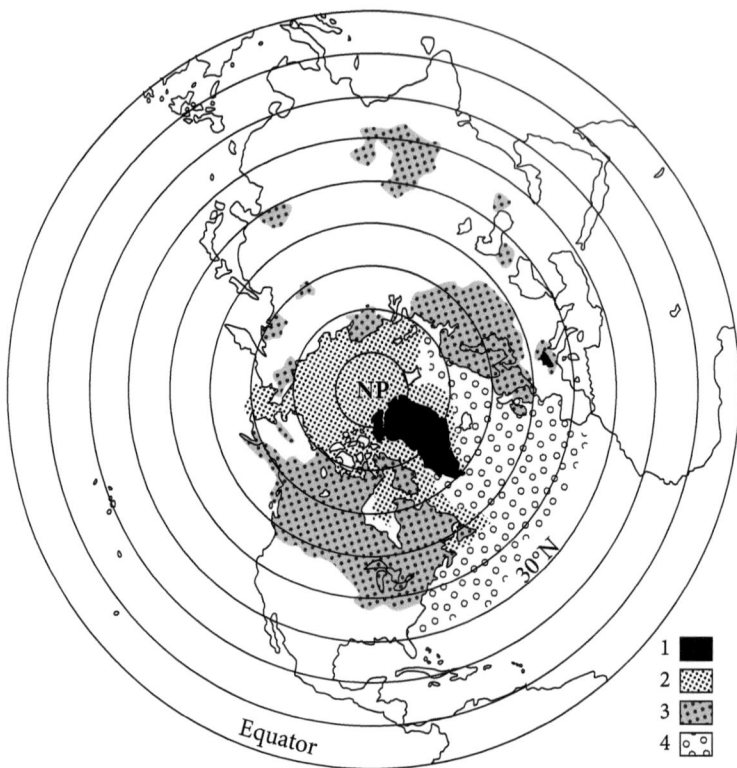

Figure 12.12 Ices in the Northern Hemisphere. This polar view shows the extent of continental ice sheets today (1) and during the Last Glacial Maximum (LGM) 21,000 years ago (3), as well as the maximum extent of the current pack ice at the end of winter (2) and the distribution of ice rafted debris in the North Atlantic during the LGM (4). *NP: North Pole.*
Source: Adapted from Harland (1965).

Sun, has steadily increased in intensity. At the end of the Precambrian, the solar constant was still 6% lower than its current value. This is not enough to trigger an ice age. What about the greenhouse gases? Atmospheric CO_2 is subject to major variations. The weathering of continental surfaces is the main sink for CO_2 and, as seen in Chapter 8, the accelerated growth of the continents led to a significant cooling of the climate at the beginning of the Proterozoic, around 2.4 Ga—another period in Earth's history known for its glaciations. Temperatures appear to have been much milder during middle part of the Proterozoic (i.e. between 1.8 and 0.8 Ga), when the supercontinents Columbia and Rodinia were generally situated in the tropics.

The intensity of continental weathering is influenced by mineral composition of the rocks that outcrop on the surface of the continents. For example, basaltic surfaces weather much more easily than granitic surfaces, which normally characterize the continental crust. Basalt is eight times more efficient as a carbon sink than granite. From 780 Ma onwards, the outpouring of large volumes of basalt due to hotspots heralded the initial break-up of Rodinia. For example, west of Laurentia, we can still observe the remains of a major basaltic province dated at 750 Ma, located at the level of the Tropic of Cancer, and which preceded the opening of the proto-Pacific Ocean (Figure 11.8). However, sub-tropical areas are always almost devoid of precipitation due to the atmospheric circulation that turns them into deserts. This is where the most extensive deserts on the planet—such as the Sahara, Kalahari, and Atacama—are located today. Thus, weathering of the large basaltic surfaces of the Laurentian LIP was delayed until they arrived at equatorial latitudes after southward rifting and drifting of North America. Another massive flood basalt, the Franklin LIP, was extruded over much of North America around 719 Ma, just a few million years before the onset of the Sturtian glacial. In 2003, Yves Goddéris, a researcher at the Geosciences and Environment Laboratory of the University of Toulouse (GET), and his colleagues at the Laboratory of Climate and Environmental Sciences (LSCE) near Paris, simulated this scenario using a climate model. The result of the calculations is clear: the catastrophic decline in pCO_2 triggers a global glaciation, even in the absence of a continent in a polar position.

The following year, researchers from LSCE and GET showed that the paleogeography of the onset of the Sturtian glaciation, with Rodinia beginning to break up into several continental fragments separated by nascent oceans, was also an important parameter that increased the intensity of chemical weathering on the continents, and therefore the decline in pCO_2 (Donnadieu et al., 2004). Indeed, the interiors of supercontinents are always arid, as seen in the widespread deposition of sand dunes and evaporites in the interior of the Phanerozoic supercontinent Pangea. A supercontinent that receives little rain will not be prone to chemical weathering. Conversely, the climate is much wetter when continents are fragmented because more of the land area is in close proximity to the oceans—the source of moisture. The causes of the first Snowball glaciation appear to be clear. All three are essential: a paleogeography favouring abundant precipitation on the continents; widespread, highly weatherable basaltic surfaces; and slightly lower solar luminosity. Because of this last point, the reader need not be afraid: a further global glaciation has been unlikely to happen since the end of the Precambrian.

The paleogeography changed prior to the second—Marinoan—global glaciation (c.640 Ma). By this time, much of Rodinia had rifted apart and several continental fragments had begun to collide again, initiating the formation of the Gondwana supercontinent and forming extensive mountain ranges. Moreover, Gondwana was then at the South Pole, like Antarctica is today. Ice caps likely nucleated on the new mountain ranges in the high southern latitudes, where snow would have progressively accumulated after Earth recovered from the previous Snowball glaciation (Figure 12.13). The high relief of the mountain ranges would have also favoured high rates of physical and chemical weathering, as was the case for the Himalayas, as discussed at the beginning of this chapter. However, solar luminosity was little changed from the beginning of the Sturtian glaciation. The Marinoan glaciation therefore appears to have had a somewhat different origin from the Sturtian glaciation, since the chemical weathering of the continents that drew down pCO_2 was due to their elevation rather than to

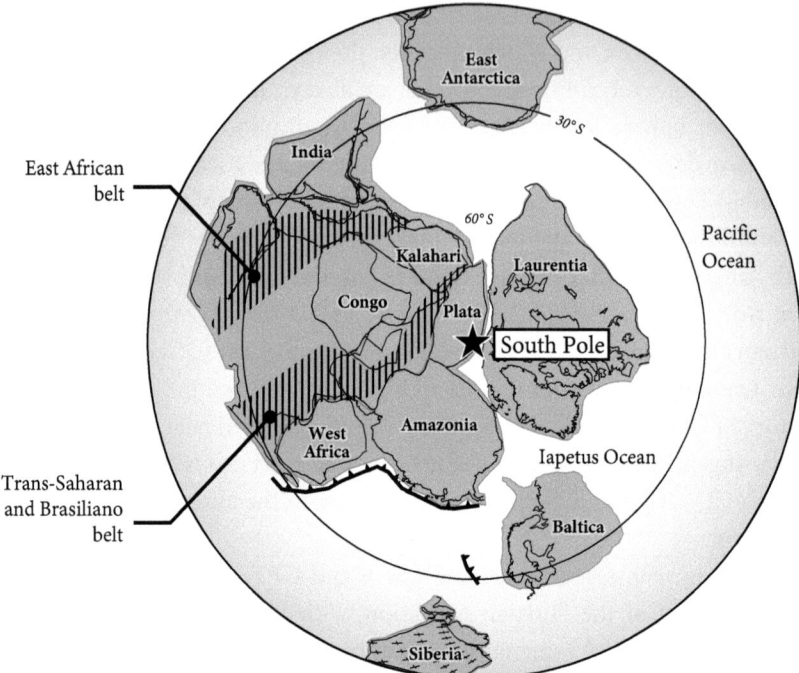

Figure 12.13 Polar view of the southern hemisphere continents during the Marinoan ice age. Ice sheets are not shown.
Source: Adapted from Meert (2003).

the nature of the rocks present. Paleogeography also played a favourable but different role in triggering Snowball glaciation, with continental masses at polar latitudes, including high mountain ranges providing an ideal setting for large ice caps to initiate and grow into continental-scale ice sheets capable of extending into the tropical latitudes.

The Marinoan glaciation was shorter-lived than the Sturtian glaciation: 5–10 million years instead of nearly 60 million. The Varangian (or Gaskiers) glaciation is even shorter—just one million years—and its nature is under debate. As in the case of the Marinoan glaciation, it was probably favoured by the tectonic context of the time, namely a new stage of the Pan-African orogeny.

During the heated debates between supporters and opponents of the Snowball glaciation hypothesis, one argument of the 'anti' side recognized local advance and retreat of continental ice, and thus the existence of climatic variations similar to those of recent glaciations. When the Snowball Earth hypothesis was first articulated, it imagined a world completely frozen under the ice, where little changed due to an acutely limited water cycle. However, sophisticated climate modelling has shown this idea to be incorrect. In fact, even in the absence of liquid water, a water cycle exists, as ice can sublimate directly when exposed to a weak sunlight. The attentive observer can see this today on some winter days. The water vapour produced by sublimation in the equatorial latitudes would yield snow precipitation at higher latitudes and elevations.

What happened to life during the Neoproterozoic ice ages?

The existence of late Precambrian global glaciations, especially one as long and severe as the Sturtian glaciation, inevitably raised questions about what happened to life when the oceans were frozen over and the average temperature of the globe was −20 °C towards the equator and around −50 °C at high and mid latitudes! It should be remembered that there was only aquatic life at that time, and in particular marine life, since complex organisms did not colonize land surfaces until the middle of Paleozoic era.

Today, we can see that microbial and even multicellular life is possible under the ice pack. A wonderful video shot by a robot under the Antarctic sea ice was put online at the end of 2016 by the Australian Antarctic Program.[5] The video shows sponges of various shapes and colours, worms, and starfishes, among other animals. Life abounds in these conditions, which are in fact relatively stable and mild compared to the surface: the water remains at −1.7 °C under the 1.5 m of sea ice that persists for three quarters of the year. However, conditions on Snowball Earth would have been much more severe, with strongly negative air temperatures all year round. Furthermore, any ice thinned in the equatorial region by sublimation would have quickly been replaced by equatorward flowing ice from higher latitudes. Therefore, limited to no sunlight would have been able to penetrate the thick sea ice. Nevertheless, at the base of the sea ice and ice shelves, the sea temperature was the same as today (i.e. close to the freezing point of sea water), hence providing conditions compatible with life. Furthermore, most of animal groups would not yet have existed at this time, with the possible exception of the sponges (Love et al., 2009), biomarkers from which have been found in sediments contemporary with the Marinoan glaciation, over 60 million years before appearance of the earliest Ediacaran fossils.

If animals were scarce, microbial life must have survived, and perhaps even thrived during the Snowball glaciation. The ferruginous brine of the Taylor Glacier represents ancient sea water trapped at the base of the glacier as seen before. This brine contains a wide variety of microorganisms. One can therefore imagine equivalent ecosystems under the sea ice and floating ice shelves of the Snowball Earth. The main problem was the lack of light where the ice was more than 100 metres thick, which was probably nearly everywhere. Yet we know that cyanobacteria and algae, both photosynthesizers, must have survived the glaciations because they were present before and after.

Researchers have reported that life is possible today on the surface of Antarctic ice in the small meltwater pits created by the presence of dark rock debris that locally reduces albedo and allows melting during the summer (Figure 12.14). These holes

---

[5] http://www.antarctica.gov.au/news/video?result_26545_result_page=4.

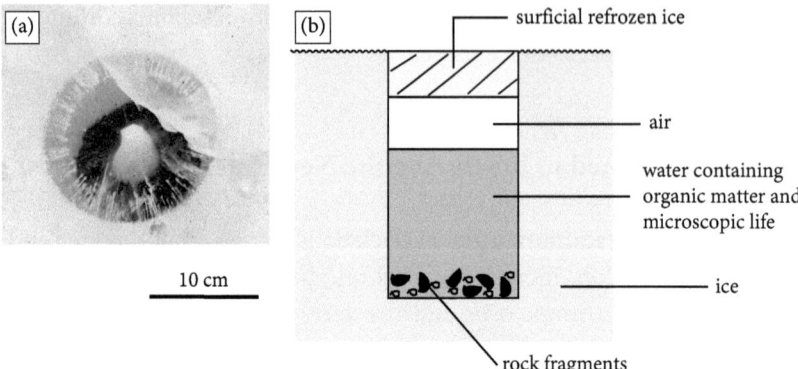

Figure 12.14 A partially melted hole on top of the Taylor Glacier in Antarctica.
(a) Top view. (b) Vertical section.
Source: Adapted from MacDonell and Fitzsimons (2008).

are commonly 10–20 cm in diameter and are inhabited by cyanobacteria and microscopic multicellular algae, among other organisms. Ice forms on the surface of the hole in winter but, at the bottom of the hole, a small pocket, rich in mucilage and other substances of bacterial origin, does not freeze, allowing its inhabitants to survive in this improbable environment. Could life have survived in similar melt pockets under the more severe conditions of Snowball Earth?

What does the fossil record say? Acritarchs declined at the end of the Precambrian. Was it because they didn't cope well with the cold climates or because they found themselves competing with, or even being consumed by, early animals represented by the Ediacara biota? Chinese researchers have published photos of small (about 1 cm long) fossil multicellular algae that are very well preserved in fine sediments of the Marinoan-aged Nantuo glacial formation (Ye et al., 2015). These findings do not reveal precisely where the refugium for these organisms was, but they do confirm that complex life survived the snowball glaciations.

> Over geological time, Earth's climate is controlled by diverse factors, including the amount of solar energy it receives, and internal factors, such as volcanism and the formation of mountain ranges that regulate the source and sinks of atmospheric carbon dioxide. The development of continental ice also depends on the geographical distribution of the continents and the ocean circulation. The combination of a lower solar luminosity at the end of the Precambrian, widespread basalt, and favourable paleogeography favoured the initiation of two global ice ages—the Neoproterozoic Snowball Earth glaciations. Life survived these glaciations but was surely strongly impacted by them.

13
Thawing of the Snowball Earth

At the end of the previous chapter, the blue planet had become the white planet, because the land and sea were frozen everywhere. Under these conditions, the high albedo reflected the Sun's rays back into space, and the Sun no longer warmed the Earth—a seemingly inescapable scenario ...

How did Earth recover from global glaciation?

Despite a seemingly static situation during the deep freeze of snowball glaciation, the atmosphere gradually changed. This change was due to the fact that the planet's internal engine remained unaffected by the ice cover. Convection in the mantle continued, and magmas were still produced. We learned earlier that the melting of mantle peridotites is possible under the lithosphere when hot mantle is dragged upwards. In the oceans, basalts continued to flow at the mid-ocean ridges. Volcanic eruptions happened here or there, in some oceanic islands or on the continents, at hotspots. A few volcanoes were high enough to emit lava, ashes, and volcanic gases above the ice, just as Mount Erebus in Antarctica, which rises to 3,794 m above the Ross Ice Shelf, does today. The gases escaping from this active volcano consist mainly of water vapour and CO_2, the second-most voluminous gas emitted.

Therefore, volcanism was the key to our planet escaping the slow, icy grip of Snowball Earth. The CO_2 emitted from the volcanoes was unable to dissolve into the ice-covered ocean. Furthermore, in the near absence of liquid water on the continents, the chemical weathering reactions that typically remove CO_2 through hydrolysis reactions with silicate minerals almost ceased. Although the massive continental-scale glaciers continued to erode the continents, this was a mechanical rather than a chemical process.

Thus, in the absence of efficient sinks, CO_2 gradually accumulated in the atmosphere where it could counteract the extraordinary radiative forcing of the high albedo of the frozen Earth. How much atmospheric CO_2 would have been required to raise air temperatures above freezing in the tropics, and thus initiate melting? It turns out that solving this problem is not trivial, and climate modellers have yet to reach a consensus. However, it seems that a CO_2 level in the atmosphere of at least 12% was needed to reach the critical threshold—an impressive figure compared to today's 0.04%! Even more impressive is the sequence of phenomena that was triggered when the fateful threshold was crossed. One can imagine a rapid disappearance (on a geological scale) of the sea ice and continental glaciers as the ice albedo feedback ran in inverse! While the melting sea ice would have had no effect on sea level, melting of glaciers would have driven a massive and rapid rise in sea level. Although the local response would have been complicated by other factors, such as isostastic uplift as the continents responded to the loss of the weight of ice, the net effect was a **transgression**, with coastlines marching up the continents globally.

During the last glacial maximum 21,000 years ago, the average sea level was about 120 metres lower than today. The low sea level allowed prehistoric men to walk on land from France to England across the Pas-de-Calais Strait, and from Siberia to Alaska across the Bering Strait. As northern hemisphere ice sheets melted, sea level rose for about 15,000 years before the coast lines stabilized. But remember that this was not the end of the Quaternary glaciation—since Greenland and Antarctic ice caps remain—but only the transition from a glacial to an interglacial stage of the Quaternary ice age. During the Neoproterozoic global glaciations, the amount of water retained in the form of ice on the continents was much greater than during the last glacial maximum, so sea level was much lower. Furthermore, the ice melted completely, because as albedo

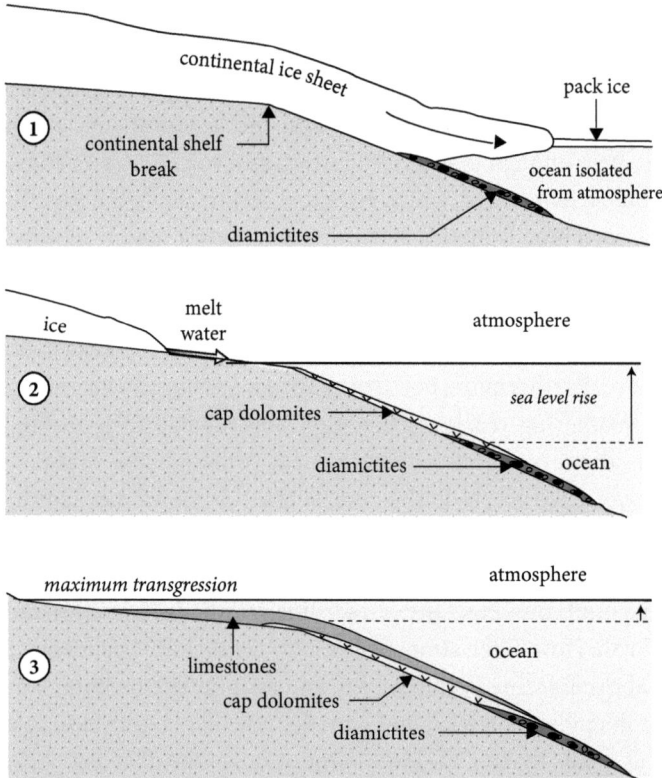

Figure 13.1 **Melting of continental ice, rising sea level, and deposition of cap carbonates during post-glacial transgression.** The cap carbonates deposited during the transgression are dolostones (2); limestones succeeded them during the maximum flooding (3) of the continental shelves. Diamictites are glaciomarine deposits (1) that contain pebbles and boulders transported at the base of the glaciers.

decreased, the extremely high CO_2 levels would have caused global temperatures to skyrocket. A rise in mean sea level of at least several hundred metres must have occurred after the Snowball Earth aftermath (Figure 13.1).

Evidence of global warming

The rapid return to warmer conditions and the transgression caused by thawing of the Snowball Earth were recorded by the deposition of unique carbonate rocks commonly referred to as 'cap carbonates'. The *cap* indicates that these rocks directly overlie the glacial deposits. However, they are far more extensive and are found far inland of glacial deposits in many regions, as they were deposited during the post-glacial transgression and the subsequent sea-level highstand.

These cap carbonates are found all over the world: in Namibia (Figure 13.2), Canada, Svalbard, China, West Africa, and Brazil, among many other places. They record a global phenomenon. Their connection with the glaciogenic diamictites is clear and unambiguous, showing that a warm climate succeeded the severe cold climate of the Snowball Earth without transition. This is an unprecedented situation in the history of the Earth. The effectively geologically instantaneous climatic shift was the consequence of the very high atmospheric CO_2 threshold required to trigger deglaciation. Some of the atmospheric CO_2 could dissolve in the ocean as soon as the sea ice disappeared. In the ocean, dissolved CO_2 disassociated into bicarbonate and carbonate ions, which combined with calcium and magnesium ions washing into the ocean to form the dolostones and limestones in the cap carbonates. Therefore, the cap carbonates represent the sequestration back into the solid Earth of the enormous reservoir of atmospheric CO_2 that accumulated in the snowball atmosphere.

As noted, the base of the cap carbonates is generally dolomitic, meaning it contains an equal balance of calcium and magnesium. These cap dolomites are one to thirty metres thick. They are light beige and laminated (Figure 13.3), and characteristically directly overlain by limestones (i.e. carbonates with no magnesium), which are generally thicker and often black in colour (where they have not been exposed to oxidative weathering on Earth's surface), due to an abundance of organic matter.

How were post-glacial cap dolostones formed?

Geologists agree that cap dolomites were formed during the post-glacial transgression. Microscopy suggests that the dolomite crystals are primary rather than replacement minerals, as is typical for most dolomite in the recent geological record. In addition, many sedimentary features that formed during deposition (i.e. before compaction and induration of the sediment) are still recognizable, highlighting the early origin of the dolomite. These sedimentary structures and textures allow us to reconstruct the depositional conditions. Thus, the trained eye of a sedimentologist can recognize laminae deposited under shallow marine conditions in some cases, with preserved

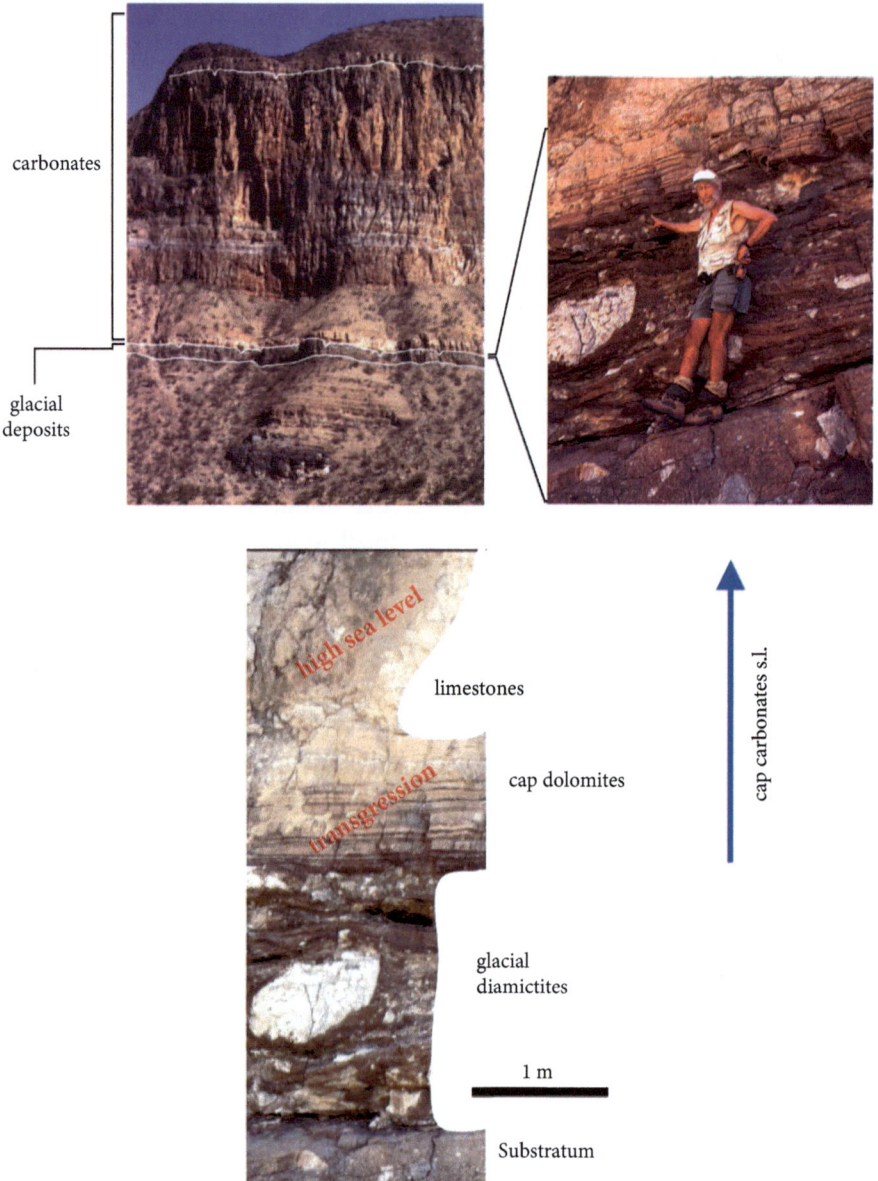

Figure 13.2 Marinoan glacial and post-glacial strata of the Otavi Group in northern Namibia. Left: view of the 'Maieberg Wall', showing pre-glacial rocks, a thin zone with Marinoan-aged glacial deposits barely resolvable at this scale, and the thick Maieberg formation cap carbonates deposited in the glacial aftermath. Right: Paul Hoffman pointing to the contact between glacial diamictites below and the cap carbonate above. Diagram below shows the interpretation of depositional conditions according to Hoffman and Schrag (2000) and Hoffman (2023).
Source: Paul Hoffman; Hoffman and Schrag (2000) and Hoffman (2023).

Figure 13.3 A cap dolomite sample from Ghana. This vertical section of a cap dolomite shows the fine bedding where remains of microbial mats were recognized under the microscope.
Source: Nédélec et al., 2007.

traces of microbial films that once covered the seafloor. Locally, these microbial textures form abundant stromatolites. Some of them are dome-like, whereas elsewhere, they are more parallel and display a continuous appearance, pierced by vertical tubes, indicating an unusual development of the surface and thickness of the microbial mats (Figure 13.4). These microbial textures provide evidence of life's recovery and proliferation as soon as the ice melted.

The formation of primary dolomite is a geological enigma that has only recently been solved. The direct precipitation of dolomite from seawater is impossible to reproduce in the laboratory. Yet seawater contains all the necessary ingredients in dissolved form, namely calcium (Ca^{2+}), magnesium (Mg^{2+}), and carbonate (CO_3^{2-}) ions. One explanation is that magnesium ions are weakly bound to sulphate ions (SO_4)$^{2-}$ that are abundant in current oxygenated seawater, and therefore cannot easily combine with calcium and carbonate to form dolomite.

Crisogono Vasconcelos, a Brazilian researcher working at the Swiss Federal Institute of Technology in Zurich (the famous ETH), solved the problem (Vasconcelos and McKenzie, 1997). He studied the conditions of dolomite formation in a lagoon not far from Rio de Janeiro, called Lagoa Vermelha (*red lagoon*) because of its waters, which are locally coloured red or pink by microorganisms that thrive in hypersaline conditions. A microbial mat several centimetres thick covers the bottom of the lagoon

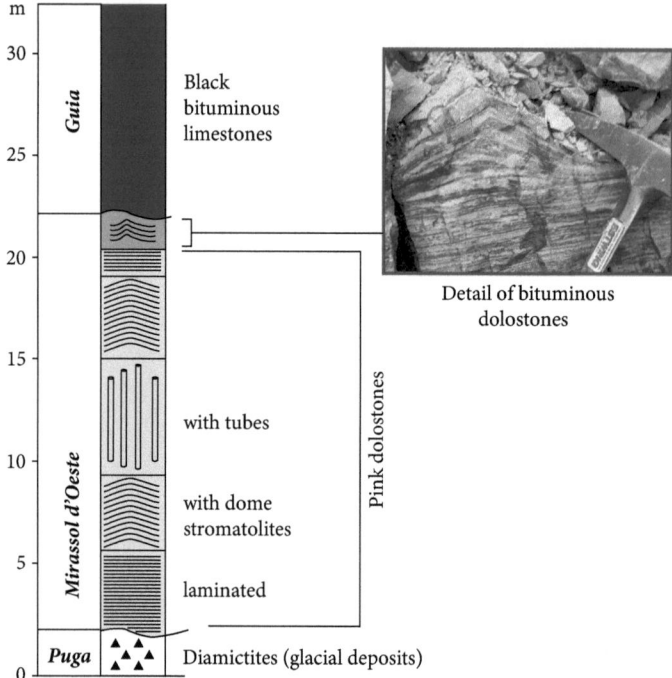

Figure 13.4 Post-glacial cap carbonate succession in Mato Grosso, Brazil. Dark, bitumen-impregnated beds can be seen at the top of the dolomites.
Source: Adapted from Font et al. (2006).

(Figure 13.5). Those who have swum in a lake or a pond may have experienced the soft, supple, but relatively firm contact of a microbial mat under their feet. The microbial carpet of Lagoa Vermelha contains sulphate-reducing bacteria, because the sediment is poor in oxygen. These bacteria reduce the sulphate ions in the seawater that soaks the sediment at the bottom of the lagoon. The magnesium ions are then free to combine with the calcium and carbonate ions to form primary dolomite crystals.

The precipitation of dolomite from seawater in the presence of sulphate-reducing bacteria was reproduced in the laboratory at ETH Zurich (Warthmann et al., 2000). Eric Font (then a PhD student, now a professor at the University of Coimbra in Portugal) and I proposed an identical mechanism for the formation of post-glacial cap dolomites in Brazil and Ghana (Font et al., 2006; Nédélec et al., 2007). This process is not unanimously accepted and formation of the Neoproterozoic cap dolostones is still debated.

The post-glacial ecosystem

The knowledge of life after Snowball Earth has been taken a step further. Deep in Brazil, in the state of Mato Grosso near the border with Colombia, rocks from the end of the Marinoan glaciation are exposed in a quarry near the small town of Mirassol

Figure 13.5 The microbial mat of Lagoa Vermelha (Brazil). View of a sample cut directly from the bottom of the lagoon with its microbial mat. The white layers consist of mineral fragments. The dark upper layer of the microbial mat is alive, while the light-brown base is about 20 years old (carbon-14 dating in Font et al., 2010). It will take much longer to compact this microbial mat, indurate it, and transform it into microbial dolomite similar to the ones found in the Neoproterozoic cap dolostones.
Source: Anne Nédélec, Font et al., 2010.

d'Oeste, where our French–Brazilian team studied in 2005.[1] These rocks are cap dolomites with all the characteristics previously mentioned, including the domed stromatolites which here reach a height of up to one metre, overlain by other massive and continuous stromatolites, with vertical tubes. Above, the rocks are impregnated with hydrocarbons, to the extent that they stain one's fingers when handling these bituminous dolomites (Figure 13.4). As the quarry was still in operation, the working face was constantly being rejuvenated and it was possible to collect fresh bitumen for biomarker analysis. These analyses were carried out in Nancy (France) by Marcel Elie, who found typical biomarkers of cyanobacteria, sulphate-reducing bacteria, other so-called 'green sulphurous' bacteria, and finally red algae, thus providing a picture of the post-glacial biological community (Figure 13.6).

During the post-glacial transgression, it is likely that there was a bloom of algae and photosynthetic bacteria in the warm coastal waters, enriched in nutrients coming from the newly ice-free continent. Such seasonal blooms are well known today. All these primary producers settled to the seafloor after their death. The organic remains were so abundant that they could not be completely decomposed. Indeed, because of

---

[1] I would like to thank our Brazilian colleagues, Afonso Nogueira from Belem University, and Ricardo Trindade from São Paulo University.

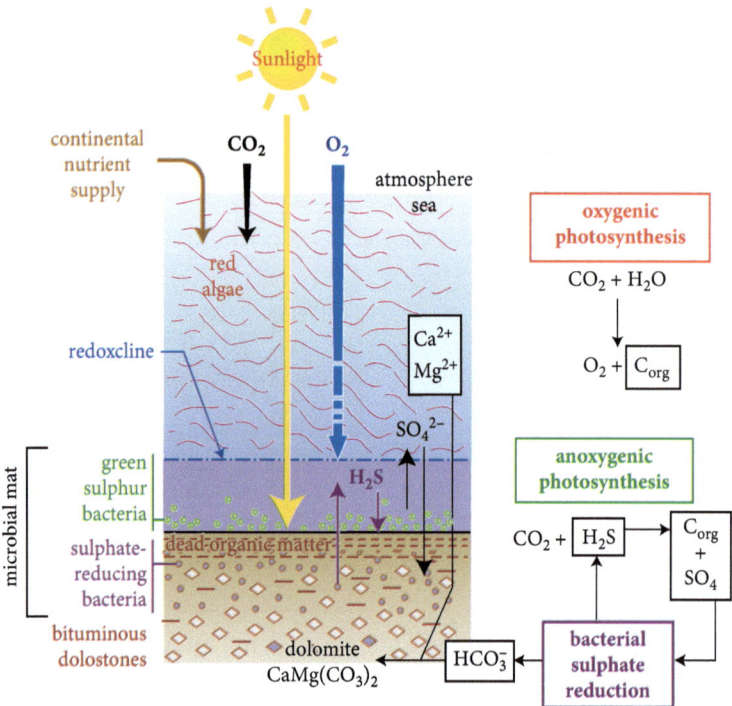

Figure 13.6 Reconstruction of the post-glacial ecosystem from biomarkers preserved in the dolomites of Mirassol d'Oeste, Brazil.
Source: Adapted from Elie et al. (2007).

this accumulation of dead organic matter, the sediment and even the base of the water column become anoxic, evidence of which is provided by the presence of green sulphur bacteria. These bacteria carried out anoxygenic photosynthesis; they therefore needed light, but they could not tolerate oxygen. In turn, the sulphate-reducing bacteria, which thrive in oxygen-free, organic-rich conditions, consumed the sulphate at the sea floor, facilitating the formation of dolomite, analogous to the modern Brazilian lagoon.

Landscapes of the Snowball Earth aftermath

Transgression and warming that followed the last glacial maximum of the Quaternary glaciation considerably changed landscapes and ecosystems due to increased temperature and moisture. For example, the reindeer and mammoth herds sought refuge in the higher latitudes. Rainforest area was enlarged by half.

There is no fossil record of life on the continents after Snowball Earth, since life at the end of the Precambrian was exclusively aquatic. However, the environmental changes must have been dramatic—much more so than during the glacial–interglacial oscillations of the Quaternary. The sea ice that covered the oceans must have disappeared quite rapidly, despite a probable thickness of several hundreds of metres across most of the oceans, because of the greenhouse effect induced by the high CO_2 content of the atmosphere. Melting of the sea ice and retreat of the floating and continental ice must have produced a layer of cold, fresh water on the surface of the ocean, which would have generated thick and long-lasting fogs due to the temperature contrast between the air and the water. The higher solubility of CO_2 in this cold surface water 'lid' may have briefly tempered the warming in a way similar to the Younger Dryas cooling event,[2] which momentarily reversed the warming trend following the last glacial maximum. Finally, the ocean surface warmed and the mists dissipated as an intensely hot climate prevailed due to the overwhelming mass of CO_2 in the atmosphere. Despite the precipitation of cap carbonates, the CO_2 content of the atmosphere must have remained high until the effects of continental weathering could moderate it. Based on modelling by Le Hir et al. (2009), this process would have taken 2–3 million years.

Can we imagine what our planet looked like then? A few parts of the world today may present analogues. The first is the Persian Gulf, more specifically the coast of the United Arab Emirates and Qatar region. The Persian Gulf is shallow. Nowhere is the depth greater than 90 m, and it is even less than 20 m deep in its very flat south-eastern part. Formation of the Persian Gulf is very recent: water only inundated the Gulf after the last glacial maximum. Furthermore, the Persian Gulf is an almost closed sea, as it only communicates with the Indian Ocean through the Strait of Hormuz. In one of the hottest climates in the world, evaporation is intense and therefore the salinity is high. Sabkhas, or natural salt pans, dot the southern coast. Islets and almost-enclosed lagoons create many sheltered areas along the coastline (Figures 13.7 and 13.8). An almost continuous microbial carpet has developed on the foreshore. Cyanobacteria abound, as do others adapted to such environments, such as sulphate-reducing bacteria. Consequently, dolomite is currently precipitating here (Bontognali et al., 2010).

Later in the post-glacial transgression, when the limestone part of the cap carbonates formed, a landscape more like that of the Bahamas likely prevailed. This archipelago country is made up of numerous limestone islets off the coast of Florida (Figure 13.9a, b). The high temperatures and relatively low rainfall allow the spontaneous formation of microscopic calcium carbonate crystals in the seawater. When this happens, the water appears to turn white: these are known as whiting episodes. The small crystals accumulate on the seafloor and form a calcareous sludge, made of **micrite.**

The Bahaman islands are also famous for their stromatolites. Large stromatolitic domes were built gradually and then flooded during the recovery from the LGM. Their impressive dimensions resemble those of Precambrian stromatolites. Finally, along the windward coast of Exuma Island, very extensive and coalescing tabular stromatolites have recently developed (Figure 13.9c, d). They are reminiscent of the thick,

---

[2] The Younger Dryas climate event occurred from c.12,900 to c.11,600 years ago. Its causes are still debated, but it appears related to meltwater discharge from the North American ice sheet.

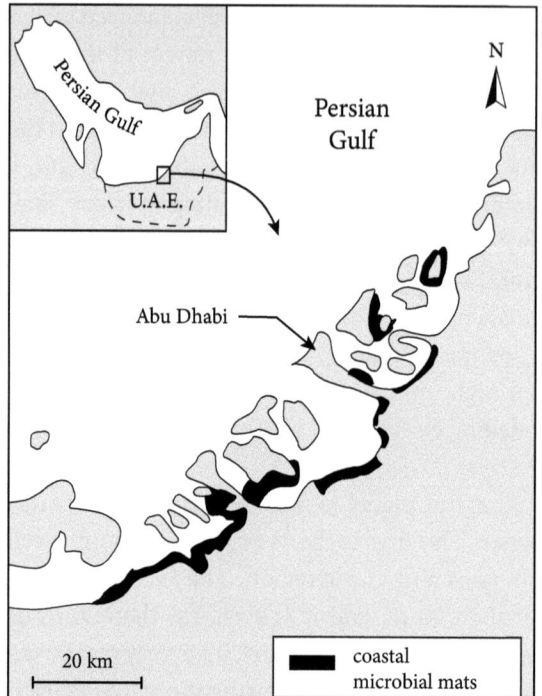

Figure 13.7 Coastal microbial mats (black) in the Abu Dhabi region (United Arab Emirates).
Source: Adapted from Kendall et al. (2003).

continuous stromatolites locally observed following the snowball glaciations. These recent analogues help to visualize the aftermath of the Neoproterozoic Snowball Earth; but to truly understand it, we must extend these environments to coastlines across the globe.

A last observation will contribute to illustrate the landscapes that followed the Snowball Earth. Indeed, at the transition between the transgressive dolomites and the highstand limestones, peculiar structures, unknown today in the marine environment, are observed in many localities. These are ancient crystals of **aragonite**—the same calcium carbonate mineral as that found in the whitings of the Bahamas today—but here growing directly on the seafloor in the form of shrubs or fans (Figure 13.10). How can we explain the growth of these clusters of large crystals?

In some sheltered settings, seawater may have reached extreme supersaturation with dissolved carbonates. This is conceivable under a very warm climate in protected areas where evaporation was intense. Indeed, during post-glacial transgression, lagoons or shallow basins could easily have formed, for example due to the shelter of residual glacial moraines, behind the coastal bulge that separated the continent, formerly collapsed under the weight of the ice cap, from the open sea, or behind reefs formed by the rapidly developing stromatolites. The post-glacial coastlines were therefore

Figure 13.8 The Khor Al Adaid lagoon in Qatar. (a) Location. (b) General view. (c) The foreshore and the microbial mat under shallow water. (d) Cross-section of the microbial mat (thickness 5 mm). (e) Seasonally dried microbial mat.

Source: (b) and (e) Tomaso Bontognali. (c) and (d) in Bontognali et al. (2010), reproduced with permission of the International Association of Sedimentology.

strongly replete with restricted basins at the time of the maximum flooding. At that time, background sediment supply was also low.

The rapid growth of aragonite in seawater subjected to 30–40% evaporation at a temperature of 40 °C has been numerically modelled. Today, although such high evaporation rates may be reached in some parts of the world, aragonite does not precipitate. The difference lies in the CO_2 content of the atmosphere. The atmospheric CO_2 content must have reached 10% for aragonite crystals to form (Figure 13.11). Even in the Snowball aftermath, these conditions may not have prevailed everywhere and all the time. In the field, growth of aragonite crystals appears periodic, possibly seasonal, and corresponding to the hottest and driest season when evaporation was at its maximum. If so, these aragonite crystal layers attest to extraordinarily fast carbonate precipitation rates.

Duration of deglaciation

Time and duration are crucial questions in geology. The end of both global glaciations is known: respectively 661 Ma for the Sturtian glaciation and 635 Ma for the Marinoan glaciation. Determining the duration of the deglaciation comes down to the question

Figure 13.9 Stromatolites of the Bahamas. (a) The Bahamas from space. (b) Map showing the shallow areas (light blue) where calcium carbonate precipitates.
(c) Coalescing tabular stromatolites on the reef fringing Exuma Island. (d) Large domed stromatolites thought to have been formed at the beginning of the transgression following the Last Glacial Maximum.
Source: (a) NASA. (b) and (c) Andres et al. (2009), reproduced with permission of the International Association of Sedimentology. (d) Adapted from Dill et al. (1986).

of the time elapsed during the deposition of the cap carbonates. Indeed, deposition of the cap dolostones was concurrent with the post-glacial transgression and therefore with melting of continental ice.

When thawing began, sea level rise was primarily due to melting of the continental ice. The thermal expansion of the ocean also played a role. Melting of the sea ice has no effect on sea level. It was initially thought that melting of the continental ice was very rapid, a few thousand years at most, due to the high temperatures caused by a strong greenhouse effect. However, paleomagnetists identified several magnetic field reversals during the deposition of the cap dolomites (Trindade et al., 2003). Magnetic

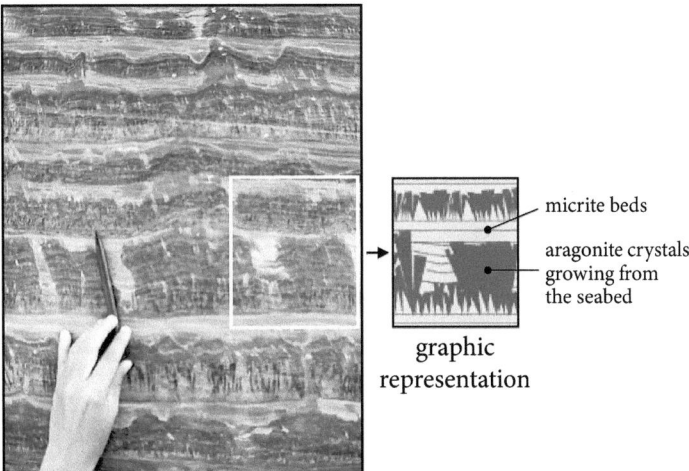

Figure 13.10 A succession of large aragonite crystals grown from the floor of a shallow basin (Sambra, Brazil). Periods of crystal growth alternate with periods of micrite precipitation.

field reversals do not occur regularly, but comparison with recent periods in the Earth's history during which reversals were very numerous suggests a depositional time for cap dolostones of at least several tens of thousands of years.

The contradiction with the assumption of relatively rapid melting of continental ice is only apparent. Indeed, sea level rise is a complex issue. It interferes with isostatic readjustments, that is, the slow ascent of continents freed from continental ice. Despite its rigidity, the continental lithosphere bends under the overload of an ice sheet, particularly in the centre of the ice sheet where the ice is thickest. The sub-glacial topography of Antarctica or present-day Greenland illustrates this phenomenon (Figure 13.12): the central part of both continents is below sea level today.

Melting of an ice cap causes an elastic rebound of the areas that supported the weight of the ice. The uplifted beaches in Canada and Scandinavia are evidence of the rebound that followed the last glacial maximum and is still ongoing. The rate of this rebound depends on the viscosity of the asthenospheric mantle, while its amplitude depends on the thickness of the continental ice that depressed the lithosphere. The result is a regression (the opposite of a transgression)—i.e. a relative drop in sea level in the previously glaciated areas. As the ice load was not the same everywhere and the Quaternary ice sheets did not melt everywhere at the same time, but retreated progressively, the post-glacial rebound did not take place everywhere at the same time, nor with the same speed. Today, it is one centimetre per year in the Gulf of Bothnia to the west of Finland, as well as in Hudson Bay in Canada, but only 0.2 cm per year to the north of Norway.

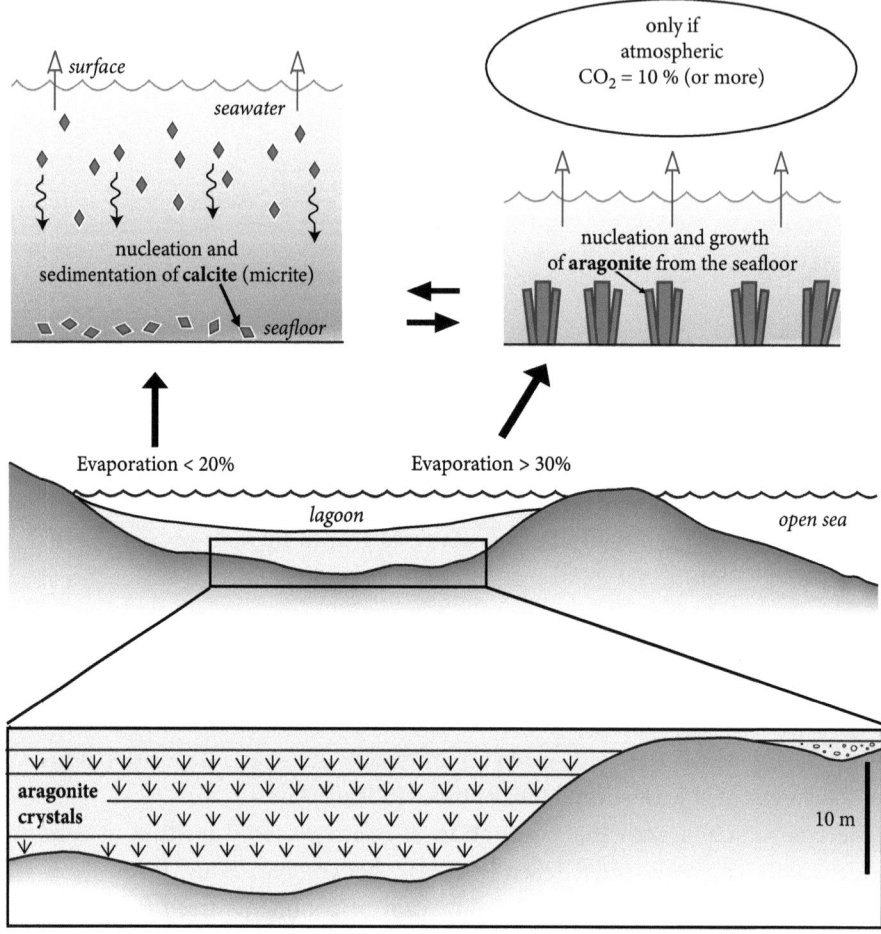

Figure 13.11 **One scenario for the formation of giant aragonite crystals** in an almost-closed basin subjected to high evaporation rates under a CO_2-rich atmosphere. *Source*: Adapted from Vieira et al. (2015).

During the thawing of the Snowball Earth, even though melting of glacial ice was extremely rapid, basins may have momentarily been open to the ocean, then closed, and then reopened again due to the competing effects on sea level. All things considered, the sea level history of the Baltic region during the retreat of the Scandinavian ice sheet is a useful analogue for understanding the complexity of sea level rise following Snowball Earth. Indeed, closed lakes and more-or-less open seas geographically linked to the Baltic Sea succeeded one another over a few thousand years in this region. The evolution of sea level following the melting of continental ice in the Snowball Earth aftermath was modelled in 2014 by a young researcher from the University of Oregon (USA), J. C. Creveling, and Professor Jerry Mitrovica from Harvard. They showed that the post-marine transgression was highly spatially variable and nowhere corresponds to the whole period of deglaciation. The competing effects of glacial melting and

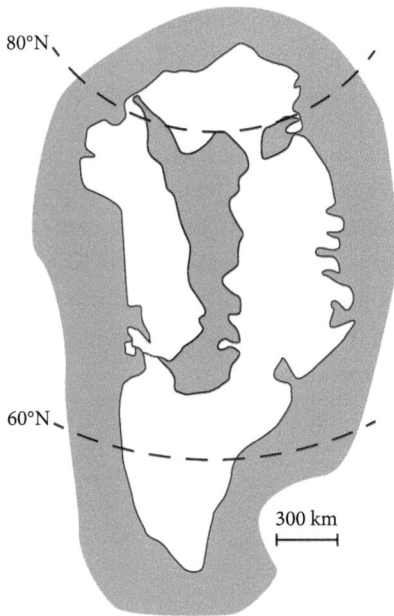

Figure 13.12 Topography of the Greenland bedrock under the present-day ice sheet. The central part of the continent (shaded area) is below sea level. This area is currently covered by a permanent ice cap with an elevation of almost 3000 m above sea level.

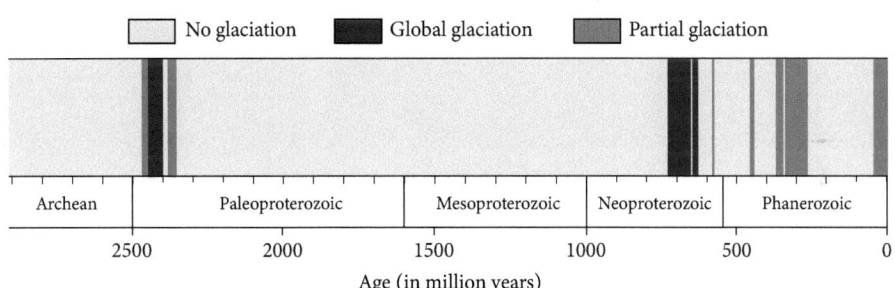

Figure 13.13 Glaciations since 2,500 Ma. The Proterozoic eon experienced multiple global glaciations, but also very long periods without any glaciation. The Phanerozoic eon has experienced multiple glacial epochs, but none of global extent.

post-glacial rebound may have lasted up to 50,000 years in some locations—long after the ice had melted and consistent with the paleomagnetic data from Brazil.

After the Cryogenian period, during which the Sturtian and Marinoan snowball glaciations occurred, followed by rapid and marked warming, Earth continued to experience climatic oscillations, but never on the same scale. The Phanerozoic eon, which began about 540 million years ago, has experienced several ice ages, including the present glaciation (Figure 13.13).

The Phanerozoic ice ages, which (fortunately!) never became global, will be discussed in the next chapter. They occurred despite the gradually increased luminosity and warming of the Sun since the end of the Precambrian.

> Global freezing of the Earth did not last indefinitely. The content of CO_2 in the atmosphere gradually increased due to volcanism, and reached a threshold where deglaciation was suddenly triggered. The transfer of CO_2 from the atmosphere to the ice-free ocean led to the precipitation of unique carbonate rocks, the cap dolomites, during the post-glacial transgression. Life recovered and proliferated during this transient swing to hot temperatures.

14
Life invades the continents

After the tectonic and climatic ups and downs of the late Proterozoic, the Phanerozoic world began with a pleasant climate and hosted a diversity of marine animals. Did this world resemble our own? Not yet, because the continents remained devoid of complex life. However, this too changed during the Paleozoic era.

The Cambrian explosion: optical illusion or reality?

Although the continents were uninhabited, the Cambrian oceans teemed with animals, as evidenced by the many kinds of fossils found in rocks from the early Cambrian, the first period of the Paleozoic era. Indeed, Cambrian strata contain so many animal fossils that an explosion of biodiversity is thought to have occurred in the early Paleozoic era. While the fauna of the late Ediacaran comprises mainly simple soft-bodied animals such as sponges, jellyfish-like species, and sea anemones, many Cambrian fossils resemble today's marine **arthropods**, the crustaceans. Their bodies, protected by a carapace, are generally segmented and have various appendages, such as legs and antennae. Their large, prominent eyes suggest animals on the lookout, ready to react quickly to the appearance of prey or predator. Trilobites are the emblematic fossils of these primitive arthropods (Figure 14.1). Their bodies were divided into three longitudinal lobes, hence their name. Trilobites ranged from a few millimetres to tens of centimetres in length.

The Burgess Shale in the Rocky Mountains of British Columbia, Canada, contains many very diverse fossils, which are certainly arthropod-like, but in some cases also

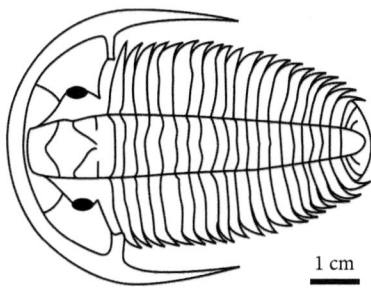

Figure 14.1 Early Cambrian trilobite fossil.
Source: Adapted from Vannier (2009).

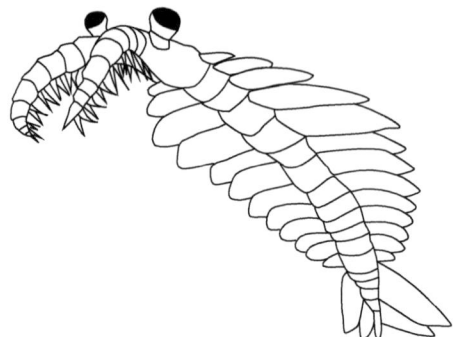

Figure 14.2 *Anomalocaris*, referred to as 'the terror of the Burgess Sea'. Apparently, a formidable hunter and measuring 80 cm to 1 m in length, *Anomalocaris* resembles no other known later arthropod and became extinct before the end of the Cambrian.
Source: Drawing by Nobu Tamura (CC-BY-SA licence).

very different from present-day arthropods (Figure 14.2). They are dated to the middle Cambrian, around 505 Ma. Though discovered just over a hundred years ago by Charles Walcott, it was only in the 1970s that the Burgess fauna was rediscovered in museum collections and its significance truly appreciated. Paleontologists from Cambridge University in England then studied it in detail and revealed its incredible diversity. Stephen Jay Gould (1941–2002), professor at Harvard University, told this story in his bestselling book *Wonderful Life: The Burgess Shale and the Nature of History* (1989), accessible to any reader and illustrated with numerous drawings.

Equivalent faunas have now been discovered in southern China and several other parts of the world. What is the significance of this exceptional Cambrian biodiversity? The Burgess Shale, as well as the fossiliferous terrains of China, represent ancient, poorly oxygenated muds that preserved carbonaceous traces of the bodies of dead animals, including their soft tissues, in exceptional detail. But do these fossils represent an actual explosion of animal diversity in the Cambrian, or is this record biased by the rare preservation conditions? We do not attempt to answer this question here, but instead refer interested readers to more detailed descriptions and discussions of the Cambrian fauna and the evolutionary problems it raises (see e.g. Erwin and Valentine, 2013). We do note, however, that many Burgess fossils do not seem to have had any descendants.

These exceptional fossil discoveries fuelled the debate between supporters of the slow and gradual evolution of species, as envisioned by Darwin, and supporters of the theory of 'punctuated equilibria' developed by Stephen Jay Gould and Niles Eldredge in 1972. According to this theory, periods of muted evolutionary changes in the fossil record are punctuated by episodic extinctions, followed by abrupt appearance of new species. For Darwin, natural selection acting on organisms can explain the gradual appearance of new species over long geological time. However, the fossil record does appear to indicate periods of stasis followed by rapid turnover, commonly linked to significant environmental changes. Of course, it is difficult to quantify the duration of such episodes, and the duration of the Cambrian explosion remains widely debated. Moreover, the fossil record is notoriously incomplete and suffers from both preservational bias, as mentioned above, and observational bias.

Initially heavily criticized, the theory of punctuated equilibria has now gained traction among paleontologists, for it is not necessarily inconsistent with Darwin's gradual evolution of species. Rather, much of this gradual evolution is masked in the geological record until major environmental catastrophes wipe out dominant species. In 2002 Stephen Jay Gould published a monumental synthesis of his hypothesis in the context of Darwinian evolution in which he did not so much challenge Darwin's theory but rather elucidated how it is expressed in the ponderously long geological timescale.

The first land plants

As a result of recent discoveries, we now know that the first land plants date back to at least 470 Ma, during the **Ordovician** period—the second period of the Paleozoic (Figure 14.3). The evidence for land plants is found in the form of **spores** (organs of dispersal and vegetative propagation) found in rocks in Argentina and Saudi Arabia (i.e. on the edge of the ancient Gondwana continent, at a paleolatitude of about 35°S). The spores probably originated from small, primitive, green plants similar to modern mosses. Fossil evidence for fungi is also abundant in the Ordovician, and molecular clock estimates[1] suggest that they may have colonized land as early as 600 Ma (Heckman et al., 2001). Fungi are neither plants nor animals. Indeed, fungi are incapable of photosynthesis, unlike plants. Rather, they feed on existing organic matter, as do animals—but they are not mobile, and their mode of reproduction is very different.

It was not until the **Silurian**, following the Ordovician period and the first Paleozoic glaciation, that the first complete fossils of primitive land plants, such as *Cooksonia*, first described in Ireland but now known from many places around the world, are found (Figure 14.4).

Plant fossils become abundant in the subsequent **Devonian** period. Some fossils are sufficiently large to be called trees, with trunks that exceed 10 cm in diameter and several metres in height (Figure 14.5). Defined as large plants with a self-supporting vertical trunk, trees have arisen independently in different plant groups. While the tree *Wattieza*, despite being 10 m high, did not yet have true leaves, *Archaeopteris*, which appeared at the end of the Devonian, had beautiful leafy branches.

These Devonian plants formed true forests. The first petrified forest in the history of the Earth was discovered in the early twentieth century during the construction of the Gilboa Dam in New York State (Goldring, 1927). Numerous fossil stumps, some still upright and up to 1.5 m tall, were uncovered in a sandstone quarry used to supply material for the dam. The stumps and plant remains were excavated and archived in

---

[1] The molecular clock uses genetic mutation rates to deduce the time when different life forms diverged. Molecular clocks require age-calibrated divergence points, and traditional molecular clock techniques assumed constant mutation rates. Modern molecular clocks, however, allow for variable mutation rates. Molecular clocks are useful in paleontology because they make predictions about when key divergences that otherwise are not captured in the fossil record should have occurred.

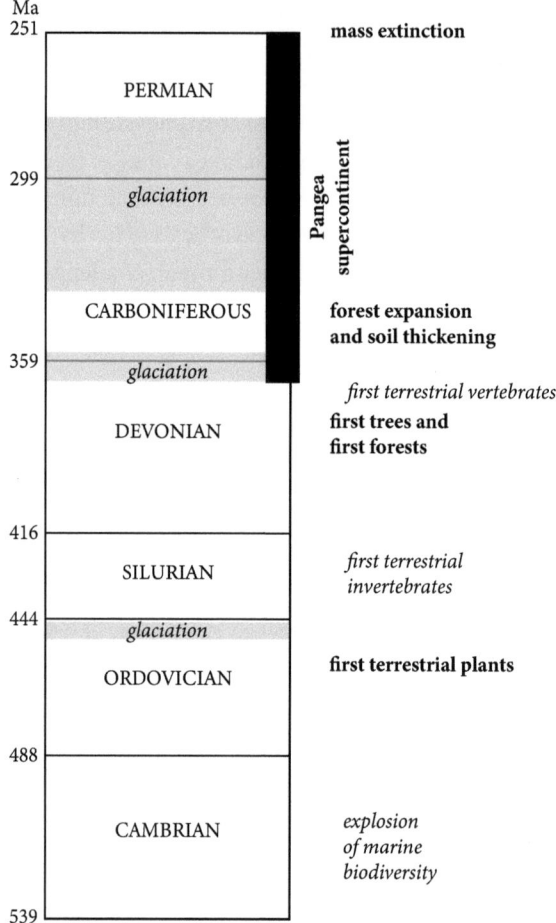

Figure 14.3 The periods of the Paleozoic era, along with major tectonic, ecological, and climatic events.

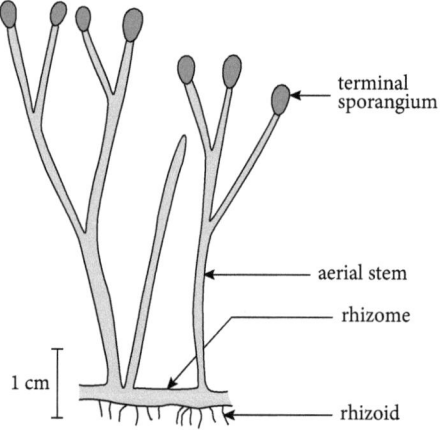

Figure 14.4 Reconstruction of Cooksonia, a plant from the Silurian of Ireland.

Figure 14.5 Devonian trees of Gilboa. (a) Discovery of a petrified trunk base at Gilboa, NY, in 1921. (b) Reconstruction of the Devonian tree *Wattieza* from Gilboa.

Source: (a) New York Public Library (Science Photo C033/4261). (b) Adapted from Meyer-Berthaud and Decombeix (2009).

the nearby museum, and the quarry was filled in after the work was completed. In 2010, the quarry was partially reopened, making it possible to reach the horizon representing the original soil, which preserved the arrangement of tree stumps, as well as the remains of trunks that reached exceptional dimensions. These fossils reveal a dense forest resembling a littoral mangrove that had developed on the southern margin of Laurentia (Figure 14.6).

The petrified forest of Gilboa dates from c.380 Ma (i.e. from the Middle Devonian). Other plants and trees then appeared and developed. The **Carboniferous**, which followed the Devonian, is well known for its forests rich in ferns and arborescent **lycophytes**, as well as giant horsetails. Lycophytes are vascular plants, that is, plants that contain vessels (the xylem) carrying sap. Whereas the lycophytes that remain today are very small, Carboniferous lycophytes included large trees, such as *Lepidodendron* (Figure 14.7). The Carboniferous world is depicted as one of giants. If dead trees are protected from oxidation and breakdown by rapid burial in an anaerobic setting, their organic matter is transformed into coal. Most of the coals in Europe and North America were formed in vast coastal swamps in the Carboniferous and give this period its name. Importantly for us, these coals are not only a source of energy, but also contain well-preserved imprints of the Carboniferous plants.

Figure 14.6 Reconstruction of the Gilboa fossil forest.
Source: Anna Kotler.

Consequences of the greening of continents

The Paleozoic Earth was very different from that of the Precambrian. Through the gradual greening of the land, it came to resemble today's Earth by the middle of the Paleozoic. The conquest of land by plants was not without consequences, as terrestrial plants are important agents in the global carbon cycle. Plants are photosynthesizers, meaning they use carbon dioxide from the atmosphere to generate their tissues and the sugars they require to function. Their roots respire, and therefore release CO_2 into the soil, as well as organic acids, creating acidic solutions in their vicinity. These acidic fluids, as well as the roots themselves, contribute to rock weathering. Therefore, the evolution of vascular land plants resulted in the deepening of soils, from a just

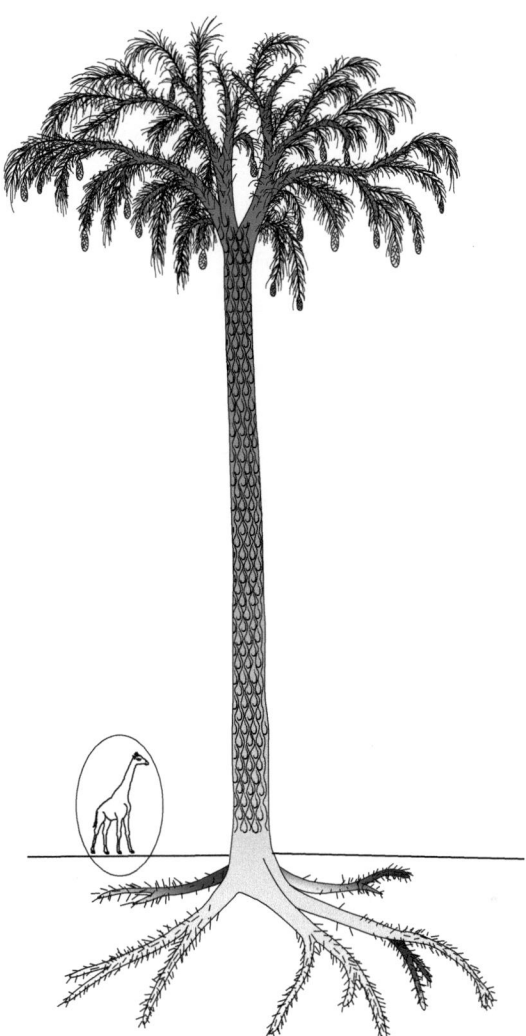

Figure 14.7 Reconstruction of *Lepidodendron*, a Carboniferous tree. The giraffe is shown to provide a sense of scale.
Source: Adapted from McGhee Jr (2018).

few centimetres thick in the late Ordovician to an average of one metre thick by the beginning of the Carboniferous. The consequence, as you might have already foreseen, is that the weathering of the silicate minerals that make up much of the continental surface was amplified. Because silicate weathering removes CO_2 from the atmosphere and ultimately stores it as carbonate minerals, the natural consequence of the evolution of land plants was a decrease in the greenhouse effect and cooling of the climate.

At the beginning of the Cambrian, the CO_2 content of the atmosphere was at least 4,000 ppm, or 0.4%—about ten times the present level. This was much lower than in the Precambrian, but high enough to induce a warm global climate. However, by the end of the Ordovician an extensive, if relatively short-lived, ice sheet developed on

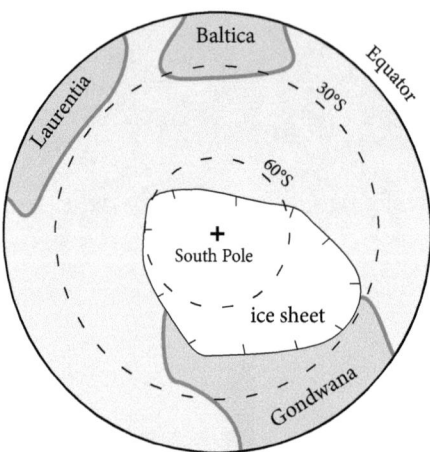

Figure 14.8 End-Ordovician paleogeographic reconstruction in polar view.

Gondwana, which then lay astride the South Pole (Figure 14.8). The centre of the ice sheet was in West Africa. The extreme aridity in the modern Sahara makes it possible to recognize unambiguously the traces left by the Ordovician ice, while cyclical sediments deposited in other parts of the world capture the sea level effects of the waxing and waning of the Ordovician ice sheet. Gondwana's austral position certainly favoured the formation of an ice sheet, but paleogeography was insufficient to counteract the strong early Paleozoic greenhouse. The additional factor might have been the role of the earliest terrestrial plants in drawing down CO_2, as suggested by Lenton et al. (2012).

A second, brief glacial epoch occurred in the late Devonian period, but the most significant cooling occurred in the latter part of the Paleozoic era: the long Permian–Carboniferous glaciation, which lasted about 70 million years (Figure 14.9). This time, the effect of vegetation seems to have been paramount. Models that account for the extent and nature of vegetation cover in the Carboniferous period indicate a drop in atmospheric CO_2 to about 400 ppm—comparable to modern levels. However, Earth did not completely freeze over as it did during the Neoproterozoic Snowball Earth events, perhaps because the Sun was a little warmer than at the end of the Precambrian. As during the Ordovician glaciation, Carboniferous ice sheets developed on Gondwana at the South Pole. This time, however, the ice sheet was centred on southern Africa and South America, as Gondwana had drifted due to plate tectonics.[2]

In the northern hemisphere, a second large continent formed at the end of the Silurian from the collision of Laurentia and Baltica: Euramerica, formerly known in Europe as the Old Red Sandstone continent. This collision gave rise to the **Caledonian** mountain chain, which extended across Scotland (whose Latin name was *Caledonia*), Norway, and Greenland. The new continent was centred on the Tropic of Cancer, resulting in widespread arid climate comparable to the Sahara today. The sands of

---

[2] The American geologist Christopher Scotese offers remarkable animations illustrating plate tectonics in the Phanerozoic: www.scotese.com.

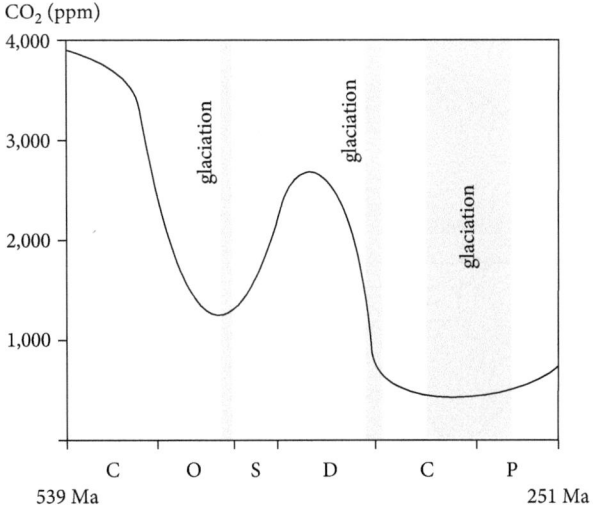

Figure 14.9 Evolution of atmospheric CO_2 content and timing of glacial epochs during the Paleozoic era. C, O, S, D, C, P: Cambrian, Ordovician, Silurian, Devonian, Carboniferous, Permian.

Source: Adapted from Myers (2016).

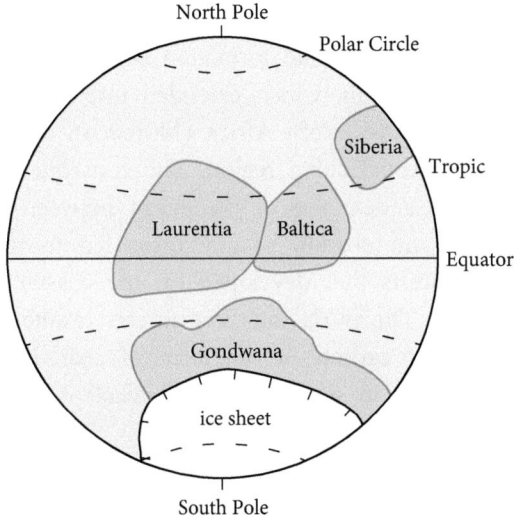

Figure 14.10 Paleogeographic reconstruction in the early Carboniferous. Laurentia and Baltica form a new, large continent in the northern hemisphere: the Old Red Sandstone continent or Euramerica.

this desert, impregnated with iron oxides, were pink or red like today's Saharan sands. They were transformed into red sandstones after induration. These are the Devonian Old Red Sandstones, common in Northern Europe. In contrast, the southern part of the Old Red Sandstone continent was in the equatorial region, where Devonian and Carboniferous forests expanded (Figure 14.10).

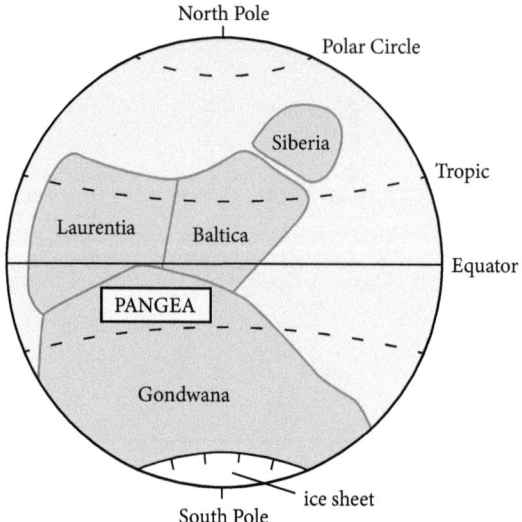

Figure 14.11 Paleogeographic reconstruction in the early Permian. Siberia is now close to Baltica and their collision will soon enlarge Pangea and give rise to the Urals.

During the Carboniferous, the narrow ocean separating Gondwana from Euramerica (the Old Red Sandstone continent) gradually closed by plate tectonics. The resulting collision led to the formation of a massive mountain chain, the Hercynian/Alleghanian chain, which extended into western Europe (Spain, France, Germany), and north-western Africa (Morocco), and along the eastern edge of Laurentia in the Appalachian region. The consequences for the vegetation were drastic. The disappearance of the ocean between the colliding continents and the genesis and erosion of landforms led to the repeated burial of the expansive, lush forests that developed on the coastal plains of the Old Red Sandstone continent. The burial of these forests resulted in the sequestration of a large amount of carbon in the form of coal, hence a decrease in atmospheric CO_2, and then in the greenhouse effect. A new cooling episode occurred.

Things changed during the **Permian** (the last period of the Primary era), by which time the continents had assembled into the supercontinent Pangea, which was centred on the equator and extended across the northern and southern tropics (Figure 14.11). The environments in the centre of this supercontinent were harsh, with a large part of its interior in the tropics far from oceanic sources of moisture. Summers were hot and arid. The global climate became drier, deserts expanded, and the southern ice sheet disappeared. The sands of the subtropics gave rise to the New Red Sandstones. Originally identified in Scotland, these sandstones are also observed in many areas of England. At that time, arrangement of land masses into a single supercontinent instead of several separate continents decreased the total length of coastlines, and so the total area of coastal platforms, which are generally rich in marine life, also decreased.

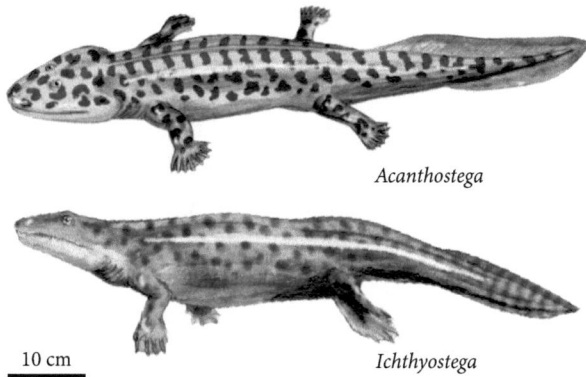

Figure 14.12 Two Devonian tetrapods.
Source: Nobu Tamura.

The first land animals

Arthropods resembling millipedes, protected by their cuticle, ventured onto dry land as early as the Silurian period, about 430 million years ago. During the Devonian and Carboniferous periods, insects proliferated. Fossils of dragonflies with a 70 cm wingspan have been found; these impressive animals buzzed around the Carboniferous marshes and swamps. Insects already made up most of the animal diversity by this time, just as they do today, where the number of species is in the millions.

Vertebrates (animals with an internal skeleton including a vertebral column) appeared at the beginning of the Paleozoic era, as witnessed by many fish fossils found as early as the Cambrian. However, it was not until the end of the Devonian, around 360 Ma, that the first fossils of tetrapod vertebrates (i.e. those with four legs) were found (Figure 14.12). These were **amphibians** resembling large newts. *Acanthostega*, the oldest of these fossils, was certainly still aquatic, as its legs were too weak to support the weight of the animal out of water. Well-preserved fossils reveal internal gills (Coates and Clack, 1991). The slightly more recent *Ichthyostega* had stronger legs, suggesting it spent at least part of its life out of water (Ahlberg et al., 2005). These early terrestrial animals still depended on the aquatic environment for their reproduction, like today's amphibians. Their evolutionary descendants the reptiles, on the other hand, developed the ability to produce hard shells for their eggs such that they could lay them out of the water. This critical adaptation allowed early vertebrates to take up permanent residence on land. The oldest fossil reptiles date back to 314 Ma, in the late Carboniferous (Modesto et al., 2014).

Under threat from volcanoes and other hazards: the great extinctions

The conquest of land was an evolutionary milestone, but it also exposed these terrestrial colonists to new risks. Atmospheric O_2 levels were high enough throughout the

Paleozoic era to maintain an ozone layer, which protected animals and plants from the mutagenic effects of ultraviolet radiation. However, other dangers came from within the Earth, as volcanic eruptions of varying degrees of violence occur periodically on the surface of the globe, releasing huge volumes of toxic gases into the atmosphere. Explosive eruptions of volcanoes such as Pinatubo (1991, Philippines) or Krakatoa (1883, Indonesia) produced clouds of stratospheric aerosols that cooled the global climate for one or two years. They caused many casualties, but no lasting global impact, as they were brief events.

The greatest volcanic threat comes from huge eruptions of a scale never witnessed by humans, evidence of which is found in the geological record. These are the eruptions that produce massive flood basalt flows (or basaltic traps) at new hotspots. They are not explosive, like Pinatubo, but they are nevertheless dangerous because of the enormous volumes of lava and gas emitted over tens to hundreds of thousands of years. The fissure eruption of Laki (1783, Iceland) provided a miniature analogue. Volcanic activity continued for eight months and emitted 15 km^3 of lava. Acid rains affected parts of Europe up to 3,000 km away and led to agricultural disasters and well-documented excess mortality (Garnier, 2011). What then of eruptions such as those that emitted basaltic flows over 1,000 km wide, as in India during the major phase of Deccan volcanism (Keller et al., 2009)? This hotspot volcanism occurred at the end of the Mesozoic era (66 Ma) and likely contributed to the extinction of dinosaurs and other animal and plant groups. In total, 70% of living species are thought to have disappeared. However, this mass extinction was also coeval with the fall of a large meteorite in the Chicxulub region of Yucatan (southern Mexico). Alvarez et al. (1980) suggested that the meteorite was around 10 kilometres in diameter. This spectacular but brief *coup de grâce* probably completed the environmental degradation by sending a huge amount of debris into the atmosphere. The corresponding impact structure is not visible on the surface. It was first recognized as circular geophysical anomalies, and was confirmed by the identification of melted rocks and shocked minerals from boreholes (Kring, 2007).

At the end of the Paleozoic era, a similar, but even more extensive volcanic episode occurred in Siberia. The Siberian traps originally covered an area of about two million square kilometres, four times the size of France (Figure 14.13). Volcanic rocks are still visible at the surface over a quarter of this area; the rest has been eroded or is buried under more recent sediments.

Based on the composition of microscopic fluid inclusions in these basaltic rocks, it has been estimated that the end-Permian Siberian volcanism released 7,000 billion tonnes of sulphur, 4,000 billion tonnes of chlorine, and 5,000 billion tonnes of fluorine into the atmosphere (Black et al., 2012). This volcanism spanned a million years at most, but the main eruptions were restricted to a few periods of intense volcanic activity that lasted around ten thousand years (Schoene et al., 2019). That's a lot of toxic gases emitted for these periods! This pattern of eruptions could help explain why the most severe mass extinction in Earth's history occurred at the end of the Permian. Nevertheless, biodiversity might have already started to decline as a result of the formation of Pangea and the decrease of shallow marine areas, but the volcanic emissions

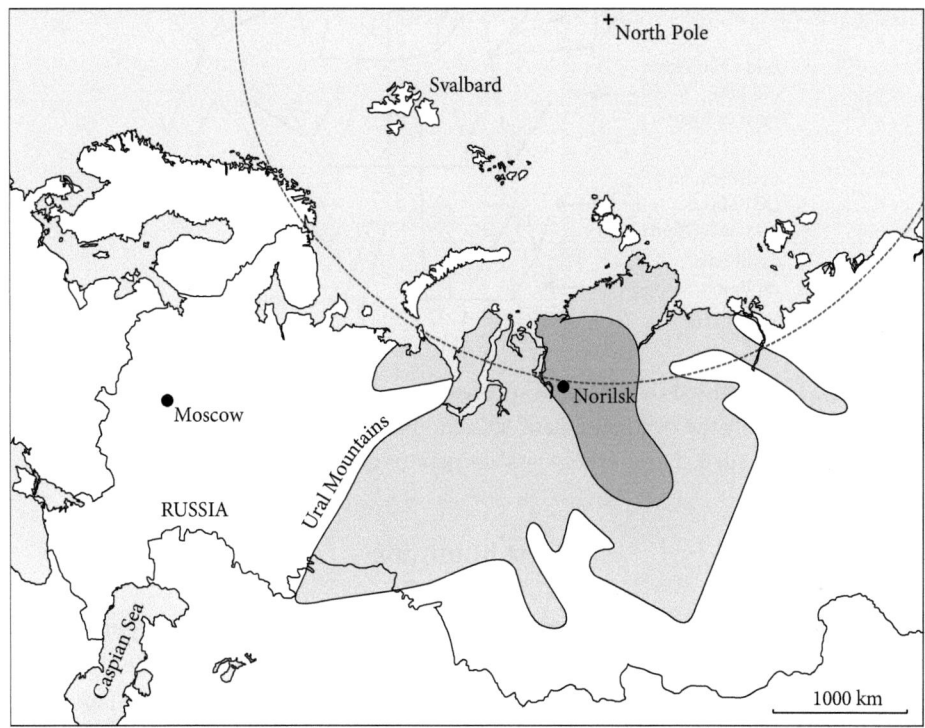

Figure 14.13 Map of Siberian traps in outcrop (dark grey) or subsurface (medium grey).

of the late Permian were catastrophic for both terrestrial and marine fauna and flora, with about 90% of species disappearing at the end of the Paleozoic era.

The biosphere was slow to recover after the end-Permian mass extinction—it took a few million years, and the recovery may have been interrupted by a series of smaller extinctions. Nevertheless, some species were able to survive and proliferate in the ecological niches opened by the catastrophe (Figure 14.14). The early Mesozoic world was very different from the late Paleozoic, and it witnessed the diversification of reptiles and the emergence of dinosaurs, as well as the growth of new types of forest dominated by conifers. Also in the Mesozoic, Pangea fragmented into several continents, while the Atlantic Ocean opened. Mesozoic climatic conditions were generally warm, though the end of the Triassic period saw another mass extinction event associated with flood basalt volcanism around 200 Ma.

With the demise of the dinosaurs at the end of the Mesozoic era, mammals quickly proliferated in the early Cenozoic. Africa, or rather its former promontory which is now Italy, and India collided with Eurasia, causing the formation of the Alps and the Himalayas. Global climate cooled, in part due to the changes in the distribution and elevation of the continents. Modifications of oceanic circulation also played a role, as already mentioned in Chapter 12. In the middle of the Cenozoic era, around 34 Ma, a new glacial epoch began with a permanent ice sheet on Antarctica. Much later in the Cenozoic, the first northern-hemisphere ice sheets formed (see Chapter 12).

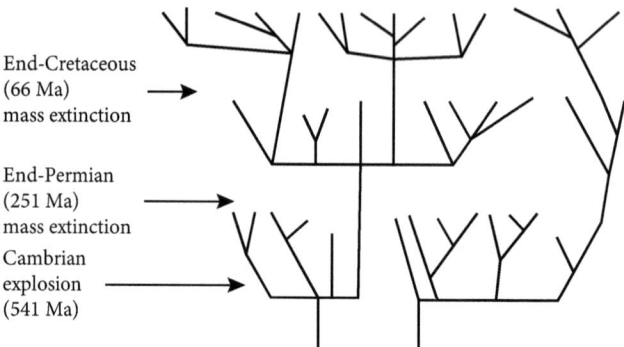

Figure 14.14 Simplified depiction of species diversity since the end of the Precambrian. Only the two main mass extinctions (end-Permian and end-Cretaceous) have been represented. Three other mass extinctions also occurred.

First hominins

A last environmental change is currently underway, related to the ingenuity and explosive growth of one living species, *Homo sapiens*. The human population recently reached 8 billion individuals. The early history of humans began in Africa. The genus *Homo* belongs to the Hominidae family together with the great apes. Chimpanzees and humans are sister species, sharing a common ancestor dating to about 7 million years. From this ancestor, two branches separated, one leading to the chimpanzees and the other leading to a multitude of now extinct species and finally to *Homo sapiens*, collectively called **hominins**. *Sahelanthropus tchadensis*, nicknamed 'Toumaï', discovered in 2001 in northern Chad, may represent this first hominin ancestor (Brunet et al., 2002). This fossil was associated with a rich fauna, suggesting a former landscape of lakes, wetlands, patches of forest, and grasslands.

South Africa delivered the first fossil of another ancestor species in 1924, the Taung Child. Other discoveries led to the definition of a new genus, **Australopithecus**, that lived between 4 and 2 million years ago. A biped, *Australopithecus* was different from both chimpanzees and the most ancient humans (Figure 14.15). The discovery of the skeleton of Lucy, a female *Australopithecus*, in 1974 by the International Afar Research Expedition, founded by the French geologist Maurice Taieb, had a worldwide impact (Johanson and Taieb, 1976). In spite of the uncertainties about the timing of the evolution of hominins, their African origins are firmly established.

The first discoveries of a fossil hominin in Europe occurred in the 19th century. A few bones and a skull very different from the modern human skull, found in Germany near Düsseldorf, were identified by Hermann Schaaffhausen as a primitive human, named 'man of Neanderthal' (Neander Valley in German) in 1857, recognized as a distinct species, *Homo neanderthalensis*, by William King in 1864. The study of ancestral hominins was invigorated by Charles Darwin's publication of the *Origin of Species* in 1859, and the subsequent application of evolution to the origin of man in 1871. New fossils discovered in Belgium definitively established the scientific ideas that

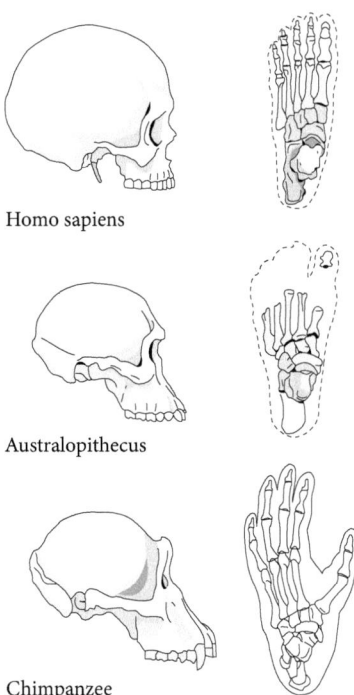

Figure 14.15 Skulls and feet of *Homo sapiens*, *Australopithecus*, and chimpanzee.
Australopithecus had an endocranial volume of c.460 cc, slightly more than the 360 cc of the chimpanzee, but only one third of *Homo sapiens*.
Source: Adapted from McCarthy and Rubidge (2005).

modern man was preceded by other different species. In Africa, Europe, and Asia, other fossils were discovered, with a larger endocranial volume than *Autstralopithecus*, but more primitive features than Neanderthals. They already belonged to the genus *Homo*, and diverse species names were proposed (*H. habilis* and *H. erectus*, among others). Another hominin genus appeared but left no known descendants: the vegetarian *Paranthropus* (Figure 14.16).

The oldest fossil *Homo* used stone tools, mainly hand axes. When did they master fire? How did they use language? These questions remain controversial. Fossil evidence suggests advances in the use of tools, along with geographic specialization, by about half a million years ago. When and where did 'modern' humans (*Homo sapiens* or their direct ancestors) appear? Recent discoveries in Morocco suggest that early *Homo sapiens* lived in Africa around 200,000 years ago (Hublin et al., 2017). They migrated from Africa in likely more than one wave, colonizing Eurasia while continuing to evolve in Africa.

Neanderthals and modern humans likely shared a common ancestor, meaning we are cousins. Other archaic humans, including the Denisovans (named after the Denisova cave in the Altai Mountains), lived in Asia. Advances in gene sequencing have yielded remarkable insight into the history of early humans and their Neanderthal and Denisovan relatives. Svante Pääbo, a Swedish geneticist, was awarded the

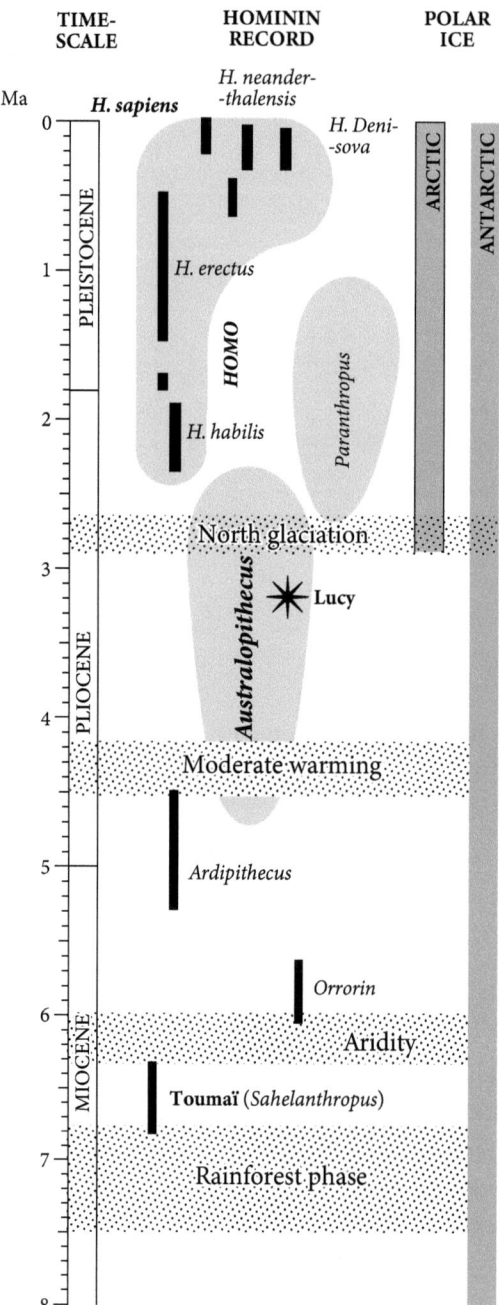

Figure 14.16 **Hominins and climate record.**
Source: Adapted from Bonnefille (2010).

Nobel Prize in Physiology and Medicine in 2022 for his work on the Neanderthal genome (Pääbo, 2014). Neanderthals and Denisovans likely split from a common archaic ancestor c.400,000 years ago. Paleogenetics also reveal that some interbreeding occurred between Neanderthals and other Eurasian humans at c.50,000 or 60,000

BP ('before present') in the Middle East. Similarly, Denisovans appear to have interbred with both modern humans and Neanderthals. Thus, despite these species having become extinct, genes of Neanderthals and Denisovans are recognized in non-African modern humans, which preserved 2% of genes from Neanderthals and up to 6% from Denisovans. Other studies have shown that the genetic variability of African humans is much larger than that of non-African humans, confirming. the 'Out of Africa' hypothesis for modern humans.

The expansion of *Homo sapiens*

Why did Neanderthals and Denisovans disappear? *Homo sapiens* arrived in Europe during a brief interval of warming about 50,000 years that preceded the descent into the Glacial Maximum 21,000 years ago (Figure 14.17). Neanderthals were robust, with an average height of 1.70 m and an average weight about one third more of that of a modern *H. Sapiens*. Thus, they are generally considered as well adapted to a cold climate. Nevertheless, the newcomer *Homo sapiens* likely had many cultural and technological advantages. *Homo sapiens* left wonderful paintings in caves in France and Spain; the oldest one is Chauvet cave in southern France, dated at 35,000 BP. Neanderthals never left any similar productions. Slimak et al. (2022) reported the first evidence of the arrival of *Homo sapiens* in south-eastern France at c.53,000 BP. Neanderthals appear to have disappeared from most areas of Europe at around 40,000 BP and to have found their last refuge in southern Spain, where they eventually succumbed to extinction around 35,000 BP. Thus, the decline and extinction of Neanderthals are, directly or not, due to *Homo sapiens*.

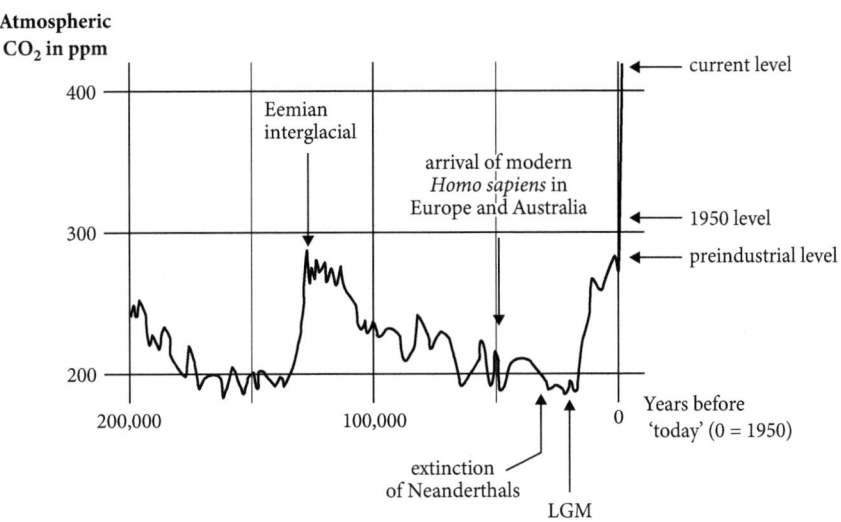

Figure 14.17 Variations of atmospheric CO_2 content and expansion of *Homo sapiens*. LGM: Last Glacial Maximum.

Homo sapiens was able to colonize fairly high latitudes just before the LGM, as evidenced by the Sungir tomb near Moscow, which is dated to c.29,000 BP. This site is unique for the abundance of ornaments, including thousands of ivory pearls worn by the buried individuals. The timing of the arrival of *Homo sapiens* in America remains heavily debated. The earliest humans came from Asia, most likely traversing the region that is now the Bering Strait but was emergent for several thousand years before and after the LGM. Humans had spread across North and South America, as indicated by the presence of human remains in southern Chile dated to c.13,000 BP. Human arrival closely coincided with the disappearance of the southern American megafauna. Was this disappearance the result of hunting, climate change, or both? Similar questions arise about the arrival of *Homo sapiens* in New Guinea and Australia, which happened at c.50,000 BP. Whereas multiple *Homo* species had coexisted in the past, the worldwide expansion of *Homo sapiens* led to the disappearance of the other human species and possibly of other living beings.

The end of the story is the technical and social evolution of *Homo sapiens*. Agriculture began in different parts of the world as early as 10,000 BP. It resulted in significant demographic and social changes. Very recently, industrial development based on the use of fossil carbon (coal and oil) has increased atmospheric CO_2 levels to over twice what they were during the LGM and to their highest levels for 3 million years. The climatic consequences of rapidly increasing CO_2 are becoming more apparent as the oceans warm, sea and glacial ice recedes, and weather events become more extreme.

The end of the Precambrian was followed by an explosive diversification of animal life in the oceans. By the middle of the Paleozoic era, plants and then animals had colonized the continents. Plants fundamentally changed the global carbon cycle, not just by directly utilizing atmospheric CO_2, but also by increasing the chemical breakdown of rocks and creating deep soils. The history of the Paleozoic era illustrates how the Earth's climate depends on both biological factors that affect the carbon cycle and geographical factors linked to the position of the continents, itself governed by plate tectonics. Late in the Cenozoic era, hominins diverged from chimpanzees. Hominin species diversified across Africa, and *Homo* species migrated out of Africa. Of these, only modern humans remain, though many of us carry genetic remnants of our Neanderthal and Denisovan cousins. Now, as the result of burning fossil fuels, humans are modifying the composition of the atmosphere and the global climate.

Epilogue

We have now reached the end of our long journey of over four billion years since the formation of the Earth. Let us recall the most striking facts. Life appeared early on, at the bottom of the oceans, and probably in the vicinity of hydrothermal sources. Then, a little over three billion years ago, the internal dynamics of the mantle triggered the onset of plate tectonics, which led to the growth of the continents and was particularly active at the end of the Archean eon. This process continues today, constantly reshaping the oceans and continents. Earthquakes witness this internal activity, sometimes devastatingly.

After the slow and gradual transformations of the Archean, the beginning of the Proterozoic was a period of major changes. Processes originating from the interior of the planet interacted with processes occurring on the surface. The production of continental crust exposed to weathering by meteoric waters provided a carbon sink that decreased the CO_2 greenhouse effect and cooled the climate. Life played its part: free oxygen, a metabolic byproduct of bacterial photosynthesis, changed the composition of the oceans and the atmosphere. The colour of the land surface was irreversibly altered due to the formation of red soils, which still characterize intertropical landscapes.

Oxygen, which was toxic to many organisms that had evolved in anoxic environments, presumably played a role in one of the great transitions in evolution: the appearance of eukaryotes, with their much more complex cells than bacteria and archea. The eukaryotic cell is a chimera that resulted from the symbiotic association of different microbes. Multiple eukaryotes evolved into multi-cellular organisms, giving rise to plants and animals.

On multiple occasions, life has disturbed the environment to the extent that it had consequences for its own evolution. This is a fascinating fact for us to remember, as we humans have now reached the point where we have, perhaps irreversibly, altered our environment. What will the future be for humans? As we are beginning to understand, the current environmental change is a human problem.

In the distant future, perhaps several hundred million years from now, the Earth's internal engine will began to weaken because of steady, inexorable cooling. Plate tectonics will progressively slow down. Oceanic ridges will no longer produce new oceanic crust because the melting temperature of the mantle will no longer be reached. By a billion years from now, the geological world will effectively freeze in place. Mars, which is a smaller planet and therefore cooled more quickly than the Earth, may offer an analogue of Earth's future. In the even more distant future, perhaps five billion years from now, the Sun will enter its final phase. From a medium-sized star, it will evolve into a red giant, consuming our planet in the process.

Will this mean the end of Earth's history? Yes, certainly; but not the end of all stories, since new exoplanets are discovered every day, revolving around other suns. What do they look like? Is there life on them? We don't know yet, but their existence offers many possibilities for both exploration and the imagination.

References

1. Formation of the Earth

Allègre, C. et al. (1995) Age of the Earth. *Geochimica Cosmochimica Acta* 10, 230–237.
Amelin, Y. et al. (2010) U–Pb chronology of the Solar System's oldest solids with variable^{238} U/^{235} U. *Earth and Planetary Science Letters* 300, 343–350.
Borg, I. E. and Carlson, R. (2023) The evolving chronology of Moon formation. *Annual Review of Earth and planetary Sciences* 51, 25–52.
Cameron, A. G. W. and Ward, W. R. (1976) The Origin of the Moon. *Lunar Science* VII, 120–122.
Canup, R. M. and Righter, K. (eds) (2000) *Origin of Earth and Moon*, University of Arizona, Tucson, 133–144.
Hartman, W. and Davis, D. (1975) Satellite-Sized Planetesimals and Lunar Origin. *Icarus* 24, 504–505.
Kleine, T. et al. (2009) Hf-W chronometry and the accretion and early evolution of asteroids and terrestrial planets. *Geochimica Cosmochimica Acta* 73, 5150–5188.
Pahlevan, K. and Stevenson, D. (2007) Equilibration in the aftermath of the lunar-forming giant impact. *Earth Planetary Science Letters* 262, 438–449.
Patterson, C. (1956) Age of meteorites and the Earth. *Geochimica Cosmochimica Acta* 10, 230–237.
Shi, C. et al. (2013) Formation of an interconnected network of iron melt at Earth's lower mantle conditions. *Nature Geoscience* 6, 971–975.
Thiemens, M. et al. (2019) Early Moon formation inferred from Hf-W systematics. *Nature Geoscience* 12, 696–700.

2. The mysteries of the first 500 millions years

Albarède, F. and Blichert-Toft, J. (2007) The split fate of the early Earth, Mars, Venus and Moon. *Geoscience Proceedings* 339, 917–927.
Belousova, E. A. et al. (2010) The growth of the continental crust: contraints from zircon–Hf isotope data. *Lithos* 119, 457–466.
Bernadet, J. et al. (2025) Making continental crust on water-bearing terrestrial planets. *Science Advances* 11, eads6746.
Blichert-Toft, J. and Albarède, F. (1997) The Lu–Hf isotope geochemistry of chondrites and the evolution of the mantle-crust system. *Earth Planetary Science Letters* 148, 243–258.
Borisova, A. Y. and Nédélec, A. (2021). A simple recipe for making the first continental crust. *Eos, Transactions American Geophysical Union* 102, 36–41.
Cavosie, A. et al. (2004) Internal zoning and U–Th–Pb chemistry of Jack Hills detrital zircons: a mineral record of early Archean to Mesoproterozoic (4348–1576 Ma) magmatism. *Precambrian Research* 135, 251–279.
Compston, W. and Pidgeon, R. (1986) Jack Hills, a further occurrence of very old zircons in Western Australia. *Nature* 321, 766–769.
Connelly, J. et al. (2011) The age of lunar ferroan anorthosite 60025 with implications for the interpretation of lunar chronology and the magma ocean model. *42nd Lunar and Planetary Science Conference*, 1171.
Froude, D. O. et al. (1983) Iron microprobe identification of 4,100–4,200 Myr-old terrestrial zircons. *Nature* 616, 175–178.
Garde, A. A. et al. (2012) Searching for giant, ancient impact structure. *Earth and Planetary Science Letters* 337–338, 197–210.
Gomes, R. et al. (2005) Origin of the cataclysmic Late Heavy Bombardment period of the terrestrial planets. *Nature* 453, 466–469.
Helz, R. (2009) Processes active in mafic magma chambers: the example of Kilauea Iki lava lake, Hawaii. *Lithos* 111, 37–46.
Holden, P. et al. (2009) Mass-spectrometric mining of Hadean zircons by automated SHRIMP multi-collector and single-collector U/Pb zircon age dating: the first 100,000 grains. *International Journal of Mass Spectrometry* 286, 53–63.
Hopkins, M. D. and Mojzsis, M. D. (2015) A protracted timeline for lunar bombardment from mineral chemistry, Ti thermometry and U–Pb geochronology of Apollo 14 melt breccia zircons. *Contributions to Mineralogy and Petrology* 169, 30.
Liu, D. et al. (2012) Comparative zircon U-Pb geochronology of impact melt breccias from Apollo 12 and lunar meteorite SaU169, and implications for the age of the Imbrium impact. *Earth and Planetary Science Letters* 319–320, 277–286.

Nemchin, A. et al. (2009) Timing of crystallization of the lunar magma ocean constrained by the oldest zircon. *Nature Geoscience* 2, 133–136.

Norman, M. and Nemchin, A. (2014) A 4.2-billion-year old impact basin on the Moon: U–Pb dating of zirconolite and apatite in lunar melt rock 67955. *Earth and Planetary Science Letters* 388, 387–398.

Reynard, B. et al. (2022) Primordial serpentinized crust on the early Earth. *Physics of the Earth and Planetary Interiors* 332, 106936.

Whitehouse, M. J. et al. (2017) What can Hadean detrital zircon really tell us? A critical evaluation of their geochronology with implications for the interpretation of oxygen and hafnium isotopes. *Gondwana Research* 51, 78–91.

Wielicki, M. M. et al. (2012) Geochemical signatures and magmatic stability of terrestrial impact produced zircons. *Earth and Planetary Science Letters* 321–322, 20–31.

Wilde, S. et al. (2001) Evidence from detrital zircons for the existence of continental crust and oceans on the Earth 4.4 Gyr ago. *Nature* 409, 175–178.

Wilhems, D. E. (1987) *The Geological history of the Moon.* US Geological Survey Professional Paper 1348.

3. Origin of the atmosphere and ocean

Carr, M. H. and Head, J. W. (2003) Oceans on Mars: an assessment of the observational evidence and possible fate. *Journal of Geophysical Research (Planets)* E5, 5042.

Elkins-Tanton, L. T. (2008) Linked magma ocean solidification and atmospheric growth for Earth and Mars. *Earth and Planetary Science Letters* 271, 181–191.

Encrenaz, T. (2010) *Searching for Water in the Universe.* Springer, New York.

Forget, F. et al. (2007) *Planet Mars: Story of Another World.* Springer Praxis Books.

Gargaud, M. (2005) *Lectures in Astrobiology.* Springer.

Hamilton, V. E. et al. (2019) Evidence for widespread hydrated minerals on asteroid (101955) Bennu. *Nature Astronomy* 3, 332–340.

Jakosky, B. M. (2021) Atmospheric loss to space and the history of water on Mars. *Annual Review of Earth and Planetary Sciences* 49, 71–93.

Lecar, M. et al. (2006) On the location of the snow line in a protoplanetary disk. *The Astrophysical Journal*, 640, 1115–1129.

Marty, B. (2020) Origins and early evolution of the atmosphere and the oceans. *Geochemical Perspectives* 9, 136–313.

McCord, T. B. et al. (2022) Ceres, a wet planet: the view after Dawn. *Geochemistry* 82, 125745.

Turbet, M. and Forget, F. (2019) The paradoxes of the Late Hesperian Mars ocean. *Scientific Reports* 9, 5717.

Wordsworth, R. D. and Pierrehumbert, R. T. (2013) Water loss from terrestrial planets with CO_2-rich atmospheres. *The Astrophysical Journal* 778, 154–173.

Zahnle, K. et al. (2010) Earth's earliest atmospheres. In *Cold Spring Harbor Perspectives in biology*, 2010.

4. The oldest rocks on Earth

Bowring, S. and Williams, I. (1999) Priscoan (4.00–4.03 Ga) orthogneisses from northwestern Canada. *Contributions to Mineralogy and Petrology* 134, 3–16.

de Wit, J. et al. (2011) Geology and tectonostratigraphy of the Onverwacht Suite, Barberton Greenstone Belt, South Africa. *Precambrian Research* 186, 1–27.

Lay, T. et al. (2008) Core–mantle boundary heat flow. *Nature Geoscience* 1, 26–32.

Martin, H. (1986) Effect of steeper Archean geothermal gradient on geochemistry of subduction-zone magmas. *Geology* 14, 753–756.

Nédélec, A. et al. (2012) TTGs in the making: natural evidence from Inyoni shear zone (Barberton, South Africa). *Lithos* 153, 25–38.

Nimmo, F. (2022) Formation, composition and evolution of the Earth's core. In *Oxford Research Encyclopedia of Planetary Science.*

Reimink, J. R. et al. (2016) The birth of a cratonic nucleus: lithogeochemical evolution of the 4.02–2.94 Ga Acasta Gneiss Complex. *Precambrian Research* 281, 453–472.

Szilas, K. et al. (2015) The petrogenesis of ultramafic rocks in the >3.7 Ga Isua supracrustal belt, southern West Greenland: geochemical evidence for two distinct magmatic cumulate trends. *Gondwana Research* 28, 565–580.

Viljoen, M. J. and Viljoen, R. P. (1969) The geology and geochemistry of the Lower Ultramafic Unit of the Onverwacht Group and a proposed new class of igneous rocks. *Upper mantle project. Special Publication of the Geological Society of South Africa* 2, 55–85.

5. Earth's internal cooling through time

Bouhallier, H. et al. (1995) Strain patterns in Archean dome-and-basin structures: the Dharwar craton. *Earth and Planetary Science Letters* 135, 57–75.

Chown, E. H. et al. (1992) Tectonic evolution of the northern volcanic zone, Abitibi belt, Quebec. *Canadian Journal of Earth Sciences* 29, 2211–2225.

Condie, K. C. (1994) *Archean Crustal Evolution*. Elsevier, Amsterdam.

French, S. W. and Romanowicz, B. (2015) Broad plumes rooted at the base of the Earth's mantle beneath major hotspots. *Nature* 525, 95–99.

Herzberg, C. (1992) Depth and degree of melting of komatiites. *Journal of Geophysical Research (B)* 97, 4521–4540.

Juteau, T. and Maury, R. (1999) *The Oceanic Crust, from Accretion to Mantle Recycling*. Springer.

Lee, C.-T. A. et al. (2010) Upside-down differentiation and generation of a 'primordial' lower mantle, *Nature*, 463, 930–933.

Moyen, J.-F. and Martin, H. (2012) Forty years of TTG research, *Lithos*, 148, 312–336.

Nédélec, A. et al. (2017) The Hadean–Archean transition at 4 Ga: from magma trapping in the mantle to volcanic resurfacing of the Earth. *Terra Nova* 29, 218–223.

Self, S. et al. (2008) Correlation of the Deccan and Rajahmundry trap lavas: are these the longest and largest lava flows on Earth? *Journal of Volcanology and Geothermal Research* 172, 3–19.

Steinberger, B. and Anstretter, M. (2006) Conduit diameter and buoyant rising speed of mantle plumes: implications for the motion of hot spots and shape of plume conduits. *Geochemistry Geophysics Geosystems* 7, Q11018.

Wyllie, P. J. (1984) Sources of granitoid magmas at convergent plate boundaries. *Physics of Earth and Planetary Interiors* 35, 12–18.

Zhao, D. et al. (2013) Global mantle heterogeneity and its influence on teleseismic regional tomography. *Gondwana Research* 23, 595–616.

6. The Archean ocean

Bekker, A. et al. (2010) Iron formation: the sedimentary product of a complex interplay among mantle, tectonic, oceanic and biospheric processes. *Economic Geology* 105, 467–508.

de Wit, M. J. and Furnes, H. (2016) 3.5 Ga hydrothermal fields and diamictites in the Barberton greenstone belt—Paleoarchean crust in cold environments. *Science Advances*, 2, e1500368.

Eriksson, K. A. and Simpson, E. L. (2000) Quantifying the oldest tidal record: the 3.2 Ga Moodies Group, Barberton greenstone belt, South Africa. *Geology*, 28, 831–834.

Konhauser, K. O. et al. (2002) Could bacteria have formed the Precambrian banded iron formations? *Geology* 30, 1079–1082.

Ledevin, M. et al. (2014) Silica precipitation triggered by clastic sedimentation in the Archean: new petrographic evidence from cherts of the Kromberg type section, South Africa. *Precambrian Research* 255, 316–334.

Rossignol, C. et al. (2023) Neoarchaean environments associated with the emplacement of a large igneous province: insights from the Carajás Basin, Amazonia Craton. *Journal of South American Earth Sciences* 130, 104574.

Shibuya, T. et al. (2007) Middle Archean ocean ridge hydrothermal metamorphism and alteration recorded in the Cleaverville area, Pilbara craton, Western Australia. *Journal of Metamorphic Geology* 25, 751–767.

Shibuya, T. et al. (2010) Highly alkaline, high-temperature hydrothermal fluids in the early Archean ocean. *Precambrian Research* 182, 230–238.

Shibuya, T. et al. (2013) Decrease of seawater CO_2 concentration in the Late Archean: an implication from 2.6 Ga seafloor hydrothermal alteration. *Precambrian Research* 236, 59–64.

Simpson, E. et al. (2012) 3.2 Ga eolian deposits from the Moodies Group, Barberton Greenstone Belt, South Africa: implications for the origin of first-cycle quartz sandstones. *Precambrian Research* 214–215, 185–191.

Tang, M. et al. (2016) Archean upper crust transition from mafic to felsic marks the onset of plate tectonics. *Science* 351, 372–375.

Trendall, A. F. et al. (2004) SHRIMP zircon ages constraining the depositional chronology of the Hamersley Group, Western Australia. *Australian Journal of Earth Sciences* 51, 621–644.

7. Origin of life

Allwood, A. C. et al. (2007) Stratigraphy and facies of the 3.43 Ga Strelley Pool chert in the southwestern North Pole dome, Pilbara craton, Western Australia. *Precambrian Research* 158, 198–227.

Altwegg, K. et al. (2007) Organics in comet 67P—a first comparative analysis of mass spectra from ROSINA-DFMS, COSAC and Ptolemy. *Monthly Notices of the Royal Astronomical Society* 469, S130–S141.

Baker, B. J. et al. (2020) Diversity, ecology and evolution of Archaea. *Nature Microbiology* 5, 887–900.

Brasier, M. et al. (2002). Questioning the evidence for Earth's oldest fossils. *Nature* 416, 76–81.

Buick, R. et al. (1984) Carbonaceous filaments from North Pole, Western Australia: are they fossil bacteria in Archean stromatolites? *Precambrian Research* 24, 157–172.

Catling, D. C. and Zahnle, K. J. (2020) The Archean atmosphere. *Science Advances* 6, eaax1420.

Dick, G. J. (2019) The microbiomes of deep-sea hydrothermal vents: distributed globally, shaped locally. *Nature reviews Microbiology* 17, 271–283.

Dodd, M. S. et al. (2017) Evidence for early life in Earth's oldest hydrothermal vent precipitates. *Nature* 543, 60–64.

Eager-Nash, J. K. et al. (2023) 3D climate simulations of the Archean find that methane has a strong cooling effect at high concentrations. *Journal of Geophysical Research (Atmospheres)* 128, e2022JD037544.

Elsila, J. E. et al. (2009) Cometary glycine detected in samples by Stardust. *Meteoritics and Planetary Science*, 44, 1323–1330.

Forterre, P. (2016) *Microbes from Hell*. University of Chicago Press.

Franklin, R. and Gosling, R. (1953) Molecular configuration in sodium thymonucleate. *Nature*, 171, 740–741.

Gilbert, G. (1986) The RNA world. *Nature* 319, 618.

Hickman-Lewis, K. et al. (2016) Carbonaceous microstructures from sedimentary laminated chert within the 3.46 Ga Apex basalt, Chinaman Creek locality, Pilbara, Western Australia. *Precambrian Research* 278, 161–178.

Homann, M. (2019) Earliest life on Earth: evidence from the Barberton Greenstone Belt, South Africa. *Earth-Science Reviews* 196, 102888.

Imachi, H. et al. (2020) Isolation of an archaeon at the prokaryote–eukaryote interface. *Nature* 577, 519–525.

Kelley, D. S. et al. (2001) An off-axis hydrothermal vent field near the Mid-Atlantic Ridge at 30°N. *Nature*, 412, 145–149.

Luisi, P. (2016) *The Emergence of Life: From Chemical Origin to Synthetic Biology*. Cambridge University Press, second edn.

Martins, Z. et al. (2008) Extraterrestrial nucleobases in the Murchison meteorite. *Earth and Planetary Sciences Letters* 270, 130–136.

Maurel, M. C. and Leclerc, F. (2016) From foundations stones to life: concepts and results. *Elements* 12, 6.

Miller, S. L. (1953) A production of amino acids under possible primitive Earth conditions, *Science*, 117, 528–529.

Mojzsis, S. J. et al. (1996) Evidence for life on Earth before 3800 million years ago. *Nature* 384, 55–59.

Rasmussen, B. et al. (2021) Apatite nanoparticles in 3.46–2.46 Ga iron formations: evidence for phosphorus-rich hydrothermal plumes on early Earth. *Geology* 49, 647–651.

Schopf, J. W. and Kudryavtsev, A. B. (2012) Biogenicity of Earth's earliest fossils: a resolution of the controversy. *Gondwana Research* 222, 761–771.

Schwartz, A. W. et al. (1982) Prebiotic adenine synthesis via HCN oligomerization in ice. *BioSystems*, 15, 101–193.

Ueno, Y. et al. (2001) Carbon isotopic signatures of individual Archean microfossils from Western Australia. *International Geology Review* 43, 196–212.

Watson, J. and Crick, F. (1953) Molecular structure of nucleic acids; a structure for desoxyribose nucleic acid. *Nature*, 171, 737–738.

Woese, C. R. et al. (1990) Towards a natural system of organisms: proposal for the domains Archaea, Bacteria and Eukaryota, *Proceedings of the National Academy of Sciences of the United States of America*, 87, 4576–4579.

Woese, C. R. and Fox, G. E. (1977). Phylogenetic structure of the prokaryotic domain: the primary kingdoms. *Proceedings of the National Academy of Sciences of the United States of America* 74, 5088–5090.

Wynn-Williams, D. D. et al. (2002) Pigmentation as a survival strategy for ancient and modern photosynthetic microbes under high ultraviolet stress on planetary surfaces. *International Journal of Astrobiology* 1, 39–49.

8. Everything changes on Earth

Antonio, P. et al. (2017) Turmoil before the boring billion: paleomagnetism of the 1880–1860 ma Uatumã event in the Amazonian craton. *Gondwana Research* 49, 106–129.

Bekker, A. et al. (2014) Iron formations: their origins and implications for ancient seawater chemistry. In: *Treatise on Geochemistry* (second edn), Elsevier, 9.18, 561–628.

Eriksson, P. G. (1999) Sea level changes and the continental freeboard concept: general principles and application to the Precambrian. *Precambrian Research* 97, 143–154.

Fabre, S. et al. (2011) Iron and sulphur isotopes from the Carajas mining province (Para, Brazil): implications for the oxidation of the ocean and the atmosphere across the Archean–Proterozoic transition. *Precambrian Research* 289, 124–139.

Farquhar, J. et al. (2000) Atmospheric influence of Earth's earliest sulfur cycle. *Science* 289, 756–758.

Hao, J. et al. (2017) A model for late Archean chemical weathering and world average river water. *Earth and Planetary Science Letters* 457, 191–203.

Lantink, M. L. et al. (2018) Fe isotopes of a 2.4 Ga hematite-rich IF constrain marine redox conditions around the GOE. *Precambrian Research* 305, 218–235.

Melezhik, V. A. et al. (2009) Petroleum surface oil seeps from a Paleoproterozoic petrified giant oil field. *Terra Nova* 21, 119–126.

Melezhik, V. A. et al. (2013) *Reading the Archive of Earth's Oxygenation*. Springer.

Mossman, D. J. et al. (2005) Black shales, organic matter, ore genesis and hydrocarbon generation in the Paleoproterozoic Franceville Series, Gabon. *Precambrian Research* 137, 253–272.

Prave, A. R. et al. (2022) The grandest of them all: the Lomagundi–Jatuli event and Earth's oxygenation. *Journal of the Geological Society*, vol. 179.

Rasmussen, B. et al. (2013) Correlation of Paleoproterozoic glaciations based on U–Pb zircon ages for tuff beds in the Transvaal and Huronian Supergroups. *Earth and Planetary Science Letters* 382, 173–180.

Strauss, H. et al. (2013) Enhanced accumulation of organic matter: the Shunga event. In Melezhik, V. A. et al. *Reading the Archive of Earth's Oxygenation*. Frontiers in earth Sciences. Springer, Berlin, pp. 1195–1273.

Taylor, S. R. and McLennan, S. M. (1985) *The Continental Crust: Its Composition and Evolution*. Blackwell.

9. Columbia: the first supercontinent

Artemieva, I. (2007) Dynamic topography of the East European craton: shedding light upon lithospheric structure, composition and mantle dynamics. *Global and Planetary Change* 58, 411–434.

Brunhes, B. (1906) Recherches sur la direction d'aimantation des roches volcaniques. *Journal de Physique Théorique et Appliquée* 5, 705–724.

Channel, J. E. T. et al. (2004) *Timescales of the Paleomagnetic Field*. AGU Geophysical Monograph Series 145.

Fabre, S. et al. (2021) Harsh or balmy weathering conditions onto the first continent surface? *Precambrian Research* 353, 106025.

Hoffman, P. F. (1988) United plates of America, the birth of a craton: Early Proterozoic assembly and growth of Laurentia. *Annual Review of Earth and Planetary Sciences* 16, 543–603.

Jefferson, C. W. et al. (2007) Unconformity-associated uranium deposits of the Athabasca Basin, Saskatchewan and Alberta. In: Goodfellow, W. D. (ed.), *Mineral Deposits of Canada: A Synthesis of Major Deposit-Types, District Metallogeny, the Evolution of Geological Provinces, and Exploration Methods*. Geological Association of Canada, Special Publication 5, 273–305.

Lenardic, A. et al. (2011) Continents, supercontinents, mantle thermal mixing, and mantle thermal isolation: theory, numerical simulations, and laboratory experiments. *Geochemistry, Geophysics, Geosystems* 12, Q10016.

Meert, J. G. and Santosh, M. (2022) The Columbia supercontinent: retrospective, status, and a statistical assessment of paleomagnetic poles used in reconstructions. *Gondwana Research* 110, 143–164.

Simpson, G. et al. (2004) Sedimentary dynamics of Precambrian eolianites. In: Eriksson, B. K. et al. (eds) *The Precambrian Earth: Tempos and Events*, Elsevier, 642–657.

Zhang, S. et al. (2012) Pre-Rodinia supercontinent Nuna shaping up: a global synthesis with new paleomagnetic results from North China. *Earth and Planetary Science Letters* 353–354, 145–155.

10. Diversification of life in the Proterozoic

Altermann, W. et al. (2004) Evolving life and its effect on Precambrian sedimentation. In: Eriksson B. K. et al. (eds), *The Precambrian Earth: Tempos and Events*, Elsevier, see pp. 539–545.

Becker-Kerber, B. et al. (2017) Ecological interactions in *Cloudina* from the Ediacarian of Brazil: implications for the rise of animal biomineralization. *Scientific Reports*, 7, 5482–5493.

Butterfield, N. J. (2015) Early evolution of the Eukaryota. *Palaeontology* 58, 5–17.

Dutkiewicz, A. et al. (2007) Oil and its biomarkers associated with the Paleoproterozoic Oklo natural fission reactors, Gabon. *Chemical Geology*, 244, 130–154.

El Albani, A. et al. (2010) Large colonial organisms with coordinated growth in oxygenated environments 2.1 Gyr ago. *Nature* 466, 100–104.

El Albani, A. et al. (2019) Large organism motility in an oxygenated shallow-marine environment 2.1 billion years ago. *Proceedings of the National Academy of Sciences* 116, 3431–3436.

Gauthier-Lafaye, F. and Weber, F. (1989) The Francevillian (Lower Proterozoic) uranium ore deposits of Gabon. *Economic Geology* 84, 2267–2285.

Gibson, T. M. et al. (2018) Precise age of *Bangiomorpha pubescens* dates the origin of eukaryotic photosynthesis. *Geology* 46, 135–138.

Gray, M. W. and Doolittle, W. F. (1982) Has the endosymbiont hypothesis been proven? *Microbiological Reviews* 46, 1–42.

Hoshino, Y. et al. (2017) Cryogenian evolution of stigmasteroid biosynthesis. *Science Advances* 3, doi 10.1126/sciadv.1700887.

Laflamme, M. et al. (2013) The end of the Ediacara biota: extinction, biotic replacement, or Cheshire Cat? *Gondwana Research* 23, 558–573.

Margulis (Sagan), L. (1967) On the origin of mitosing cells. *Journal of Theoretical Biology* 14, 225–274.

Maynard Smith, J. and Szathmary, E. (2000) *The Origins of life: from the Birth of Life to the Origin of Language*, Oxford University Press.

Moreira, D. and López-García, P. (1998) Symbiosis between methanogenic archaea and δ-proteobacteria as the origin of eukaryotes: the syntrophic hypothesis. *Journal of Molecular Evolution* 47, 517–530.

Moussavou, M. et al. (2015) Multicellular consortia preserved in biogenic ductile-plastic nodules of Okondja Basin (Gabon) by 2.1 Ga. *Journal of Geology and Geosciences* 4, 100095.

Neuilly, M. et al. (1972) Sur l'existence dans un passé reculé d'une réaction en chaîne naturelle de fission, dans le gisement d'uranium d'Oklo (Gabon). *Comptes rendus de l'Académie des Sciences de Paris* 275, 1847–1849.

Yin, L. M. (1997) Acanthomorphic acritarchs from Meso-Neoproterozoic shales of the Ruyan Group, Shanxi, China. *Review of Palaeobotany and Palynology* 98, 15–25.

11. From Columbia to Gondwana: the ballet of the continents

Baragar, W. R. A. et al. (1996) Longitudinal petrochemical variation in the Mackenzie dyke swarm, Northwestern Canadian Shield. *Journal of Petrology* 37, 317–359.

Boullier, A. M. (1991) The Pan-African trans-Saharan belt in the Hoggar shield (Algeria, Mali, Niger): a review. In: Dallmeyer, R. D. and Lécorché, I. P. (eds), *The West-African Orogens and Circum-Atlantic Correlatives*, Springer.

Bradley, D. C. (2008) Passive margins through Earth history. *Earth-Science Reviews* 91, 1–26.

Brito-Neves, B. B. et al. (1999) From Rodinia to Western Gondwana: an approach to the Brasiliano–Pan-African cycle and orogenic collage, *Episodes* 22, 3, 155–166.

Durand-Delga, M. and Seidl, J. (2007) Eduard Suess (1831–1914) and his global geological synthesis 'Das Antlitz der Erde', Comptes Rendus Geoscience 339, 1, 85–99.

Evans, D. and Mitchell, R. (2011) Assembly and breakup of the core of Paleoproterozoic–Mesoproterozoic supercontinent Nuna. *Geology* 39, 443–446.

Kennedy, W. Q. (1964) The structural differentiation of Africa in the Pan-African (± 500 m.y.) tectonic episode. *Annual Report of the Research Institute of African Geology*, University of Leeds, 8, 48–49.

Li, Z. X. et al. (2013) Neoproterozoic glaciations in a revised global paleogeography from the breakup of Rodinia to the assembly of Gondwanaland. *Sedimentary Geology* 294, 219–232.

Meert, J. G. and Torsvik, T. H. (2003) The making and unmaking of a supercontinent: Rodinia revisited. *Tectonophysics* 375, 261–268.

Nédélec, A. et al. (2016) A-type stratoid granites of Madagascar revisited: Age, source and links with the breakup of Rodinia. *Precambrian Research* 280, 231–248.

Wegener, A. (1919) *Die Entstehung der Kontinente und Ozeane*, Vieweg und Sohn, Braunschweig.

Wilson, J. T. (1966) Did the Atlantic close and then re-open? *Nature* 211, 676–681.

12. The Snowball Earth

Bjørlykke, K. et al. (1967) The Eocambrian Reusch moraine at Bigganjargga and the geology around Varangerfjord Northern Norway. *Studies on the latest Precambrian and Eocambrian Rocks in Norway no. 4, Norges geologiske undersøkelse* 251, 18–44.

Donnadieu, Y. et al. (2004) A 'Snowball Earth' climate triggered by continental break-up through changes in runoff. *Nature* 428, 303–306.

Emiliani, C. (1955) Pleistocene temperatures. *The Journal of Geology* 63, 538–578.

EPICA (2004) Eight glacial cycles from an Antarctic ice core. *Nature* 429, 623–628.

Goddéris, Y. et al. (2003) The Sturtian 'snowball' glaciation: fire and ice. *Earth and Planetary Science Letters* 211, 1–12.

Halverson, G. P. et al. (2011) Fe isotope and trace element geochemistry of the Neoproterozoic syn-glacial Rapitan iron-formations. *Earth and Planetary Research Letters* 309, 100–112.

Harland, W. B. (1965) Critical evidence for a great infra-Cambrian glaciation. *Geologische Rundschau* 54, 45–61.

Hoffman, P. et al. (1998) A Neoproterozoic snowball Earth. *Science* 281, 1342–1346.

Jouzel, J. et al. (1987). Vostok ice core: a continuous isotope temperature record over the last climatic cycle (160,000 years). *Nature* 329, 403–408.

Kirschvink, J. L. (1992) Late-Proterozoic low-latitude global glaciation: the snowball Earth. In: *The Proterozoic Biosphere*, Cambridge University Press, 51–52.

Laskar, J. et al. (1993) Stabilization of the Earth's obliquity by the Moon. *Nature*, 361, 615–617.
Love, G. D. et al. (2009) Fossil steroids record the appearance of Demospongiae during the Cryogenian period. *Nature* 457, 718–721.
MacDonell, S. and Fitzsimons S. (2008) The formation and hydrological significance of cryoconite holes. *Progress in Physical Geography* 32, 595–610.
Meert, J. G. (2003) A synopsis of events related to the assembly of eastern Gondwana. *Tectonophysics* 362, 1–40.
Mikucki, J. A. et al. (2009) A contemporary microbially maintained subglacial ferrous ocean. *Science* 324, 397–400.
Pu, J. P. et al. (2016) Dodging snowballs: geochronology of the Gaskiers glaciation and the first appearance of the Ediacaran biota. *Geology* 44, 955–958.
Railsback, L. B. et al. (2015) An optimized scheme of lettered marine isotope substages for the last 1.0 million years, and the climatostratigraphic nature of isotope stages and substages. *Quaternary Science Review* 111, 94–106.
Rooney, A. D. et al. (2015) Cryogenian chronology: two long-lasting synchronous Neoproterozoic glaciations. *Geology* 43, 459–462.
Ye, Q. et al. (2015) The survival of benthic macroscopic phototrophs on a Neoproterozoic snowball Earth. *Geology* 43, 507–510.

13. Thawing of the Snowball Earth

Andres, J. et al. (2009) Microbes versus metazoans as dominant reef builders: insights from modern marine environments in the Exuma Cays, Bahamas. *Special Publication of the International Association of Sedimentology* 41, 149–165.
Bontognali, T. R. (2019) Anoxygenic phototrophs and the forgotten art of making dolomite. *Geology* 47, 591–592.
Bontognali, T. R. et al. (2010). Dolomite formation within microbial mats in the coastal sabkha of Abu Dhabi (United Arab Emirates). *Sedimentology* 57, 824–844.
Creveling, J. R. and Mitrovica, J. X. (2014) The sea-level finger print of a Snowball Earth deglaciation. *Earth and Planetary Science Letters* 309, 74–85.
Dill, R. F. et al. (1986) Giant subtidal stromatolithes forming in normal salinity waters. *Nature* 324, 55–58.
Élie, M. et al. (2007) A red algal bloom in the aftermath of the Marinoan Snowball Earth. *Terra Nova* 19 303–308.
Font, E. et al. (2006) Chemostratigraphy of the Neoproterozoic Mirassol d'Oeste cap dolostones (Mato Grosso, Brazil): an alternative model for Marinoan cap dolostone formation. *Earth and Planetary Science Letters* 250, 89–103.
Font, E. et al. (2010) Fast or slow melting of the Marinoan snowball Earth? The cap dolostone record. *Palaeogeography, Palaeoclimatology, Palaeoecology* 295, 215–225.
Hoffman, P. F. (2023) Snowball Earth: The African legacy? *Journal of African Earth Sciences* 205, 104976.
Hoffman, P. F. and Schrag, D. P. (2000) The Snowball Earth hypothesis: testing the limits of global change. *Terra Nova* 14, 129–155.
Kendall, C. G. S. C. et al. (2003). Changes in microclimate tracked by the evolving vegetation cover of the Holocene beach ridges of the United Arab Emirates. In: Alsharan, A. S. et al., *Desertification in the Third Millenium*, Swets and Zeitlinger B.V., Lisse, the Netherlands, 91–98.
Le Hir, G. et al. (2009) The Snowball Earth aftermath: exploring of continental weathering processes. *Earth and Planetary Science Letters* 277, 453–463.
Nédélec, A. et al. (2007). Sedimentology and chemostratigraphy of the Bwipe Neoproterozoic cap dolostones (Ghana, Volta Basin): a record of microbial activity in a peritidal environment. *Comptes Rendus Geoscience* 339, 223–239.
Trindade, R. I. F. et al. (2003) Low-latitude and multiple geomagnetic reversals in the Neoproterozoic Puga cap carbonates. *Terra Nova* 15, 441–446.
Vasconcelos, C. and McKenzie, J. (1997) Microbial mediation of modern dolomites precipitation and diagenesis under euxinic conditions (Lagoa Vermelha, Rio de Janeiro, Brazil). *Journal of Sedimentological Research* 67, 378–390.
Vieira, L. et al. (2015). Aragonite crystal fans in Neoproterozoic cap carbonates: a case study from Brazil and implications for the post-snowball earth coastal environment. *Journal of Sedimentary Research* 85, 285–300.
Warthmann, R. et al. (2000) Bacterially induced dolomite precipitation in anoxic culture experiments. *Geology* 28, 1091–1094.

14. Life invades the continents

Ahlberg, P. E. et al. (2005) The axial skeleton of the Devonian tetrapod *Ichthyostega*. *Nature* 437, 137–140.
Alvarez, L. W. et al. (1980) Extraterrestrial cause for the Cretaceous–Tertiary extinction. *Science* 208, 1095–1108.

Black, B. A. et al. (2012) Magnitude and consequences of volatile release from the Siberian traps. *Earth and Planetary Science Letters* 317–318, 363–373.

Bonnefille, R. (2010) Cenozoic vegetation, climate changes and hominid evolution in tropical Africa. *Global and Planetary Change* 72, 390–411.

Brunet, M. et al. (2002) A new hominid from the Upper Miocene of Chad, Central Africa. *Nature* 418, 152–155.

Coates, M. I. and Clack, J. A. (1991) Fish-like gills and breathing in the earliest known tetrapod. *Nature* 352, 234–236.

Darwin, C. (1859) *On the Origin of Species by Means of natural Selection*, John Murray, London.

Darwin, C. (1871) *The Descent of Man, and Selection in Relation to Sex*, John Murray, London.

Erwin, D. H. and Valentine, J. W. (2013) *The Cambrian Explosion: The Construction of Animal Biodiversity*, Bedford/Saint Martin's.

Garnier, E. (2011) Les brouillards du Laki en 1783: volcanisme et crise sanitaire en Europe. *Bulletin de l'Académie nationale de médecine* 195, 1043–1055.

Goldring, W. (1927) The oldest known petrified forest. *The Scientific Monthly* 24, 514–529.

Gould, S. J. (1989) *Wonderful Life: The Burgess Shale and the Nature of History*. Norton.

Gould, S. J. (2002) *The Structure of Evolutionary Theory*. Harvard University Press.

Gould, S. J. and Eldredge, N. (1972). Punctuated equilibria: an alternative to phyletic gradualism. *Models in Paleobiology*, 82–115.

Heckman, D. S. et al. (2001) Molecular evidence for the early colonization of land by fungi and plants. *Science* 293, 1129-1133.

Hublin, J. J. et al. (2017) New fossils from Jebel Irhoud, Morocco and the pan-African origin of Homo sapiens. *Nature* 546, 289–292.

Johanson, D. C. and Taieb, M. (1976) Plio-Pleistocene hominid discoveries in Hadar, Ethiopia. *Nature* 260, 293–297.

Keller, G. et al. (2009) Deccan volcanism, the KT extinction and dinosaurs. *Journal of Biosciences* 34, 709–728.

King, W. (1864) *The Reputed Fossil Man of Neanderthal*, Longmans Green and Company.

Kring, D. A. (2007) The Chicxulub impact event and its environmental consequences at the Cretaceous–Tertiary boundary. *Paleogeography, Paleoclimatology, Paleoecology* 255, 4–21.

Lenton, T. M. et al. (2012) First plants cooled the Ordovician. *Nature Geoscience* 5, 86–89.

McCarthy, T. and Rubidge B. (2005) *The story of Earth and Life: a southern African Perspective on a 4.6-billion-year Journey*, Struik Publishers, Cape Town.

McGhee Jr., G. (2018) *Carboniferous Giants and Mass Extinction*, Columbia University Press.

Meyer-Berthaud, B. and Decombeix A. L. (2009) Evolution of earliest trees: The Devonian strategies. *Comptes Rendus Palevol* 8, 155–165.

Modesto, S. P. et al. (2014) The oldest parareptile and the early diversification of reptiles. *Proceedings of the Royal Society of London B* 282, 20141912.

Myers, T. S. (2016) CO_2 and late Palaeozoic glaciation, *Nature* 9, 803–804.

Pääbo, S. (2014) *Neanderthal Man: In Search of Lost Genomes*. Basic Books, New York.

Rubinstein, C. V. et al. (2010) Early middle Ordovician evidence for land plants in Argentina (eastern Gondwana). *New Phytologist* 188, 365–369.

Schoene, B. et al. (2019) U–Pb constraints on pulsed eruption of the Deccan Traps across the end-Cretaceous mass extinction. *Science* 363, 862–866.

Slimak, L. et al. (2022) Modern human incursion into Neanderthal territories 54,000 years ago at Mandrin, France. *Science Advances* 8, eabj9496.

Vannier, J. (2009) L'explosion cambrienne ou l'émergence des écosystèmes modernes. *Comptes Rendus Palevol* 8, 133–154.

Glossary

Acritarch: an organic-walled microfossil of uncertain biological affinity, hence the name derived from the Greek *akritos*, 'uncertain'. Many acritarchs were likely eukaryotic. Many of them have external ornaments and prolongations suggesting planktonic life.

Albedo: a dimensionless quantity (between 0 and 1), expressing the reflectivity of a surface defined as the ratio of reflected to incident light. The albedo of fresh snow is as high as 0.9, whereas most land areas have an albedo around 0.3 and the albedo of the ocean is a maximum of 0.1.

Alkalinity of water: capacity of water to absorb H^+ ions by the presence of anions such as carbonate (CO_3^{2-}) and bicarbonate (HCO_3^-), which are weak bases. Other ions, such as borates, also contribute to the alkalinity of seawater, though in practice only carbonate alkalinity is usually considered.

Amino acids: organic compounds with an amino group ($-NH_2$) which are the basic building blocks of proteins. The first discovery of an amino acid (asparagine) was made by French chemists Louis-Nicolas Vauquelin and Pierre Robiquet in 1806.

Amphibolite: a metamorphic rock containing the minerals amphibole and plagioclase that forms when basaltic rocks are subjected to temperatures between 500 and 750 °C and pressures between 2 and 12 kilobars.

Andesite: a volcanic rock of intermediate composition produced by volcanoes above a subduction zone, such as in the Andes, whence the name is derived.

Anorthosite: a light-coloured intrusive magmatic rock consisting mainly of calcic plagioclase feldspar. Anorthosites are common on the Moon, but rare on Earth.

Anoxic: devoid of oxygen.

Aragonite: calcium carbonate mineral that has the same formula ($CaCO_3$) as calcite but differs in crystal structure.

Archean (or Archaean): An eon in the Precambrian spanning from 4 to 2.5 Ga (i.e. between the Hadean and Proterozoic eons).

Argillite: a fine-grained, low-permeability sedimentary rock, mainly made of clay minerals as a result of the lithification of muds. *Synonyms*: pelite, mudstone.

Argon: a noble gas found in the Earth's atmosphere, of which it makes up nearly 1%. Its name derives from the Greek *argos*, meaning 'inactive'. Most atmospheric argon is argon-40, derived from the decay of potassium-40 in the Earth's crust, whereas the most common argon isotope in the universe is argon-36.

Arthropods: a phylum of invertebrate animals characterized by a segmented body with paired articulated appendages, and covered by a protective cuticle that forms an exoskeleton. Living arthropods include crustaceans, myriapods (millipedes), arachnids, and insects. Trilobites are extinct arthropods.

Australopithecus: a genus of early hominins that lived in eastern and southern Africa; named after the Latin *australis* ('southern') and the Greek *pithekos* ('ape'). *Australopithecus* was bipedal and had a brain one third the volume of current *Homo sapiens*.

Basalt: a dark volcanic rock resulting from partial melting of the Earth's mantle.

Bedding: The arrangement of layers (or strata) of sedimentary or volcanic rocks.

BIF: acronym derived from 'banded iron formation', referring to a sedimentary rock formed by the chemical precipitation of iron oxides and quartz.

Biomarker: a geological molecule of biological origin.

BP: abbreviation of 'before present', where 'present' is conventionally taken to be 1950. The BP timescale is currently used in archaeology and prehistorical studies.

Breccia: rock containing more than 50% angular grains or clasts set in a natural matrix or cement. Breccia can be of volcanic (from an explosive eruption) or detrital origin (from the erosion of pre-existing rocks).

Cambrian: first period of the Paleozoic era, between 539 and 485 Ma.

Cap carbonates: carbonate layers on top of glacial deposits. Cap carbonates are made of dolostone and limestone and were deposited during the post-Snowball Earth transgression.

Carboniferous: penultimate period of the Paleozoic Era, between 359 and 299 Ma. Its name means 'coal-bearing', and refers to the fact that large deposits of coal were formed during that time.

Chert: a hard, siliceous sedimentary rock made of microcrystalline or cryptocrystalline quartz formed by chemical precipitation.

Chlorite: a greenish mineral, belonging to the phyllosilicate group, which includes clays and micas. Chlorite is characteristic of metamorphism under so-called greenschist conditions, corresponding to temperatures between 250 and 500 °C, and pressures between 2 and 10 kilobars. Chlorite contains 14% water by weight. It often results from hydrothermal alteration, for instance in the oceanic crust.

Chloroplast: intracellular organelle containing chlorophyll. Chloroplasts are present in photosynthetic eukaryotic cells, such as in algae and terrestrial plants. They originated as an endosymbiotic cyanobacteria within a eukaryotic cell.

Chondrite: a stony meteorite containing mostly silicates, but also 20–30% metallic iron. The name comes from the chondrules, or silicate spherules, that are found in chondrites. Chondrites are primitive (undifferentiated) meteorites. Carbonaceous chondrites are distinguished from the more common ordinary chondrites by the presence of carbon (1–5%).

Climate forcing: a factor that influences the climate by modifying the planet's radiation balance. Natural forcings include variations of the solar constant due to astronomical parameters and variations of atmospheric greenhouse gases due to volcanic gas eruptions.

Conduction: a method of heat transfer without the displacement of matter, thus contrasting with convection.

Conglomerate: detrital sedimentary rock containing at least 50% rounded pebbles or blocks, resulting from the erosion of other rocks.

Convection: a mode of heat transfer involving the flow of material due to temperature-induced differences in density. Convection occurs in fluids and in some low-viscosity solids.

Craton: a continental domain that has been stable since (at least) the end of the Precambrian. Named derives from the Greek *kratos*, meaning 'strength'.

Crustal: referring to the Earth's crust.

Crustal recycling: a process that refers to the cycling of rocks in the continental crust by erosion, sedimentation, metamorphism, and (eventually) partial melting.

Cyanobacteria: Photosynthetic bacteria, formerly referred to, inaccurately, as 'blue-green algae'.

Declination: the angle formed between the direction of the geographic pole and the magnetic pole.

Deuterium: a naturally occurring stable isotope of hydrogen, denoted ^{2}H or D.

Devonian: period of the Paleozoic era that precedes the Carboniferous.

Diamictite: sedimentary rock containing unsorted clasts in a fine matrix. Some diamictites that show flattened faces or striations on the clasts can be identified as glacial in origin.

Dip: tilt of a plane or layer with respect to horizontal; the dip is defined by an angle (between 0 and 90°) and by the azimuth of the line of steepest slope. Example: a dip of 30° towards the north-east.

Dipole: a combination of two magnetic poles between which a magnetic field loop is established. The Earth's magnetic field is regarded as a mainly dipolar field such as would be created by a bar magnet in the centre of the Earth.

DNA: deoxyribonucleic acid, a molecule that encodes genetic information. The DNA molecule is made of two chains of nucleotides coiled around each other to form a double helix. Each nucleotide is composed of a nucleobase (adenine, cytosine, guanine, or thymine), a deoxyribose glucide (carbohydrate), and a phosphate group.

Dolomite: carbonate mineral containing both calcium and magnesium with the formula $CaMg(CO_3)_2$.

Dolostone: sedimentary rock mainly composed of dolomite, a calcium and magnesium carbonate.

Dyke (or dike): a vertical magma vein or sheet filling a fracture. Basaltic dykes sometimes occur in swarms.

Eon: the longest subdivision of time in the geologic timescale: an eon comprises several eras.

Eukaryotes: organisms, unicellular or multicellular, whose cells have a nucleus.

Evaporites: sedimentary rocks formed by the precipitation of salts dissolved in water. The minerals that precipitate include sulphates (e.g. gypsum), chlorides (e.g. halite), and carbonates (e.g. calcite, dolomite).

Feldspars: common aluminium silicate minerals made of two groups: alkali feldspars (potassic to sodic) and plagioclase feldspars (sodic to calcic).

Fractional crystallization: a mode of crystallization of a magma which is accompanied by the extraction of the first formed crystals, typically by settling a cumulate on the floor of the magma chamber.

Ga: abbreviation for billion years (for geological dates); for time duration 'Gyr' is used.

Gabbro: a plutonic rock of basaltic composition.

Garnet: aluminosilicate mineral crystallizing in the cubic system; garnet is variably rich in iron, magnesium, calcium, or manganese.

Geodynamics: study of the evolution of the Earth under the effect of internal forces (plate tectonics, magma migration) or external forces (e.g. erosion).

Geotherm: curve representing the inner temperature of the Earth as a function of depth.

Geothermal gradient: rate of temperature change with increasing depth in the Earth.

GIF: acronym derived from 'granular iron formation'; a variant of BIF (see above) in which iron oxides have precipitated or been otherwise deposited as spherical granules 0.2 to 10 mm in size. The granular texture implies deposition in wave-agitated, shallow-marine settings.

Glaciogenic: coming from a glacier or related to glaciation.

Gneiss: a foliated metamorphic rock composed of quartz, feldspars, and other minerals such as mica or garnet.

Granite: plutonic magmatic rock characteristic of the continental crust.

Granodiorite: a plutonic magmatic rock resembling a granite, but with a higher proportion of plagioclase feldspar to alkali feldspar.

Gravitation: interaction between massive bodies in the universe. Gravitation is responsible for the Earth's attraction, or gravity, which keeps us grounded.

GOE: great oxidation (or oxygenation) event of the Earth's atmosphere that occurred around 2.3 Ga.

Greenhouse effect: effect of heat-trapping gases, which influence the radiation balance of the planet.

Hadean: the first subdivision of the Earth's geological timescale, between 4.5 and 4 Ga.

Hafnium: a rare chemical element, with symbol Hf, chemically homologous to zirconium (Zr) and therefore commonly replacing Zr in minerals. Zircon is the main host of hafnium on Earth. Hafnium has multiple naturally occurring isotopes, all but one of which are stable.

Half-life: the time required for a population of radioactive isotopes to decrease by half as a result of its exponential decay.

Harzburgite: rock belonging to the peridotite group, containing olivine and orthopyroxene and distinguished from lherzolite by its more refractory composition.

Interglacial: a relatively warm period, lasting only a few thousand years, during an ice age, when continental ice retreats without disappearing completely.

Isotopes: different species of the same chemical element that differ only in the number of neutrons contained in the atomic nucleus (i.e. in their atomic mass).

Isotopic differentiation: a process by which isotopes of the same element are fractionated into reservoirs with different isotopic ratios.

Isotopic signature: ratio of two stable isotopes of the same chemical element measured in a material. The value of this ratio can be characteristic of specific processes or sources. For example, the oxygen isotopic signature of a granitic rock makes it possible to distinguish between a mantle, crustal, or mixed source.

Kilobar (kb or kbar): A unit of pressure equivalent to about one thousand times the atmospheric pressure at sea level, or one hundred million pascals (100 MPa).

Kimberlite: a rare volcanic rock formed by a small percentage of partial melting of the Earth's mantle; kimberlite magmas are generated deep in the continental lithosphere and may contain diamonds.

Komatiite: a magnesium-rich volcanic rock, corresponding to a high percentage of partial melting of the Earth's mantle and therefore requiring higher temperatures than found in the Earth's interior today.

LGM (Last Glacial Maximum): the coldest period of the current ice age occurring 21,000 years before present. Ice sheets were at their maximum and sea level was about 125 metres lower than now.

Lherzolite: a typical rock of the Earth's mantle, belonging to the group of peridotites and made of olivine, orthopyroxene, and clinopyroxene. Named after Lherz Lake in the French Pyrenees.

Liquidus: limit separating the completely molten (liquid) domain from the domain where liquid and solid coexist in the case of a mixture.

Lithosphere: the solid envelope of the Earth that includes the crust and the rigid upper part of the mantle. The lithosphere is segmented into different plates, which are mobile in relation to each other.

Lithospheric recycling: a process by which lithospheric plates are subducted into the mantle where they slowly (over hundreds of millions of years) heat up and eventually mix into the mantle. Only oceanic plates can be dragged into the mantle.

Ma: abbreviation meaning million years (for geological dates).

Magma: silicate liquid resulting from the melting of pre-existing rocks. Magmatic rocks are formed by the crystallization of a magma, either on the surface (volcanic rocks) or at depth (plutonic rocks). A magma may contain solid crystals, which influence its viscosity.

Magnetic inclination (magnetic dip): the angle of the magnetic field with respect to Earth's surface.

Mantle plume: an upwelling of Earth's mantle that results from a small difference in density with the surrounding mantle. The material that makes up a plume only begins to melt as the plume approaches the surface. Plumes are therefore made of a low-viscosity solid.

Metabolism: the set of reactions that take place in a living cell and which include both molecular degradation (catabolism, e.g. to produce energy as in the case of respiration) and molecular synthesis (anabolism, e.g. to build molecules required for cell function).

Metamorphism: mineralogical and structural transformations that affect a rock submitted to conditions of temperature and pressure different from those in which it was formed.

Methanogen: microbial producer of methane.

Methanotroph: a microbe (bacterium or archaeon) that uses methane as a source of carbon and energy.

Micrite: a carbonate rock made of very small crystals (of the order of a few micrometres).

Mitochondrion: intracellular organelle responsible for energy production in a eukaryotic cell; mitochondria contain their own DNA and are therefore interpreted to represent a former endosymbiotic bacterium.

Moraine: A mixture of mud and rock fragments that accumulates at the front and sides of the glacier, forming the so-called frontal and lateral moraines.

Nebula: Latin word meaning 'cloud', referring to a cloud of gas and dust from which stars commonly form.

Noble gases: colourless, odourless, and inert or poorly reactive gases, grouped in the last column of the periodic table of elements.

Nuclear fission: the process by which a heavy atomic nucleus (such as uranium) is split into two or three smaller nuclei. Fission can be spontaneous (as in the case of radioactive decay) or induced (e.g. by a collision with a neutron in a nuclear device). Fission releases a large amount of heat or energy.

Nucleic acids: macromolecules present in all cells and comprising two main classes: DNA and RNA.

Nucleosynthesis: synthesis of atomic nuclei in nuclear reactions. Primordial nucleosynthesis followed the Big Bang and provided the light elements (up to lithium); the next elements (up to iron) are produced by nucleosynthesis in stars, and the heaviest elements (beyond iron) form during the explosion of supernovas.

Oceanic accretion: formation of oceanic crust on either side of a mid-oceanic ridge.

Oceanic plateau: a region of thick oceanic crust resulting from the magmatic activity of a hotspot, for example the Kerguelen plateau in the Indian Ocean.

Oceanic ridge: the boundary between two divergent plates, corresponding to the production of new oceanic crust from basaltic magma resulting from the partial melting of the underlying mantle.

Olivine: a silicate mineral also called peridot, whose composition varies between an iron endmember, with the formula Fe_2SiO_4, and a magnesium endmember, with the formula Mg_2SiO_4. The olivine in the Earth's mantle comprises dominantly (90%) the Mg endmember.

Ophiolites: rocks originating as oceanic lithosphere that escaped subduction and are now preserved on the continents. Named after the Greek *ophis* ('snake'), due to their lustrous sheen resembling a snakeskin, because of the presence of the alteration minerals chlorite or serpentine.

Ordovician: second period of the Paleozoic era following the Cambrian.

Orogenesis: formation of mountains (from the Greek *oros*, 'mountain').

Paleolatitude: the latitude at which a rock was formed (often different from the latitude at which it can be collected today).

Paleomagnetism: study of the Earth's past magnetic field, based on the record of an old magnetization in minerals. Paleomagnetists determine the orientation of former magnetizations.

Paleozoic era: earliest era of the Phanerozoic eon.

Pallasite: a type of meteorite consisting of olivine crystals that appear to float in metallic iron.

Pangea: the supercontinent that formed at the end of the Paleozoic era.

Peridotite: rock formed mostly of the mineral olivine (also known as peridot).

Permian: last period of the Paleozoic era, between 299 and 251 Ma.

pH (or potential of hydrogen): a measure of the activity of hydrogen ions (H^+) or the acidity of a solution. The lower the pH (< 7), the more acidic the solution; a pH of 7 corresponds to neutrality and a pH above 7 indicates a basic solution.

Phanerozoic: the last eon in the geologic timescale, beginning 541 million years ago and comprising the Paleozoic, Mesozoic, and Cenozoic eras. From the Greek *phaneros*, meaning 'visible', and *zôon*, 'animal'.

Photosynthesis: from the Greek *phos*, 'light'. The process by which certain organisms harvest energy from light.

Phylogenetics: study of the evolution of living species and their relationships, resulting in a phylogeny or phylogenetic tree.

Pitchblende: the main uranium ore mineral, with the formula UO_2.

Plagioclase: *see* feldspars.

Planetary accretion: the formation of a planet by the incremental capture of matter under the effect of gravitational forces.

Planetary differentiation: the process leading to the separation of an iron core and a silicate mantle in a planetary body.

Plate: a fragment of lithosphere which is bounded by a ridge, a subduction zone, a transform fault, or a collisional mountain chain.

Plutonic rock: an intrusive magmatic rock that crystallized slowly in the crust, thus acquiring a coarse-grained texture with crystals visible to the naked eye.

Precambrian: The interval comprising all three eons prior to the Cambrian period (i.e. the Hadean, Archean, and Proterozoic eons). The Precambrian has a very long duration of four billion years, hence representing about 90% of Earth's history.

Prokaryotes: Single-celled organisms lacking a nucleus; prokaryotes are subdivided into the two domains Bacteria and Archea.

Proterozoic: from the Greek *proteros* meaning 'first' or 'earlier', and *zôon*, 'animal'. Subdivision of the Precambrian time spanning from 2,500 to 541 Ma.

Protists: single-celled eukaryotes.

Protolith: original or source rock before metamorphism or partial melting.

Pyrite: iron sulphide with the formula FeS_2, which when unaltered has golden lustre, hence its common nickname 'fool's gold'.

Pyroclastic: from the Greek *pyr*, 'fire' and *klastos*, 'fragment'. Refers to rocks formed during an explosive volcanic eruption. Pyroclastic deposits vary in size from fine ashes to bombs metres in diameter. When transported en masse by volcanic gases and heat-welded, they are called pyroclastic flows.

Pyroxenes: Ferro-magnesian silicate minerals, common in peridotites and basaltic rocks.

Quartzite: a metamorphic rock composed mainly of quartz and resulting from the metamorphism of a sandstone or a chert.

Radiogenic: an isotope resulting from the decay of a radioactive element.

Refractory: resistant to melting under high temperatures.

Remanent magnetization: persistent magnetization carried by minerals such as magnetite. It records the orientation of the magnetic field at the time magnetization was acquired, either during cooling (magmatic rock) or deposition (sedimentary rock).

Rheology: named from the Greek *rheo*, 'to flow'. Rheology is the study of how materials flow and deform under the effect of a stress.

Rhyolite: volcanic rock resulting from the eruption of a granitic magma.

Ribosome: an organelle made of RNA (known as 'ribosomal RNA') and proteins, where proteins are made.

Ribozyme: RNA molecules that serve as catalysts, similar to enzymes, during biochemical reactions.

Sagduction: a tectonic process that deforms and drives crustal rocks downwards due to their density. This process, envisioned by some as a precursor to plate tectonics, is impossible where the lithosphere is rigid.

Serpentinite: a rock composed mainly of serpentine, a mineral resulting from the transformation of olivine by hydrothermal alteration.

Shear waves (or transverse waves): seismic waves characterized by displacement of material perpendicular to the direction of wave propagation. After an earthquake, shear waves arrive after the faster primary waves, hence their designations as 'secondary' (S) waves. Shear waves cannot travel through a fluid.

Silurian: the shortest period of the Primary era, between the Ordovician and Devonian.

Solar constant: the amount of solar energy incident upon 1 m^2 of Earth's atmosphere or surface.

Solidus: limit separating the solid domain from the domain where liquid and solid coexist. The melting of a mixture begins as soon as the solidus conditions are reached. The liquidus must be reached for complete melting.

Sterane: a molecule of organic origin, derived from the geological transformation of a sterol or a steroid, which can be used as a biomarker in sedimentary rocks. Steranes are considered markers for eukaryotes.

Stratosphere: the layer of the Earth's atmosphere above the troposphere; it lies on average between 10 and 50 km above sea level.

Stratum (pl. strata): a layer of sedimentary (or volcanic) rock.

Stromatolite: a laminated and lithified structure formed by the interaction of microbial mats at the sediment–water interface. Stromatolites most commonly comprise carbonate cements and grains, but may also be formed by other precipitated minerals.

Subduction: tectonic process by which a lithospheric plate descends into the Earth's mantle.

Sulphate reduction: a bacterial metabolic process that utilizes the sulphate ion (SO_4^{2-}) to metabolize organic carbon, and produces hydrogen sulphide as a byproduct.

Sulphur bacteria: bacteria that use sulphur species in their metabolisms, including the green and purple sulphur bacteria, which photosynthesize in anaerobic environments.

Supercontinent: a continental agglomeration comprising more than half of the world's landmass and resulting from the assembly of several cratons.

Supracrustal: the character of a rock that has been deposited on the surface of the Earth's crust, either by a sedimentary process or as a result of a volcanic eruption.

Tectonics: from the Greek *tecton* ('carpenter'), the study of major geological structures and the processes responsible for them. Since the end of the 1960s, these processes have been interpreted in relation to the movements of the lithospheric plates or **plate tectonics**. Early in Earth's history, probably before 3.2 Ga, other mechanisms likely occurred, such as **gravity tectonics**, which involved only vertical movements under the effect of density contrasts.

Tidalite: sediment deposited by bimodal tidal currents.

Tillite: glaciogenic sedimentary rock that formed as a moraine.

Tonalite: an intrusive rock of the granite family containing plagioclase feldspar but no alkali feldspar (unlike granite).

Trace elements: chemical elements contained in very small quantities in rocks (measured in ppm, i.e. parts per million, or μg/g).

Transgression: a rise in sea level, resulting in the retreat of shorelines.

Traps: basaltic flows forming stepped cliffs, the total height of which is generally greater than 1,000 m. Traps, also known as flood basalts, correspond to the eruption of huge volumes of magma at a hotspot in the continental domain.

Trilobites: fossil marine arthropods from the Paleozoic era.

Trondhjemite: a kind of light-coloured tonalite, rich in quartz and poor in dark ferromagnesian minerals. The composition of a trondhjemite is very close to that of a granite, but is rich in sodium rather than potassium.

Troposphere: the lower layer of the Earth's atmosphere, whose thickness varies with latitude (thicker at the equator than at the poles) and averages 10 km.

TTGs: an association of tonalites, trondhjemites, and granodiorites; very common in the Archean continental crust.

Unconformity: An erosional surface separating sedimentary layers and corresponding to missing geologic time. Where the layers below and above the unconformity are not parallel, it is known as an **angular unconformity** and identifies a tectonic event.

Upwelling: upward movement of deep, nutrient-rich oceanic water along continental margins. Today, significant upwellings occur along the western coasts of Africa and South America.

Volatile: easily converted to a gaseous state.

Zircon: a common accessory mineral in magmatic rocks, with the formula $ZrSiO_4$. Because zircon commonly contains uranium and is resistant to alteration, it is widely used to date rocks using the U–Pb decay system.

Index

For the benefit of digital users, indexed terms that span two pages (e.g., 52–53) may, on occasion, appear on only one of those pages.

Figures are indicated in italic

Abitibi 78, 79*f*, 88*f*, 89
Acasta 55–56, 55*f*, 56*f*
Acritarch 158–159, 159*f*
Albedo 52, 197, 203–204
Algae 163–165, 164*f*, 209–210
Amino acid 101*f*, 102–104, 104*f*, 105*f*
Amphibians 229
Amphibolite 63–64
Andesite 75–76
Anorthosite 29–32, 31*f*
Anoxic 91, 95, 195–196, 209–210
Antarctica 189*f*, 197*f*, 203, 231
Apex chert 108–109, 110*f*
Apollo 20, 30–31, 30*f*, 31*f*
Aragonite 212–213, 215*f*, 216*f*
Archea 114–118, 114*f*, 117*f*, 162*f*, 163, 163*f*
Archean 2, 3*f*, 55, 55*f*, 76–77, 80*f*, 82–85
Argon 48, 48*f*
Asteroid 5*f*, 7*f*, 8*f*, 45–46
Athabasca 148–149, 149*f*
Atmosphere 41–43, 49–51, 125*f*, 129–132
Australia 25–26, 26*f*, 111*f*, 141*f*
Australopithecus 232–235

Baltica 139–140, 139*f*
Barberton 55*f*, 57–60
Basalt 57, 70, 71*f*, 73–75, 198–199, 230–231
BIF 95–99, 127–130, 195–196
Biomarker 154, 158, 162
Bitumen 135–136, 208*f*
Brazil 97–98, 98*f*, 145*f*, 207–209, 209*f*

Cambrian 1–2, 3*f*, 219–221
Cap carbonate 204*f*, 205–208, 210–211
Carboniferous 222*f*, 223, 227*f*, 228
Carbon dioxide 49*f*, 50*f*, 52, 96, 125*f*, 203–205
Carbon cycle 125*f*, 133–135
Carbon isotopes 12–13, 106, 110–112
Carbonaceous chondrite 9*f*, 44–45, 45*f*, 106
Chert 60, 87*f*, 89–93, 109
Chlorite 58, 90
Chloroplast 160–162, 162*f*, 164
Chondrite 8, 9*f*, 16, 44–45
Cloudina 166, 166*f*
Columbia 137, 174–176

Comet 44–45, 45*f*, 106–108
Conduction 67
Continent 72*f*, 123, 124*f*
    Continental collision 176–177, 179, 179*f*, 226–228
    Continental crust 72*f*, 87–89
    Continental weathering 124, 198–199, 211
Convection 67–68, 69*f*, 73–74, 147*f*
Core 11, 16–20, 22*f*, 69*f*
Craton 137–141
Crustal recycling 28, 29*f*
Cyanobacteria 97, 109–112, 111*f*, 162*f*, 208–209

D" layer 69*f*
Degassing 46–48
Density of rocks 73, 87–89, 88*f*, 138
Deuterium 45, 45*f*
Devonian 221–223, 222*f*, 223*f*, 226–227
Diamond 138*f*, 138
Diamictite 125, 191–192
Dinosaur 1, 231
Dolomite 86–87, 87*f*, 204*f*, 205–208
DNA 101–103, 103*f*, 104*f*, 159–161, 160*f*
Dyke 174*f*, 175, 175*f*

Eccentricity 185, 186*f*
Ecosystem 210*f*, 210
Ediacaran 3*f*, 165*f*, 165–166
Energy's budget 52*f*
Eukaryote 113–114, 114*f*, 158–163

Farallon plate 73, 73*f*
Fossil 108–112, 154–159, 164*f*, 165*f*
Francevillian 154–158

Gabbro 76, 116
Genetic code 102–103
Geomagnetic reversal 143, 214–215
Geotherm 69*f*, 69
    Geothermal gradient 67
GIF 127–129, 130*f*
Gilboa 221–223, 223*f*, 224*f*
Glaciation 190–195, 217*f*, 222*f*, 227*f*, 234*f*
Glacier 191*f*, 197*f*
Gneiss 56*f*
GOE 128*f*, 129–132, 133*f*, 163
Gondwana 178–181, 179*f*, 225–226, 226*f*, 227*f*, 228*f*
Granite 63, 73, 75*f*, 78–80, 79*f*, 87–89, 148

Granodiorite 63
Gravity 6, 44
   Gravity tectonics 81
Greenhouse effect 51–53, 124, 214–215
Greenland 183–184, 185f, 188, 196, 217f
Greenstones 57f, 58–60, 61f, 80f, 81, 93
Grenville orogeny 130f, 176–177

Hadean 2, 3f, 25, 82–84
Hafnium isotopes 18–19, 24, 28, 29f
Harzburgite 138
Hawaii 33f, 69–70, 73f, 169
Heat Flux 64–66
Heat transfer 67–69
Helium 43, 47–48
Hematite 95, 96f, 98, 113f
Homo 232–236
Hot spot 78, 169, 171–175, 175f, 177, 178f
Hydrogen 12, 43, 45f, 45, 116
Hydrothermalism 89–94, 115–118
   Hydrothermal alteration 93, 116

Iapetus 169–170
Ice core 188–190
Ice line 44
Ice sheet 189f, 204f, 217f, 225–228
Idiwhaa 56
Impacts 20–22, 23f, 23–24, 30, 34–39, 230
Interglacial 187–190, 187f, 190f
Iron isotope 127–128, 128f
Iron mines 97–99
Isochron 14–15, 15f
Isostasy 87–88, 88f
Isotherm 67–68, 72, 72f, 75f, 171f
Isotope 12–13
Isua 55f, 56–57

Kimberlite 138
Komatiite 59f, 60f, 61–62, 76, 79f

Laurentia 140f, 145f, 172f, 175f, 176, 176f, 178f, 226f
LGM 188–190, 190f, 193, 198f, 204, 235f, 236
LHB 38
Lherzolite 11–12
Lipid 101, 119, 120f
Liquidus 17–18, 19f, 70f, 77f, 82f
Lithophile 12f, 18–19
Lithosphere 67–68, 72f, 74f, 75f, 138f, 139f, 147
Lost city 115, 116f, 116
LUCA 114f, 118–119, 162f

Magma 17, 26–27, 33, 63, 74–77, 81f
   Magma density trap 82–83, 82f, 83f
   Magma ocean 17, 18f, 22–24, 34f
Magnetic field 141–143
Manganese 132–133
Mantle 16–20, 69f, 72f, 73f, 74–75, 83f, 88f, 147f
Marinoan 199–200, 200f, 201, 217

Mars 5f, 35f, 50f, 51
Metamorphism 55–56, 58, 173–174
Meteorite 8–10, 14–15, 15f, 230
Methane 105, 114–118
Microbial mat 109–112, 110f, 205–208, 207f, 209f, 212f, 213f
Middle Marker 60f, 111
Mitochondria 159–163, 160f, 161f, 162f
Moon 20–24, 29–32, 36–38, 37f, 85–86, 197
Moraine 190–191, 191f, 192f

Nitrogen 42f, 43–44, 49f, 50f, 51, 101
Noble gas 41, 47–48
Nucleic acid 101–103, 120
Nucleobase 101f, 103f
Nucleotide 102

Obliquity 185–186, 186f, 197
Oceanic crust 75f, 77f, 78, 92f, 116, 173
Oceanic ridge 71–72, 72f, 74
Oklo 155–156
Olivine 9f, 34f, 61–62, 116
Onverwacht 59–61, 59f, 60f
Ophiolite 78, 173
Ordovician 221, 222f, 224–226, 226f, 227f
Orogen (mountain belt) 173, 176, 176f, 179–181, 199–200, 200f, 226–227
Oxygen 42f, 127–133, 134f, 186–188, 187f, 188f
Oxygen isotope 21f, 27–28, 186–188, 187f, 188f
Ozone 42f

Pacific Ocean 170, 177–178, 178f
Paleomagnetism 141–144, 193–194
Pan-African 178–180, 200
Pangea 172f, 222f, 228f, 231
Peridotite 11–12, 74–75, 116f, 138
Permian 228f, 230–231
Persian Gulf 211, 212f, 213f
Petroleum (oil) 135–136
Photosynthesis 96, 110, 127, 162, 164
Phylogeny 113–115
Pilbara craton 141f, 141
Pillow lava 61f
Planetary accretion 6
Planetary differentiation 16–20
Plate 71–76, 78–81
Plume 69–71, 73f, 76, 77f, 83f, 171, 171f, 173
Plutonism 75f, 76
Protein 101, 104f
Proterozoic 2, 3f, 123–169
Pyrite 89, 92, 94, 95f

Radioactivity 13–14
Refractory mantle 82–83, 83f, 138, 139f
Ribosome 102–103, 104f, 160f
Rodinia 176–178
RNA 101–103, 104f, 113, 120

Sagduction 81, 81*f*
Sandstone 59*f*, 85, 87*f*, 144–147, 145*f*, 149
Serpentine 32–34, 34*f*, 116*f*, 116
Shale 87*f*
SHRIMP 25–26, 26*f*
Siderophile 12*f*, 39
Silica 92*f*, 93
Silicification 92*f*
Smoker 90*f*, 92*f*
Snowball Earth 183, 203
Solar constant 52–53
Solidus 17–18, 19*f*, 70*f*, 77*f*, 82*f*
Stromatolite 109–112, 111*f*, 112*f*, 151–153, 207, 208*f*, 214*f*
Sturtian 196*f*, 199
Subduction 74, 74*f*, 75*f*, 76–77, 77*f*, 80*f*
Sulphur 94, 230–231
Supercontinent 137, 170*f*, 172*f*, 176, 180, 228
Superior 78, 79*f*, 129, 130*f*, 140*f*
Symbiosis 161–163

Tectonics 71–76, 78–81
Thermal boundary layer 69*f*, 69
Thermophile 114*f*, 118

Tidalite 85–86, 86*f*
Tillite 190
Tomography 73*f*
Tonalite 56, 62
Transgression 204, 204*f*, 214–215
Traps 70, 71*f*, 230, 231*f*
Trondhjemite 62–63
TTG 57–58, 59*f*, 62–64, 76–81, 80*f*, 81*f*

Uranium 12*f*, 149–150
  Uranium isotopes 12–14, 56*f*, 65*f*, 155–156
  Uranium deposits 148–150, 156*f*

Venus 5*f*, 49–50, 50*f*
Vertebrates 229
Vesta 8*f*, 45
Virus 120
Volcanism 47, 70–71, 75*f*, 75–76, 148, 203, 229–230
Vredefort 35–36

Water saturation 46, 46*f*
Wilson cycle 169–171

Zircon 25–27, 35